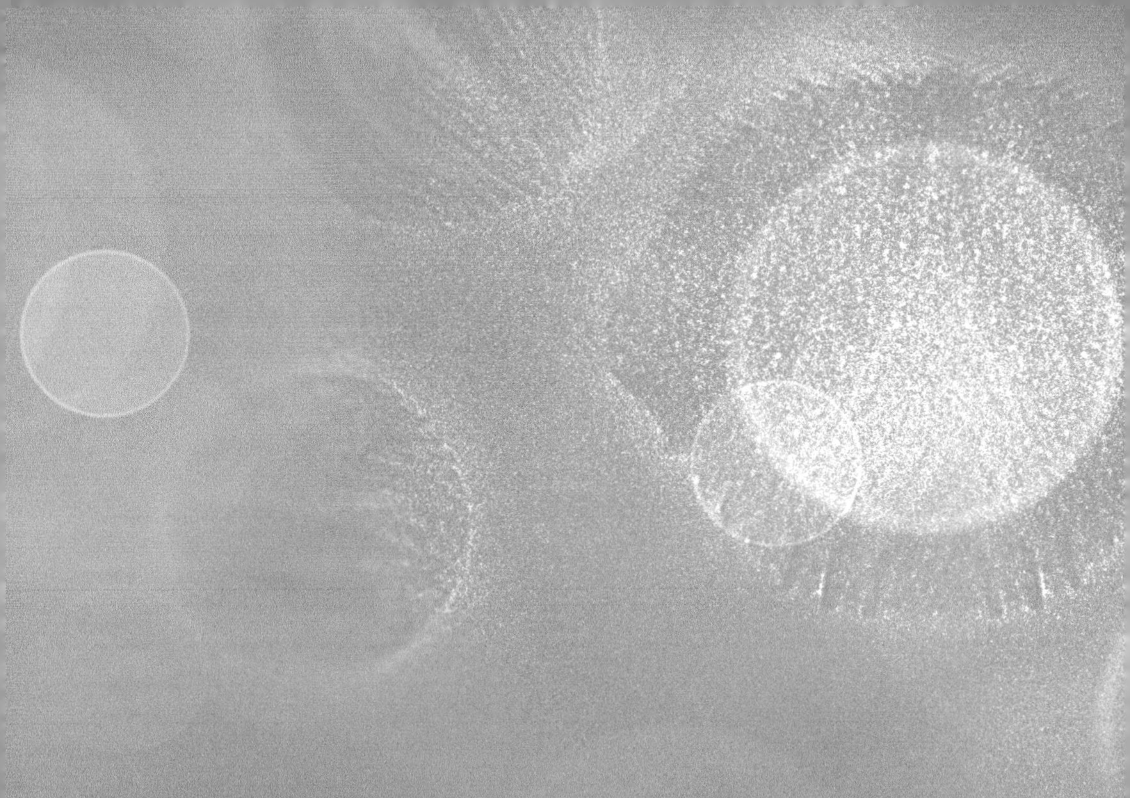

Nanotherapeutics for Infectious Diseases

Nanotherapeutics for Infectious Diseases

edited by
Íris Luz Batalha

Jenny Stanford
Publishing

Published by

Jenny Stanford Publishing Pte. Ltd.
101 Thomson Road
#06-01, United Square
Singapore 307591

Email: editorial@jennystanford.com
Web: www.jennystanford.com

British Library Cataloguing-in-Publication Data
A catalogue record for this book is available from the British Library.

Nanotherapeutics for Infectious Diseases

Microscopy image on the cover courtesy of Dr Pablo Castro-Hartmann, Cambridge Pharmaceutical Cryo-EM Consortium, Thermo Fisher Scientific, Cambridge, UK.

ISBN 978-981-5129-36-6 (Hardcover)
ISBN 978-1-003-56678-6 (eBook)

To my mom and mentor,

Fátima Luz,

who taught me how to dream

Contents

Preface

Small-molecule drugs (<900 Da), such as chemotherapeutic agents, antibiotics, and corticosteroids, have played a pivotal role in the development of modern medicine, dominating the pharmaceutical industry throughout the 19th and 20th centuries. Yet, a vast majority of preclinical small-molecule candidates exhibit low solubility and poor bioavailability, limiting their effectiveness. By the 1990s, considerable efforts were put into the development of nanoparticle-based delivery systems, which demonstrated remarkable potential in enhancing drug solubility and permeability, leading to improved pharmacokinetics and reduced toxicity but limited therapeutic benefits. As the therapeutic landscape evolved to include other modalities, such as nucleic acids, peptides, antibodies, and other proteins, the nanomedicine field had to adapt and address new challenges, particularly related to stability, immunogenicity, *in vivo* delivery, and crossing/avoidance of a variety of biological barriers.

Despite intensive academic research, the number of nanomedicines available to patients remained drastically below the initial projections, largely due to the use of a "one-size-fits-all" approach toward drug delivery. It is now known that this lack of clinical progress was influenced by the translational gap between animal preclinical studies and human clinical trials as well as heterogeneity within patients and disease states. Moreover, the complex and innovative nature of nanomedicines introduced additional manufacturing and regulatory hurdles as well as financial risks, further complicating the process of translation.

The advent of the COVID-19 pandemic has changed the face of nanomedicine and vice versa, with the first-ever mRNA vaccines being brought to market. This was only possible due to previous developments in lipid nanoparticle (LNP) technology for nucleic acid delivery, such as the one seen with the approval of Patisiran (Onpattro)—the first LNP-based siRNA drug—in 2018. These advancements have effectively paved the way for renewed investment in the field, with several hundreds of nanomedicines currently undergoing clinical testing.

With the halt in the development of effective antimicrobial drugs and the alarming rise of antimicrobial resistance, the need for alternative approaches to infection control has never been greater. In addition to providing a means to deliver combination therapies in a controlled and targeted manner, nanomaterials can have intrinsic therapeutic, imaging, and diagnostic properties. Contrary to small-molecule antimicrobials, nanotherapeutics have the potential to effectively eradicate pathogens by simultaneously using multiple modes of action, greatly reducing the emergence of drug-resistant strains.

If the nanomedicine field faced a certain degree of skepticism, the last 5–10 years have shown that it can bring novel clinical solutions and tangible medical breakthroughs, making a true difference in public health. It is certain, however, that collaboration between different stakeholders, including academia, industry, healthcare providers, regulatory agencies, patient associations, and policymakers, is a must for success. Advancements in synthetic chemistry and analytical methods have enabled the creation of sophisticated nanoparticle designs that incorporate bio-responsive and targeting moieties to overcome biological barriers in specific patient populations. Amidst the rapid rise of machine learning and artificial intelligence, large collections of nanomedicine data are expected to be generated and used to establish design-function relationship trends—that can help build platform technologies instead of universal solutions—and move a growing number of nanomedicines into clinical translation. In combination with other key enabling technologies, there are no bounds to the transformative impact nanomedicine can have on patients' lives.

Íris Luz Batalha

Autumn 2024

Chapter 1

Nanomedicine: A Historical Perspective

William Etheridge,[a] Badri Parshad,[b] Ishtiaq Ahmed,[a] Cristina Gonzalez-Lopez,[a] and Ljiljana Fruk[a]
[a]*Department of Chemical Engineering and Biotechnology, University of Cambridge, Philippa Fawcett Dr, Cambridge CB3 0AS, UK*
[b]*Wellman Center for Photomedicine, Massachusetts General Hospital, Harvard Medical School, Charlestown, MA 02129, USA*
lf389@cam.ac.uk

Although it is a very wild idea, it would be interesting in surgery if you could swallow the surgeon.

—Richard Feynman, 1959

1.1 Introduction

Throughout history, humans have relied upon macroscopic observations of symptoms for the diagnosis and assessment of the efficacy of plant- and animal-based therapeutics. As our scientific understanding of disease progressed, so did our grasp of underlying biological processes that often occur at the nanoscale—the scale

Nanotherapeutics for Infectious Diseases
Edited by Íris Luz Batalha

ISBN 978-981-5129-36-6 (Hardcover), 978-1-003-56678-6 (eBook)
www.jennystanford.com

of proteins and DNA, allowing for the creation of nanoscale tools tackling both diagnostics and therapy.

The term nanoscience refers to the study of structures at the nanoscale, usually from 1 to 100 nm, although some systems such as liposomes or polymeric nanocarriers can be larger than 100 nm. Considering that 1 nm is only 10 times larger than the size of an atom, the development of the field has largely been driven by the advent of microscopy and spectroscopy techniques capable of resolving such small lengths. This does not mean, however, that humans have not inadvertently used nanostructures to their benefit in the absence of understanding the science behind their unique properties. For example, centuries ago, master glassblowers produced colored glasses doped with noble metals to give them special hues and properties. Unknown to them, they often produced nanoparticles in the process, which resulted in unusual interaction with light. A famous example of Roman glassblowing, the Lycurgus Cup dating from the fourth century AD on display in The British Museum, is made of dichromic glass that appears green under reflected and red under transmitted light. This optical effect is due to the presence of gold–silver alloy nanoparticles embedded within the glass that scatter and absorb different wavelengths of visible light. Various other examples ranging from Mayan blue—a pigment fabricated by the Mayas in the early first millennium AD by mixing organic dyes with nanosized clays—to Damascus swords doped with carbon nanotubes have been shown to contain nanostructures [1, 2].

However, nanotechnology established itself as a field when we learned to apply scientific knowledge and method to manufacture and manipulate nanostructures that can later be used for various applications. The emergence of the concept of nanotechnology is often attributed to Richard Feynman and his famous lecture "There's Plenty of Room at the Bottom" given at Caltech in 1959, although the term "nanotechnology" was first coined 15 years later by Japanese scientist Norio Taniguchi [3].

Feynman discussed the possibilities of creating tiny machines that could create even smaller machines and thus manipulate material on the atomic scale, paving the way for finer microscopes, electronics, and computers. In the years that followed, some of his initial ideas became reality, and scientists developed not only a range of nanomaterials but also continuously expanded their applications.

Although nanomaterials first found application in electronics and material design, nanomedicine soon emerged as a discipline that utilizes these structures as diagnostics and therapeutic tools. The term "nanomedicine" was first coined in 1999 by American scientist Robert A. Freitas Jr in his textbook "Nanomedicine: Basic Capabilities" [4]. Although the field has gained traction over the last 30 years, numerous key discoveries were enabling its development in previous decades. For example, in 1912, metal colloid solutions were first studied using Zsigmondy's immersion microscope, enabling the rapid development of colloid science, principles of which are largely employed in nanomedicine today. The discovery of the structure of DNA and RNA in the 1950s allowed for a better understanding of the role of genetics in the development of diseases (Fig. 1.1). However, arguably the most impactful discovery was the invention of the scanning tunnelling microscope (STM) at the IBM in 1981; the first microscope capable of real-time visualization of nanometer-sized structures and features.

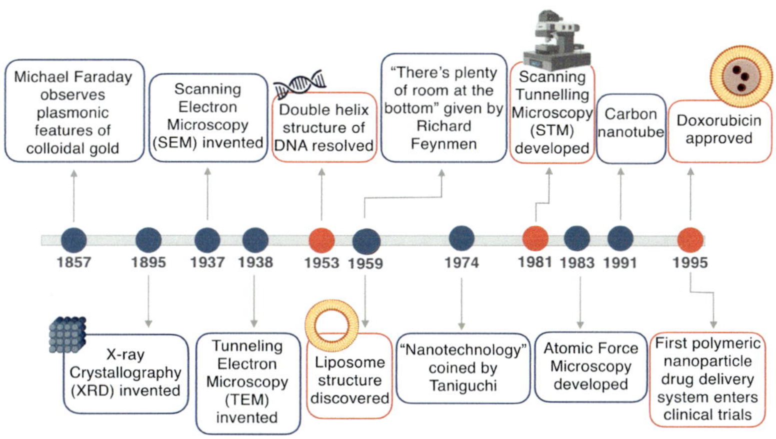

Figure 1.1 Timeline of some landmark discoveries in nanoscience (blue) and nanomedicine (red), which led to further developments in both fields.

Modern-day nanomedicine is concerned with the development of novel nanostructures for applications in targeted drug delivery, disease detection, and medical imaging. Targeted drug delivery draws from Feynman's idea that one can swallow the doctor who can perform targeted surgery and aims to design novel nanocarriers to transport and deliver therapeutic agents specifically to the diseased cells. For example, nanoscale drug carriers can be formulated to be sensitive to physiological changes such as pH or increased level of a particular enzyme, or functionalized with targeting molecules such as antibodies, to differentiate disease cells (*e.g.,* cancer cells) from healthy cells.

More than 70 therapeutic formulations based on the use of nanomaterials have been approved by regulatory agencies and used in the clinic and, judging from the number of published papers and ongoing clinical trials, there are many more to come. Most of the nanomaterials used in the clinic to date are lipid-, polymer-, or protein-based structures, with only a few inorganic materials being used as contrast agents rather than drug delivery systems. However, inorganic materials such as gold and silver nanoparticles have been employed as excellent probes for disease sensing and extensively used as components of biosensors. Functionalized with antibodies, they are an essential part of lateral flow tests for pathogens and metabolites, as their plasmonic properties make them sensitive to surface changes resulting from molecular binding. Many examples will be given throughout this book on the use of nanomaterials in sensing and therapy. Within this chapter, we will focus on the historical aspects of nanotechnology and nanomedicine, to set the stage for both the exploration of current developments and the future outlook discussed in later chapters.

1.2 Short History of Nano-biosensors

Historically, nanomaterials were first knowingly applied within biosensors, devices aimed to provide information on biochemical interactions and, in such a way, aid the assessment of the health/disease states of a patient. The first biosensor is largely accepted to be an electrochemical glucose sensor reported by Leyland Clarke in 1962 [5]. Clarke and his team immobilized the enzyme glucose oxidase onto an electrode, which could then be used as a glucose biosensor based on the oxidation of glucose in the presence of molecular oxygen.

Developments in the field of electrochemical biosensing of enzymatic activity progressed incrementally over the next decades, with a significant technological advance achieved with the development of nanomaterials, in particular plasmonic and magnetic nanoparticles, carbon nanotubes, and quantum dots. The unique properties of these nanoscale materials are today employed to design sensing platforms for pathogens and biologically relevant molecules in a sensitive and rapid fashion. Nanomaterials can be used both as enhancers of the readout signal or as an integral part of the transducers, which convert a bio-interaction event into a readable signal. For example, gold nanoparticles have been a very useful signal-enhancing element enabling an optical readout (often by eye) of antibody–antigen interactions, whereas conductive carbon nanotubes have often been employed to build new-generation sensitive electrochemical transducers. The next sections will introduce some of the most widely used nanomaterials used for biosensor design.

1.2.1 Plasmonic Nanoparticles in Biosensor Design

Nanoparticles such as gold and silver have a large size-dependent absorption cross section giving rise to colors not observed in the bulk material. For example, gold presents various shades of red for particle sizes between 10 and 100 nm and appears blue when larger than 100 nm. The aggregation of plasmonic nanoparticles into larger clusters, nanoparticles being brought close to a reflective conductive surface, or a change in local refractive index can all cause a change in the intensity and wavelength of the surface plasmon resonance (SPR) and be detected either by eye or with simple optical readout systems.

Although today it is hard to imagine the field of nano-biosensors without plasmonic nanoparticles, it was not until the 19th century that scientists began to understand the origin of their optical effects. The first reported study on the distinct behavior of noble metal nanomaterials was conducted by Michael Faraday in 1857 and focused on the unique colors of colloidal gold, which was ruby red as opposed to the golden yellow of the bulk gold. The explanation of the red color of gold nanoparticles was given by Gustav Mie in 1908 as he provided a qualitative solution to Maxwell's equations. The Mie solution explored the interaction between the electric field of a wave and the surface electrons of the sphere representing a gold nanoparticle. In short, when a photon comes into proximity to the sphere, the electric field of the photon causes the sphere's free electrons to oscillate at the same frequency and the photon is absorbed. The free electrons, or plasmons, on the surface of a metallic nano-sphere have a unique resonance frequency in which light of this frequency will be strongly coupled to the electrons causing them to oscillate, a phenomenon called surface plasmon resonance (SPR).

Faraday's ruby red colloidal gold, prepared by the reduction of chloroauric acid, has the maximum absorption of 520 nm and, therefore, when irradiated with ambient light, blue wavelengths are absorbed and the solution appears red. Faraday could not determine the size of his nanoparticles at the time, but today, even by only looking at the color of the solution still on display in the Royal Society, we can estimate that they are around 10–50 nm. Larger (50–100 nm) nanoparticles would have been purple or blue due to the size dependence of SPR frequency.

SPR is a crucial property of noble metal nanoparticles that enabled their widespread use in nano-sensor design. Studies throughout the 20th and 21st centuries found that the intensity and frequency of the SPR bands in metallic nanoparticles could be tuned by the shape and size of the particles, proximity to other particles or a conductive surface, or the local refractive index of the surface. This meant that any interactions of molecules attached to the surface of plasmonic nanoparticles that lead to changes in size or shape will result in altered plasmons interaction with light and, ultimately, differences in the optical signal. Making use of color change, Chad Mirkin and collaborators designed the first gold-based nano-biosensor for the detection of DNA mutations in 1996 [6]. Single-stranded DNAs were immobilized onto gold nanoparticles, and the sequences were chosen in such a way that the hybridization with target DNA led to the controlled aggregation of gold nanoparticles and detectable color change (Fig. 1.2). Although originally used to detect mutations within DNA, the strategy was quickly adapted to the detection of unknown bacterial and viral DNA, biomolecular barcoding [7], and even as a CRISPR system probe [8].

Soon after the first demonstration of the plasmonic gold nano-sensing platform, Bauer et al. proposed a system for the detection of various biomarkers consisting of both a metallic mirror and plasmonic nanoparticles functionalized with a linker that binds selectively to a target analyte [9]. When the analyte is present, it provides a bridge between the two linkers bringing the nanoparticles close to the surface. The oscillating plasmons in particles are coupled to the electrons in the mirror, causing an increase in the absorption cross section that can be easily detected. By changing the linkers, Bauer demonstrates the detection of antibodies, binding proteins, and DNA.

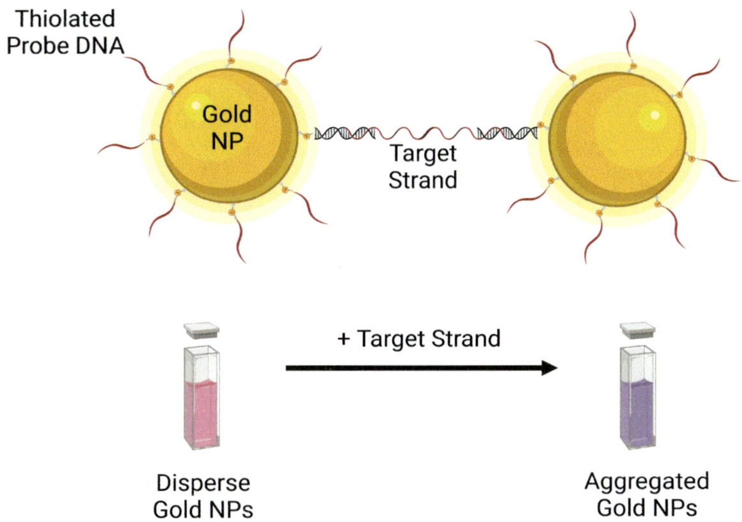

Figure 1.2 DNA-driven nanoparticle aggregation through hybridization is used to design a colorimetric detection system for a target strand. As gold nanoparticles aggregate in the presence of the target, their color will change from red to purple/blue due to the change in surface plasmons.

Often plasmonic nanoparticles are used to further advance the already existing biosensors by improving their detection limits and decreasing their production costs, such as in-home pregnancy testing. Pregnancy testing is based on the detection of the human chorionic gonadotropin (hCG) hormone in urine using a lateral flow assay. Initially, these assays employed hCG antibodies labeled with a fluorescent dye for the detection of the hormone in the sample. If the hormone is present, it binds to the fluorescent antibody, and the mixture then flows to the test line consisting of immobilized antibodies that can capture the hCG–antibody pair, which leads to the appearance of a red line. If hCG is not present in the sample, the labeled antibodies flow directly to the control line. These first test systems typically had a high false negative rate due to the limitations of the used fluorescent dyes. However, this issue was resolved in 2008, when Rojanathanes et al. demonstrated an *in vitro* biosensor for hCG based on the selective and controlled aggregation of gold nanoparticles, which then led to the development of antibody-coated gold nanoparticle tools beyond pregnancy test applications (Fig. 1.3) [10].

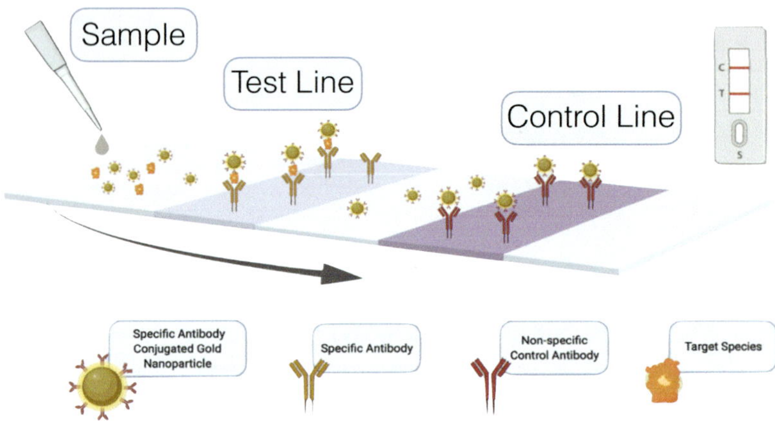

Figure 1.3 Lateral flow antibody test that uses plasmonic antibody-coated gold nanoparticles for the detection of biomolecules, hormones, or pathogens.

More recently, lateral flow tests have played a huge role in the COVID-19 pandemic, enabling the fast detection of the virus, and lots of effort went also into the design of gold nanoparticle–based systems for the fast detection of viral nucleic acids. Some of the examples have been driven by previous work on the detection of influenza virus or bacteria, which employed antibody-coated gold nanoparticles that assemble on viral or bacterial membranes resulting in color change due to the plasmon interaction of closely packed nanoparticles.

1.2.2 Carbon Nanotubes in Biosensor Design

A carbon nanotube (CNT) is another material that exhibits unique properties owing to its nanoscale size. This nanometer-scale cylindrical structure made entirely of carbon exhibits exceptional durability and unique conductive and optical properties [11]. Depending on the orientation of the atoms on the seam, CNTs can behave like metallic conductors or semiconductors. Optically, CNTs can be fluorescent and have excellent Raman scattering properties that may be tuned through size, seam orientation, and surface functionalization. Such optical and electrical properties of CNTs, as well as their large surface area, make them an appealing scaffold for protein and DNA biosensing, or the monitoring of enzymatic activity, mainly through electrochemical readout strategies.

CNTs were first reported in 1991 after the discovery of double-walled CNTs by a team of Japanese scientists led by Sumio Iijima [12]. The discovery of single-walled CNTs followed shortly after in 1993, with their benefits being widely recognized and resulting in a large number of publications describing CNT-based biosensors. The majority of the designs utilized the excellent conductive properties of CNTs, enhancing the electronic interaction between enzymes and electrodes within electrochemical sensors. The underlying design of these sensors did not differ that much from Clarke's landmark glucose electrode biosensor; however, the fact that the large surface area enabled the immobilization of a significant number of enzymes, together with the excellent conductive properties of the material, resulted in unprecedented sensitivity. The CNTs could also be "plugged" into the enzymes, giving electrons a direct, low-resistance route to the electrode, with numerous variations of this principle being utilized for the detection of metabolites, cancer biomarkers, and whole cells. [13].

Besides enzyme activity monitoring or the use of enzyme–CNT hybrids as biosensing elements, since 2007 CNTs have also been employed in sensitive electrode design for new-generation immunoassays [14]. More studies followed from this initial work, ultimately resulting in affordable FET (field effect transistor) and other electrochemical biosensors. In the meanwhile, CNT nano-sensors have been developed for specific applications in medicine through the sensitive electrochemical detection of biomarkers associated with cardiovascular diseases [15], infections, and fetal defects [16].

1.2.3 Other Nanomaterials Used in Biosensing

Although this section is non-exhaustive, it is meant to provide a snapshot of two more classes of materials that played a huge role in the development of powerful nano-sensors. For further reading on the topic, two excellent reviews by Mehrotra and Naresh et al. are suggested with the former giving an insight into areas aside from medicine in which these technologies are used [17, 18].

1.2.3.1 Quantum dots

Quantum dots are nanosized crystals of semiconductor materials such as cadmium sulfide, cadmium telluride, indium arsenide, and indium phosphide that exhibit unique optical and electrical properties owing to their small size. Quantum dots present optical features akin to single atoms due to the dominance of the quantum mechanical phenomena. Most remarkably, their interaction with light, absorption, and emission profiles depends on their size, providing us with a library of distinct, intense, and photobleaching-resistant colors. Since their simultaneous discovery in the United States and Russia during the 1980s, quantum dots have been used in the design of biosensors, due to their much higher extinction coefficient (brightness), resistance to photobleaching, and their ability to carry numerous reporter molecules on the surface. For example, they have been utilized in the design of a fluorescence resonance energy transfer (FRET) biosensor for the detection of *Helicobacter pylori's* DNA, a bacteria implicated in a range of gastric diseases, including cancer [19]. In the presence of the target strand, a complex is formed when a fluorescent probe comes in the proximity of the quantum dot, allowing for FRET to occur between the two species. FRET can be detected using a simple and affordable spectroscopic setup, making the strategy applicable to low-resource settings. Often due to their remarkable properties and high-intensity colors, quantum dots can decrease the cost of biosensing platforms and lead to simpler and more sensitive designs. An excellent review by Ma et al. summarizes the applications of quantum dots in biosensing platforms [20].

1.2.3.2 DNA and protein nanostructures

Techniques allowing the manipulation and construction of nanostructures using DNA and proteins have also mediated advances in biosensor design over the last 20 years. DNA origami, first demonstrated in 2006 by Paul Rothemund, is the process of creating two- and three-dimensional structures from long single-stranded DNA held together using complementary base pair interactions [21]. These scaffolds have since been shown to aid the design of surface-enhanced Ramen scattering (SERS) and fluorescence-based biosensors. For example, SERS sensitivity can be vastly improved by the presence of plasmonic nanoparticles due to the amplification of the local electromagnetic field excited by the SPR; however, this effect is largely dependent upon the proximity and arrangement of the nanoparticles on the surface. This requires assembly strategies that are difficult to achieve using conventional supramolecular chemistry. However, several groups have reported successful patterning of plasmonic nanostructures on surfaces using DNA origami scaffolds, which resulted in excellent enhancements in signal and sensitivity. Similar self-assembled DNA nanostructures were also used for the detection of antibodies by employing the attachment of antibody recognition elements and utilizing precisely controlled distances achievable with origami, to allow for detection based on fluorescence quenching [22].

Not only DNA but also protein nanostructures had a significant impact on the field of nanomedicine. The prime example is nanopore technology, which employs protein nanopore channels to sequence a DNA structure passing through. First conceived as an idea in the 1990s and first published in 1999, nanopore sequencing developed significantly in the past two decades into a commercial product launched by Oxford Nanopore Technology in 2014 [23]. Nanopore sequencing is an excellent example of nano-structuring, which enables further developments in molecular biology and provides powerful tools for next-generation nanomedical interventions.

A wide range of hybrid strategies involving different materials and readout systems in development allows us to design more sensitive and cheaper biosensing platforms, which will have implications for early diagnostics and monitoring of diseases. These developments are possible as fewer obstacles need to be overcome when nanomaterials are employed in devices for external use (such as current biosensors) when compared to the requirements and regulations for *in vivo* applications. The design of *in vivo* biosensors that are ingested or implanted—a field witnessing lots of development at the moment—requires a new set of design and validation guidelines before reaching the commercial space. But these guidelines and strategies will benefit from the last decade's advances that have been made in the field of drug delivery using nanocarriers.

1.3 Drug Nanocarriers in Cancer Therapy

The application of nanomaterials in drug delivery and formulation aims to improve the selectivity and specificity of commonly used small molecule drugs, as well as therapeutic antibodies and nucleic acids, but also exploits the intrinsic properties of the nanomaterial to achieve enhanced therapeutic activity. Nanocarriers—nanostructured materials that help deliver drugs—have initially been focused on the formulation and delivery of chemotherapeutics, but it should be noted that nanomaterials themselves can act as therapeutic agents either through their antipathogenic properties or the ability to heat up and release damaging reactive oxygen species. In addition to offering protection to the drug from biological removal machinery (enzymes, white blood cells), thereby extending its half-life in the bloodstream, nanocarriers can help solubilize hydrophobic drugs, minimize the amount of drug needed by improving the accumulation in diseased tissue, and offer the possibility of more precise targeting either through size/charge effect or by providing a surface to which various targeting molecules can be attached.

Almost every nanostructured material produced was at some point exploited for drug delivery or treatment. However, considering the complexity of the biological systems, it is not surprising that only a few classes are usually taken from the lab bench to the clinic (Fig. 1.4). The following section will give an overview of developments of the most commonly used drug nanocarriers.

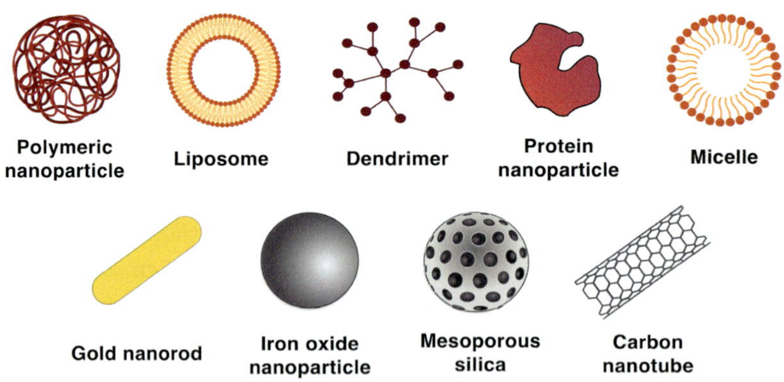

Figure 1.4 Types of the most used nanostructures employed in drug nanocarrier design.

1.3.1 Liposomes

Before the 1950s, drugs and active pharmaceuticals were formulated as capsules or tablets using various additives and released immediately upon contact with the biological medium with limited control over the drug release kinetics. In 1952, a sustained release formulation was reported by Smith, Klein, and French that allowed the controlled release of the drug Dexedrine (dextroamphetamine, used for the treatment of attention deficit hyperactivity disorder and narcolepsy) and showed efficacy over 12 hours. Based on the wax-fat-coated tablets, the technology allowed sustained release kinetics at a predetermined rate achieved by a simple change in the thickness of the wax-fat coating [24]. A few years later, in 1955, Horst Jatzkewitz reported the preparation of one of the first polymer-drug conjugates based on polyvinylpyrrolidone polymer linked with the drug mescaline (a protoalkaloid hallucinogenic drug) through a dipeptide spacer [25]. Within the same decade, Mathe et al. pioneered the conjugation of drugs to immunoglobulins (antibody-drug conjugates; ADCs), making use of their ability to target specific cells and setting the stage for targeted delivery [26]. In the 1960s, lipid vesicles, later known as liposomes, were discovered and described by Alec D. Bangham at the Babraham Institute in Cambridge and became the first nanoparticle-based drug delivery system approved in the mid-1990s [27, 28].

Liposomes are usually composed of natural, synthetic, or semi-synthetic phospholipids, which can carry both hydrophobic and hydrophilic drugs. Semi-synthetic phospholipids are particularly interesting as they offer the possibility of including additional functionalities such as polyethylene glycol (PEG), which improves blood circulation time, as well as stability and solubility of the formulations. The concept of PEG addition or PEGylation was introduced by Frank Davis in the late 1960s and aimed to enhance the stability and systemic circulation of PEG-attached protein drugs developed around that time. PEGylation soon became one of the most widely used strategies to stabilize various biomedically useful molecules and materials.

Indeed, the first U.S. Food and Drug Administration (FDA)-approved drug nanocarrier "Doxil," approved in 1995, was a PEGylated liposome carrying the anticancer drug doxorubicin (Fig. 1.5).

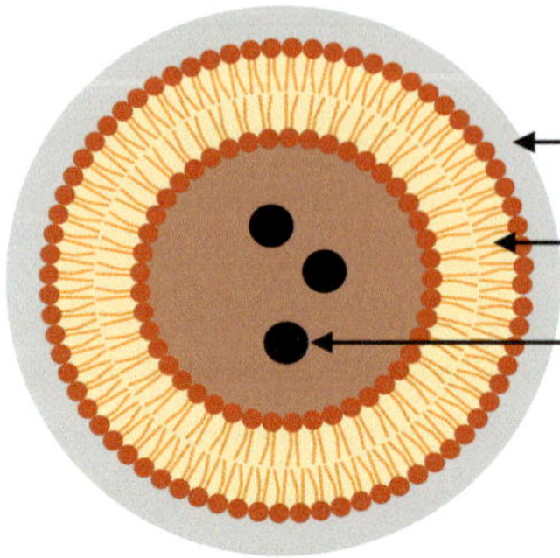

Figure 1.5 Structure of Doxil, the first nanoparticle (liposome) drug delivery system approved for medical use in 1995.

Doxil was initially produced by Liposome Technology Inc. (LTI) which was later bought by ALZA Corporation. In 1968, when it was founded, ALZA Corporation was unique as it did not focus on the development of new drugs but aimed to improve the therapy and efficacy by targeting the drugs to the right place and at the right time (targeted and controlled drug delivery). The development of targeted drug delivery systems is today one of the main objectives of nanomedical strategies.

Initially, Doxil was approved for Kaposi's sarcoma, ovarian cancer, and metastatic cancer, as well as for multiple myeloma, and has resulted in a 4- to 16-fold increase in drug levels in comparison to bare doxorubicin. After the approval of Doxil in 1995, several other liposome-based drug delivery systems, including DepoCyt® (cytarabine, 1996), DauniXome® (daunorubicin, 1996), Marqibo® (vincristine, 2012), and Onivyde® (irinotecan, 2012), were reported to improve efficacy, reduce toxicity, and extend the lifespan of cancer patients. In the meanwhile, liposomal formulations are still considered a benchmark for other delivery systems, despite suffering from some drawbacks, such as high production cost, concentration-dependent self-assembly, and leakage of encapsulated molecules.

1.3.2 Micelles

Despite the drawbacks, self-assembled structures such as liposomes are, nevertheless, desirable drug delivery systems due to the similarities to our biological structures and ease of manufacturing. Similar to liposomes, self-assembled micelles—mono-layered supramolecular nanostructures with a hydrophobic core and hydrophilic surface—have also been investigated as drug carriers for water-insoluble drugs [29, 30]. Polymeric and amphiphilic (containing hydrophilic and hydrophobic units) micelles are more suitable systems for the encapsulation of hydrophobic drugs since their hydrophobic core can accommodate the drugs, whereas hydrophilic shells increase the stability in biological fluids and improve circulation time [31, 32]. In the last few years, a new generation of polymeric micelles, which accomplish high loading and drug retention while maintaining prolonged circulation after intravenous administration, have been developed and are suitable for both active and passive drug targeting. The polymers are usually made up of biodegradable or biocompatible hydrophobic components linked to the hydrophilic shell of the micelles. One such clinically interesting polymeric micelle—NK012 (size ~20 nm)—was developed in 2006 by the self-assembly of a block copolymer composed of PEG and polyglutamate and was recently linked to the chemotherapeutic drug SN-38. The ester linkage between SN-38 and glutamic acid is gradually hydrolyzed under physiologic conditions, causing the release of SN-38 in various types of cancer (lung cancer, pancreatic cancer, renal cancer, and gastric cancer) and is currently undergoing Phase II clinical trials [33–35]. Some of the micelle-formulated drugs that are presently in clinical trials for cancer therapy are shown in Table 1.1.

Table 1.1 Examples of clinically relevant micelle-based drug nanocarriers

Micelles	Loaded drug	Applications	Status	Ref.
NK105	Paclitaxel	Metastatic or recurrent breast cancer	Phase III	[36]
Genexol-PM	Paclitaxel	Breast cancer and NSCLC	Approved in Korea	[37]
NK012	SN38	Small cell lung cancer, metastatic breast cancer, and refractory solid tumors	Phase II	[33]
NK911	Doxorubicin	Solid tumors	Phase II	[38]
SP1049-C	Doxorubicin	Advanced adenocarcinoma of the esophagus	Phase II	[39]
BIND-014	Docetaxel	Metastatic prostate cancer	Phase II	[40]

Note: NSCLC is non-small-cell lung carcinoma.

As can be seen from the large number of papers on micelles, but a low number of micelle formulations in clinical trials, some difficulties still need to be addressed in terms of the assessment of the *in vivo* toxicological profile and general bioavailability. But considering that this has become a very active research field, more developments are expected in the next decade.

1.3.3 Protein Nanostructures

Some of the challenges encountered by semi-synthetic and synthetic nanocarriers can be overcome by protein nanostructures, particularly those made of proteins commonly present in biological systems such as albumin. Abraxane, an albumin-bound paclitaxel nano-formulation, was the first protein nanocarrier approved by FDA in 2005 and is used as an efficient treatment for breast cancer [41]. It contains paclitaxel and human albumin in the ratio of 1:9 (wt/wt), where 6 or 7 molecules of paclitaxel can bind non-covalently to a single molecule of albumin, forming a paclitaxel–albumin primary complex of 4–14 nm in size. These protein–drug conjugates further aggregate to form a paclitaxel–albumin particle of around 130 nm in diameter. Moreover, Abraxane can be used for the treatment of head, neck, pancreatic, and melanoma cancers [42].

In the meanwhile, there have been many other albumin-based nanoparticles approved for the treatment of different types of diseases, including diabetes (Levemir® and Victoza®), rheumatoid arthritis (Ozoralizumab®), and hepatitis infection (Albuferon®) [43].

1.3.4 Inorganic Nanomaterials

Although the majority of FDA-approved nanoformulations used for cancer treatment are made using liposomes, polymeric materials, and protein nanocarriers, some inorganic nanoparticles have slowly entered the field of drug delivery (Table 1.2). Despite being extensively exploited in biosensors, the major obstacle to their use for drug delivery is their toxicological profile. Some nanomaterials, such as gold nanoparticles or carbon nanotubes, have biomedically useful properties, but there are contradictory reports on their toxicological profiles, which limits their applications. Other inorganic nanoparticles, such as iron oxide nanostructures, have been more successfully applied *in vivo*. For example, NanoTherm, a 15 nm superparamagnetic iron oxide nanoparticle formulation coated with aminosilane, was approved for localized ablation of glioblastoma in 2011 in Europe, and for the selective ablation of prostate cancer in 2018 in the United States [44]. These approvals were preceded by extensive studies and use of iron oxides as contrast agents, which provided more information on their toxicological profile and enabled the translation into drug delivery applications.

NanoTherm's action is based on the effect of hyperthermia, whereby after the injection into the tumor, the nanoparticles are heated to produce a locally elevated temperature of 40–45°C by applying an alternate magnetic field, which results in cell death. The major advantage of this strategy is that the external magnetic field can be applied at the localized tumor site to avoid harming the healthy cells, which might also accumulate the magnetic nanoparticles. Such thermal ablation of tumor cells is also the principle behind AuroLase, which entered clinical trials in 2014 for the treatment of metastatic lung tumors. AuroLase is composed of a PEGylated gold-silica nanoshell, which is optically active in the NIR (near infrared) region owing to the SPR of gold nanostructures [45]. The absorption of NIR light during the therapy produces rapid localized heat to selectively kill tumor cells, avoiding, at the same time, the systemic side effects that are normally associated with traditional, non-selective anti-tumor therapy.

There are a few other inorganic systems based on iron oxide or gold nanoparticles in various stages of clinical development, mainly relying on hyperthermia induced by a magnetic field or irradiation, but some also carry additional drug cargo (Table 1.2). An example is CYT-6091, a 27 nm PEGylated gold nanoparticle functionalized with recombinant human tumor necrosis factor (rhTNF), a vascular disrupting agent that preferentially and specifically targets tumors [46]. CYT-6091 was demonstrated to effectively target tumor tissue upon intravenous administration. As TNF is functionalized on the surface of nanoparticles, the nanoparticle system takes advantage of both the enhanced permeability and retention (EPR) effect and the TNF's effect on blood vessels, which improves its biodistribution, specifically within the tumor microenvironment.

Table 1.2 Therapeutic inorganic nanoparticles approved for clinical use and in various stages of clinical development

Name	Composition	Applications	Status	Ref.
NanoTherm	Fe_2O_3 NPs	Solid tumor	Approved	[44]
Auritol (CYT-21001)	PEGylated colloidal AuNPs	Solid tumor	Preclinical	[47]
Aurimune (CYT-6091)	Colloidal Au	Solid tumor	Phase II	[46]
NBTXR3	Hafnium oxide NPs	Solid tumor	Approved	[48]
Actinium-225-HuM195	Actinium-225-Labeled Humanized Anti-CD33 Monoclonal Antibody HuM195	Leukemia	Phase I	[48]

Other, more recently approved therapeutic strategies rely on radiation-induced cell death and the use of inorganic oxides or radioactive materials as radiation relays. For example, hafnium oxide nanoparticles (NBTXR3) were approved in Europe in 2019 and used for the treatment of squamous cell carcinoma [49]. When hafnium oxide nanoparticles are irradiated by X-rays in an aqueous and oxygen-rich environment, highly toxic reactive oxygen species (ROS) are generated, which induce cancer cell death [50]. Several other nanostructures, which are either made of active materials or doped with radioactive species, such as actinium-labeled antibodies, are currently in clinical trials. Multiple other inorganic nanoparticles are in clinical trials for cancer treatment [51]. However, the main challenge of the limited understanding of long-term effects and accumulation in tissue remains. Until now the lack of data on these issues has hindered the broader clinical translation of most inorganic nanostructures, and it remains to be seen if future studies will lead to the application of other materials besides the commonly used gold and iron oxide.

1.3.5 Nanogels for Drug Delivery Applications

Unlike inorganic nanomaterials, nanogels are far more promising systems for therapeutic applications. These three-dimensional, cross-linked polymeric networks, characterized by mechanical properties similar to the extracellular matrix, have a high capacity for water retention. They are also characterized by high porosity and a high degree of swelling, which makes them suitable for the encapsulation and loading of therapeutics, ranging from small drug molecules to nucleic acids and large proteins.

The term "NanoGel" was introduced in 1999 by Vinogradov et al. to represent particles of a hydrophilic polymeric network, which were synthesized by the cross-linking of PEG and polyethyleneimine [52]. These nanogels were developed as nanocarriers for the delivery of antisense oligonucleotides specific to the human *mdr1* gene and were able to inhibit the expression of the P-glycoprotein efflux pump in multidrug-resistant cell lines. The first physically cross-linked gel nanoparticles, reported in 1993 by Akiyoshi et al., were produced by the self-assembly of cholesterol-functionalized polysaccharides (pullulan) in water and used as carriers for various hydrophobic compounds [53]. The last two decades have seen the design of a large variety of polymeric systems using versatile cross-linking strategies [54, 55]. In one of the most recent studies, pH-responsive hyaluronic acid nanogels loaded with cisplatin and doxorubicin were used as a drug delivery system for osteosarcoma. The nanogel exhibited significant stability at physiological pH, while its size and swelling behavior changed at a lower pH and caused the release of the encapsulated drug at pH 6.8 and 5.5.

The ability of nanogels to be embedded within soft and hard tissues and enable the controlled release of drugs is of significant advantage for the treatment of diseases that require prolonged drug administration. New developments in the field are moving toward the design of nanogels sensitive to internal, tissue-specific molecules such as enzymes, or external stimuli such as ultrasound or magnetic field, to allow for more control over drug release. The major advantage of nanogels, such as in the case of hyaluronic acid-based gels, lies in their ability to be degraded and removed by intrinsic biomolecular machinery.

1.4 Nanotherapeutics for Infectious Diseases

As we have seen, both drug delivery systems and therapeutic nanomaterials (nanotherapeutics) have been employed to deal with the limitations of cancer chemotherapeutics. The same is true for the applications of nanomaterials in the treatment of infectious diseases, although in this field, nanotherapeutics have probably been explored more commonly than drug nanocarriers for cancer therapy.

In general, nanotherapeutics have offered solutions and new possibilities to improve the efficacy and safety aspects of conventional therapeutic systems used to deal with pathogens, particularly considering the emergence of antibiotic-resistant bacteria. Various metal (Ag, Au, Cu, Pt, Fe), metal oxide (Ag_2O, Fe_2O_3, TiO_2, SiO_2, ZnO, CuO), and hybrid organic (lipid, protein, polymer, dendrimer) nanostructures have been explored for their antimicrobial/antiviral properties over the years and became even more relevant during the recent COVID-19 pandemic. In the section that follows, we will give a short historic overview of the most important developments.

1.4.1 Antibacterial Nanotherapeutics

Antibacterial activity describes compounds capable of killing bacteria or slowing down their growth. History has seen a number of different approaches to dealing with bacterial infections, many of which have been extensively used by Indian ayurvedic Bhasma and traditional Chinese medicine [56]. However, despite some excellent early medical developments, the dark Middle age in Europe led to the development of strange and often entirely counterintuitive strategies unfounded in any empirical observation. For example, inspired by an American First Nations custom, tobacco smoke enemas were administered by medical practitioners to treat everything from colds and headaches to cholera and typhoid. Luckily, scattered islands of wisdom and scientific thought could be found where not everything was based on alchemy but rather empirical science, and plant extracts were employed as antibacterial and antiviral agents. Although the true revolution in terms of antibacterial compounds came with the discovery of penicillin by Alexander Fleming in 1928, it is a lesser-known fact that doctors in ancient Egypt were already using moldy bread to treat infections and formulated antiviral tinctures with plant extracts.

Although civilizations relied on traditional medicine for centuries, the first half of the 20th Century brought strategies for rational drug discovery. In the early 1900s, German scientist Paul Ehrlich, the father of antimicrobial chemotherapy, discovered chemical compounds capable of killing infectious microbes without harming patients. Arsphenamine, an arsenic-based chemical discovered by Paul Ehrlich and his team in 1909, proved to be an effective treatment for syphilis and was probably the first modern antimicrobial agent. Ehrlich was also the first to use the term "chemotherapy" to describe the use of chemicals to treat diseases.

In the 1930s, German biochemist Gerhard Domagk developed the first commercially available systemically active sulfonamide drug—Prontosil. Although the beginning of the "antibiotics era" is usually associated with Paul Ehrlich, the landmark contribution was given by Alexander Fleming who discovered bacteria-killing mold and described its active agent, penicillin (Fig. 1.6). However, what Fleming also predicted during his Nobel Prize speech in 1945 was that bacteria may become resistant to antibiotics, the challenge which we are currently facing. Different strategies have been explored as alternatives to molecular antibiotics, and nanomaterials have been considered to be useful alternatives, either to be employed directly as active agents or used as carriers of antibacterial compounds. Particularly interesting are nanostructured materials made of or in combination with silver.

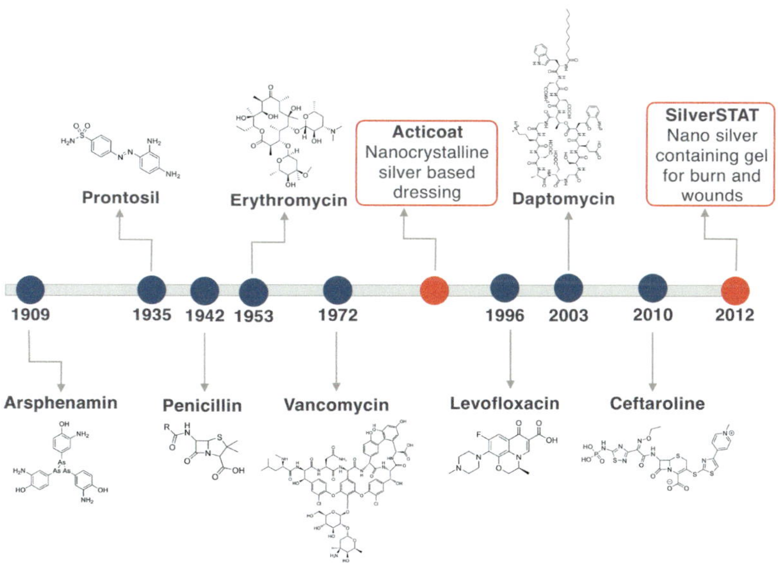

Figure 1.6 Timeline of some landmark developments in the field of antibacterial therapeutics. Nano-based strategies are marked in red.

Silver in different forms has been historically used as an antimicrobial agent by itself or combined with other materials. The antimicrobial properties of silver ions have been known for centuries, and ancient Greeks used them both for stomach pain and wound healing. When first settlers came to North America, it was common practice to drop silver coins into the water and milk transport vessels to prevent spoilage caused by bacteria. In the early 1800s, silver wires were used to suture surgical wounds, and during World War I, silver leaves were applied to the wounds of the soldiers to prevent infections and promote healing. In the 1920s, the FDA approved the use of silver solution as an antibacterial agent. Since then, the use of silver has established itself as a well-known and effective treatment strategy for the prevention of infections and has been formulated into commercial products ranging from surgical and food-handling tools, clothing, cosmetics, dental products, catheters, and wound dressings [57, 58]. For example, both silver-impregnated dressings and silver-containing biocomposites such as Bactigras (Smith & Nephew), PolyMem Silver (Aspen), Aquacel (ConvaTec), and Tegaderm (3M) have been approved by the FDA for wound-dressing applications [59].

Besides silver, copper has also been historically used to treat infections. Egyptians and Indians used copper vessels to sanitize drinking water to treat wounds and help digestion. Recent studies explored the antibacterial properties of water stored in copper pots and experimentally confirmed that there was a good rationale behind those early practices [60]. Namely, after 16 h of storage time, no bacteria such as *Salmonella* or *Escherichia coli* were found in water, indicating improved antibacterial action of copper pot–stored water. Interestingly, there have been reports that during the two 19th-century European cholera epidemics, the outbreak of cholera was suppressed among workers in the copper industry [61]. Despite the proven benefits, the use of copper in modern medical practices is limited due to the lack of understanding of the mechanism of antimicrobial activity and the general toxicity of copper ions.

Both silver and copper nanostructures have been prepared, with silver nanoparticles being used in numerous antibacterial products. However, the specific mechanism of their activity has not been completely elucidated, but it was shown that there is a range of potential routes that interfere with respiration, cell permeation, and growth of bacteria (Fig. 1.7). For example, it was shown that silver nanoparticles interfere with cell membrane components and can react with sulfur-mediated proteins and phosphorus-containing biomolecules such as DNA [62, 63].

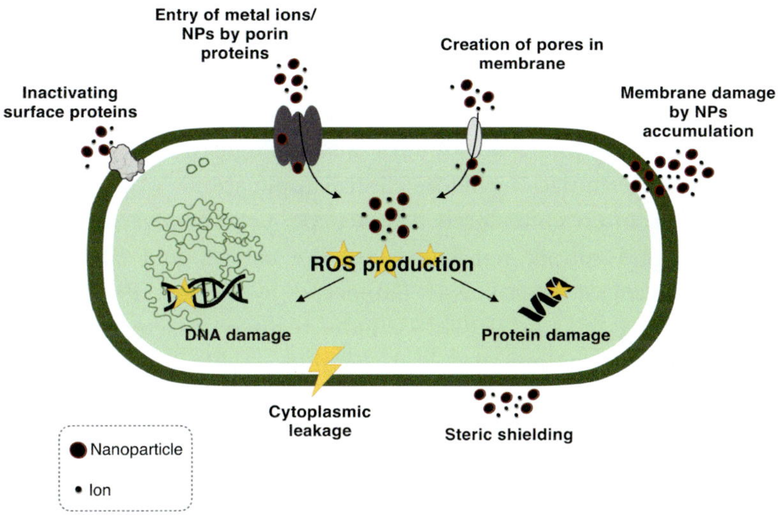

Figure 1.7 Potential pathways of antimicrobial colloidal metal nanoparticle action. ROS: reactive oxygen species.

The particle properties such as shape, size, surface charge, and diffusion influence their antibacterial properties, and silver nanoparticles smaller than 10 nm could directly penetrate the bacterial cells and initiate cell lysis [64]. Other reported mechanisms include the interaction of Ag^+ ions with the biological macromolecules (enzymes and DNA), and electron-discharge or free radical generation [65]. The latter is particularly interesting as it can result in the formation of ROS, which are believed to cause bacterial death by inducing intracellular oxidation, membrane potential changes, and the release of cellular contents. It was also demonstrated that the gram-negative bacteria are more sensitive to the effects of silver nanoparticles than the gram-positive due to the slowed or hampered diffusion of particles through a thick cell wall [66].

Interesting new approaches to dealing with antibiotic-resistant bacteria involve the use of antibiotics together with silver nanoparticles, which have been shown to improve the overall antibacterial activity [67, 68]. The strategies to render silver more effective but less toxic to normal cells have prompted the development of silver nanocomposites whereby silver nanoparticles are embedded within polymer matrices such as chitosan or different types of biocompatible hydrogels [69]. However, it should be stressed that despite numerous demonstrations of the potential of silver nanoparticles, one needs to critically inspect the underlying biodistribution, pharmacokinetics, and biological effects of silver-based nanoformulations before clinical application.

Gold is often considered a less toxic, yet effective alternative to silver and copper nanoparticles. The use of gold compounds as antimicrobial agents was pioneered by Robert Koch in the 19th century, who used gold cyanide in his antibacterial studies. Commonly, gold nanomaterials are combined with a photosensitizer to kill pathogens by photodynamic antimicrobial therapy [70]. In fact, photodynamic therapy has been particularly useful in destroying a wide range of microorganisms in antibacterial surfaces or wound dressings, and there are also some inorganic oxide nanoparticles such as TiO_2 that can be activated by light to produce ROS and effectively kill bacteria. Similarly, the activity of ZnO nanoparticles is governed by their ability to produce ROS and accumulate on the outer bacterial membrane [71]. ZnO is a GRAS (generally recognized as safe) substance by the FDA owing to its outstanding biocompatibility and low toxicity, with several ongoing clinical trials using ZnO alone or in combination with other compounds showing effectiveness in the treatment of antibiotic-resistant strains such as *Staphylococcus aureus* [72].

Apart from metal and metal oxide nanoparticles, polymeric, self-assembled amphiphilic nanoparticles and DNA nanostructures (origami) have also shown antibacterial activity as a consequence of either their inherent chemical structure (cationic polylysine) or their ability to carry antibacterial compounds [73–75]. For example, aptamer-functionalized DNA origami was recently employed to deliver antibacterial lysozyme to several gram-positive and gram-negative bacteria, including *Bacillus subtilis* and *E. coli* [76], and a tetrahedral DNA nanostructure was employed for the efficient delivery of an antisense peptide nucleic acid to inhibit the growth of methicillin-resistant *S. aureus* [77].

1.4.2 Antiviral Nanotherapeutics

Vaccination remains the most efficient strategy to prevent the spread of viral diseases. However, the long time required for the development of vaccines and the rapid evolution of viruses through mutations and genetic recombination, together with the lack of effective vaccines against some viruses, mean that there is a huge demand for the development of new strategies to deal with viral infections. Currently, several molecular drugs are used as a first-line treatment of various viral diseases, but unlike antibiotics, the development of antiviral drugs remains slow. In fact, most of the early antiviral compounds were discovered serendipitously or repurposed from other therapeutic applications. For example, thiosemicarbazones, introduced in 1951 and used against vaccinia virus, were first employed to treat tuberculosis bacilli, and the antiviral drug idoxuridine (IDU) used to treat Herpex virus in the 1960s was initially designed as an anticancer drug (Fig. 1.8).

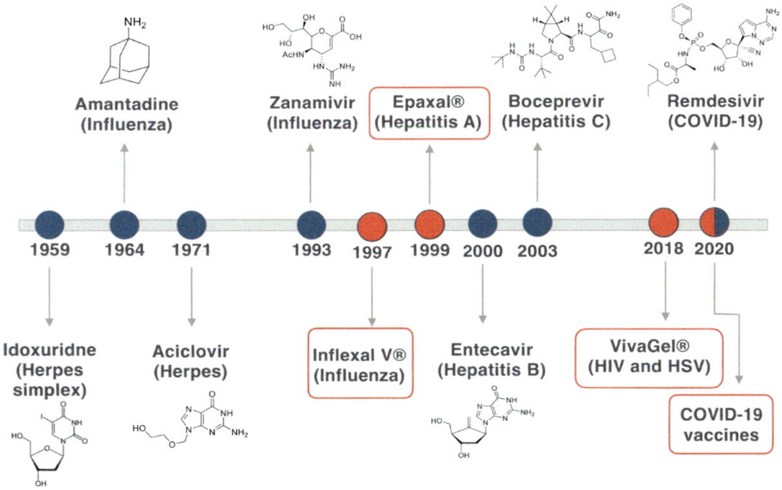

Figure 1.8 Timeline for the development of some important antiviral drugs/treatments and vaccines.

The antiviral application of idoxuridine was a landmark in the antiviral drug discovery, and many antiviral agents for the treatment of various viruses soon followed. For example, amantadine, the first antiviral drug for influenza, was reported in 1964. In the 1970s, several broad-spectrum drugs were developed, such as ribavirin and acyclovir, used as effective treatments for most known species of the herpesvirus family such as herpes simplex and varicella zoster viruses. Acyclovir has, in the meanwhile, been used for the treatment of chickenpox, hepatitis B, and hepatitis C viruses. The recent pandemic prompted the repurposing of huge libraries of existing molecular drugs to coronaviruses, with some potent hits being identified [78].

Besides small molecular drugs, antiviral polymers have been explored as antiviral agents. Starting from the late 1940s until the 1960s, several polysaccharides, such as heparin, dextran sulfate, and sulfated polysaccharides have been tested against a range of viruses [79]. In the 1990s, the concept of multivalency introduced by Whitesides et al. allowed for more potent designs whereby several copies of the active functional group added to the polymeric carriers significantly lowered the required amount of the agent needed to inhibit the virus.

This led to an interest in using nanoparticles, which have a large surface area and can incorporate a larger number of functional groups. The first nanoparticle-based strategy was approved in 1997, and it was a liposome-based vaccine Inflexal®V in which 150 nm liposomes mimicked the native virus structure. A few other vaccines based on the same principle followed (Table 1.3), leading to further development of other nano-based formulations. For example, a dendrimer-based nano-microbicide was developed for the prevention of HIV and HSV infections and approved in the EU in 2018, while lipid nanoparticle constructs were used to deliver siRNAs and inhibit the production of hepatitis B virus proteins effectively preventing the reproduction of viral particles.

In addition to nanoparticle-based vaccines and nanocarriers of antiviral agents, nanoparticle-based virus inhibitors also offer a great possibility to deal with viral infections. These inhibitors rely on the intrinsic properties of nanomaterials such as singlet oxygen production or interaction with various biomolecular building blocks of the virus, which effectively decreases the infectivity and reproduction potential of viruses (Fig. 1.9).

Table 1.3 Examples of nanoparticle-based treatments for viral infections in clinical use

Name	Composition	Applications	Status (Company)	Ref.
Inflexal V®	Liposomal (150 nm) vaccine	Influenza virus	Approved in 1997 (Crucell)	[80]
Epaxal®	Liposomal vaccine	Hepatitis A virus	Approved in 1999 (Crucell)	[81]
VivaGel® (SPL7013)	Dendrimer	HIV and HSV	Approved in 2018 (Starpharma)	[82]
ARB-001467 TKM-HBV	Lipid nanoparticles	Hepatitis B virus	Phase II (Arbutus Biopharma)	[83]

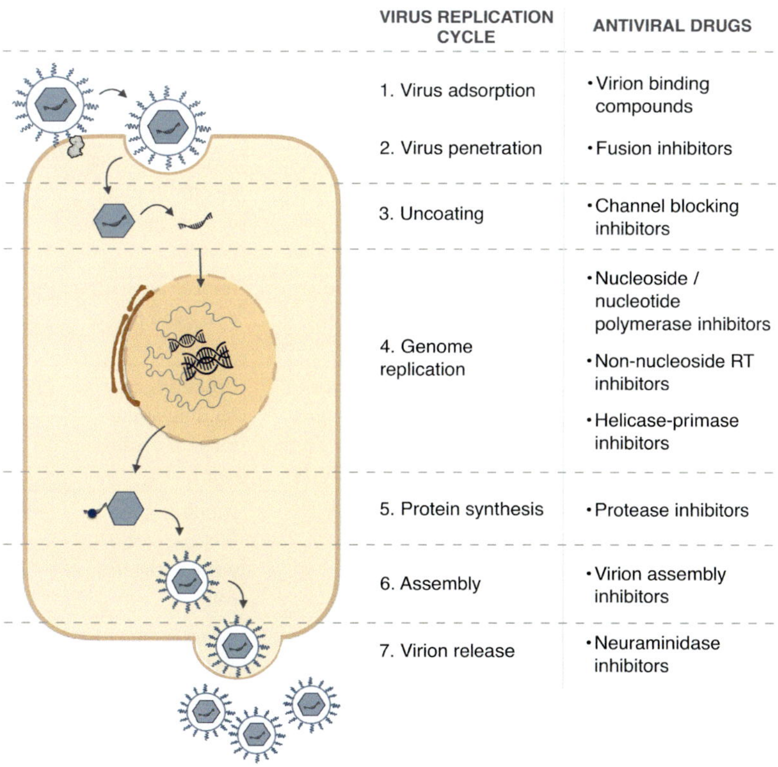

Figure 1.9 Antiviral mechanism of nanoparticles.

Metal nanoparticles, such as silver nanoparticles, have also shown the ability to destroy hepatitis B virus [84], simplex virus type 1 [85], respiratory syncytial virus, and monkeypox virus [86]. They have also been used against HIV-1 at non-cytotoxic concentrations. Overall, HIV has been extensively studied in terms of antiviral agents due to its huge impact on global health. In addition to silver nanoparticles, various gold-based nanosystems, including multivalent gold nanoparticles and porous gold structures, were developed to address both the burden of HIV and influenza viruses (Table 1.4).

Table 1.4 Some recently reported nanoparticles against different virus targets

Nanoparticles	Surface Ligands	Virus	Mechanism of Action	Ref.
Gold nanoparticles (AuNPs)	Sialic acid	Influenza virus	Inhibition of virus binding to host cell	[87]
	Virus-like spiky surface	Influenza virus	Inactivate the virus by cleaving disulfide linkage in hemagglutinin protein	[88]
	Mercaptosuccinic acid, mercaptoethanesulfonate	Influenza virus	Interfering with entry and fusion steps	[89]
	Mercaptoethanesulfonate	Herpes simplex virus type-1 (HSV-1)	Hindering the binding of the virus to the host cell	[90]
Silver nanoparticles (AgNPs)	Mercaptoethanesulfonate	HSV-1	Hindering the binding of the virus to the host cell	[91]
	Polyvinylpyrrolidone (PVP) coating	HIV-1	Interaction with gp120	[92]
	—	Hepatitis B virus	Interaction with DNA and/or binding with viral particles	[93]
	Polysaccharide coating	Monkeypox virus	Interference with viral attachment	[94]

Nanoparticles	Surface Ligands	Virus	Mechanism of Action	Ref.
Graphene	Curcumin	Respiratory syncytial virus	Inactivate virus and inhibit attachment	[95]
	Sulfated polyglycerol	Herpesvirus	Attachment inhibition	[96]
	Sialic acid	Influenza virus	Wrap around the virus and inhibit attachment	[97]
Carbon-based nanoparticles	Red blood cell membrane and zanamivir	Influenza	Inhibit binding and budding of virus	[98]
Silica nanoparticles	Virus-like spiky surface with sialyllactose and zanamivir	Influenza	Inhibit binding, budding and release of virus	[99]

It was also observed that the size of the nano-inhibitor influences the virus binding, and when sialic acid–modified gold nanoparticles were employed for influenza virus binding, 14 nm gold nanoparticles demonstrated strong inhibitory effects compared to 2 nm gold nanoparticles [87]. The rationale behind this is that the smaller nanoparticles lacked the ability to hinder the virus from binding to the host cell membrane. These observations are providing more insight into the rational design of antiviral nanostructures in the future.

Meanwhile, organic nanoparticles such as polymers, dendrimers, and nanogels have been investigated for the inhibition of different viruses by blocking viral attachment to cells using their flexible and adaptive nature together with multivalency. Various modified organic nanoparticles such as sialylated and sulfated nanogels [100, 101], sialyl lactose-functionalized polymers and dendrimers [102, 103], and sialyl lactose-modified DNA three-way junctions have been prepared to mitigate viral infection [104].

As the field progresses more toward the rational design of nanostructures and there is more understanding of the underlying cytotoxicity of molecular structures, more systems enabling precise targeting and efficient inhibition of virus will continue to appear. It remains to be seen if the field is going to move more toward the design of nano-based antiviral agents or immune system modulation (vaccines).

1.4.3 Antifungal Nanotherapeutics

Apart from bacteria and viruses, fungal infections have also been known to cause serious health problems in immunocompromised patients. Unlike antiviral drugs, the development of antifungal drugs is very slow because of the eukaryotic nature of the target cells. One of the first antifungal drugs, Nystatin, was discovered in 1950, almost 30 years later than antibiotics. However, the first orally active antifungal drug Griseofulvin was used in the early 1960s against dermatomycosis. Then came the development of the first triazole drug Terconazole, which was approved by the FDA in 1987, followed by fluconazole in 1990, and itraconazole in 1992. Fluconazole has various advantages over other antifungal agents, including the option of oral administration together with minimal side effects. However, like antibiotics, antifungal resistance and host-specific side effects have limited their widespread use.

This prompted the development of metal nanoparticles with fungicidal and fungistatic effects. For example, silver nanoparticles showed significant antifungal activity against *Candida albicans* [105], *Penicillium citrinum* [106], *Trichophyton mentagrophytes* [107], and *Aspergillus niger* [108]. Most of the nanoparticles showed antifungal activity at a minimum inhibitory concentration of 2.0 μg/mL, similar to the activity of the small molecular drug amphotericin B. As with bacteria and viruses, the antifungal effects of nanoparticles are mainly caused by the disruption of the cell membrane and inhibition of the common budding process [109]. The antifungal activity of gold nanoparticles [110], zinc oxide nanoparticles, copper–polymer nanocomposites [111], silica nanoparticles, and aluminum oxide nanoparticles [112] have been demonstrated against different fungal strains over the last 20 years.

Several drug delivery systems have also been reported for the oral and topical administration of antifungal drugs. Various liposomes, polymeric amphiphiles and micelles, solid lipid nanoparticles, microemulsions, and microsponges have all been reported for the topical, oral, and ocular delivery of fluconazole, nystatin, itraconazole, econazole, and ketoconazole [113]. Nanoformulations have been used owing to their ability to improve the pharmacokinetic and pharmacodynamic properties of drugs, such as their solubility and stability, but can also be used for specific targeting [114].

1.5 Conclusion and Future Outlook

In the past 30 years, nanotechnology has made a significant impact on medicine, and it will continue to do so as we embrace the rational design of nanostructures adapted to the latest discoveries in disease biology. Commercial biosensors that employ nanoparticles are already part of daily life and as we go toward early diagnostics and disease prevention, it could be expected that their market share will continue to increase. The use of nanomaterials in the design of therapeutic interventions, in particular drug delivery, has been slower, mainly due to the lack of comprehensive toxicological data but also the lack clear guidelines from regulatory agencies. The development of mRNA vaccines for SARS-CoV-2 is a definite game changer in terms of regulation and clinical application of nanostructures. Considering the huge need for antibacterial, antiviral, but also anticancer therapies, nanomaterials could help us address the current limits of small molecular drugs, in particular the emergence of resistance and significant side effects. Now, in the 21st century, we are well placed to exploit the efforts and knowledge accumulated by scientists over the past century and bring medicine into a new era of personalized medicine, but also an era of disease prevention and early treatment.

References

1. G. Chiari, R. Giustetto, J. Druzik, E. Doehne, and G. Ricchiardi, Pre-Columbian nanotechnology: Reconciling the mysteries of the Maya blue pigment, *Appl. Phys. A* **90**, 3 (2007).
2. M. Reibold, P. Paufler, A. A. Levin, W. Kochmann, N. Pätzke, and D. C. Meyer, Carbon nanotubes in an ancient Damascus Sabre, *Nature* **444**, 286 (2006).
3. R. P. Feynman, There's plenty of room at the bottom, *Resonance* **16**, 890 (2011).
4. R. A. Freitas, *Nanomedicine, Volume I: Basic Capabilities*, 1st ed. (CRC Press, 1999).
5. W. R. Heineman and W. B. Jensen, Leland C. Clark Jr. (1918–2005), *Biosens. Bioelectron.* **21**, 1403 (2006).

6. C. A. Mirkin, R. L. Letsinger, R. C. Mucic, and J. J. Storhoff, A DNA-based method for rationally assembling nanoparticles into macroscopic materials, *Nature* **382**, 607 (1996).
7. K. Paunovska, C. D. Sago, C. M. Monaco, W. H. Hudson, M. G. Castro, T. G. Rudoltz, S. Kalathoor, D. A. Vanover, P. J. Santangelo, R. Ahmed, A. V. Bryksin, and J. E. Dahlman, A direct comparison of *in vitro* and *in vivo* nucleic acid delivery mediated by hundreds of nanoparticles reveals a weak correlation, *Nano Lett.* **18**, 2148 (2018).
8. L. Ma, L. Peng, L. Yin, G. Liu, and S. Man, CRISPR-Cas12a-powered dual-mode biosensor for ultrasensitive and cross-validating detection of pathogenic bacteria, *ACS Sens.* **6**, 2920 (2021).
9. G. Bauer, F. Pittner, and T. Schalkhammer, Metal nano-cluster biosensors, *Microchim. Acta* **131**, 107 (1999).
10. R. Rojanathanes, A. Sereemaspun, N. Pimpha, V. Buasorn, P. Ekawong, and V. Wiwanitkit, Gold nanoparticle as an alternative tool for a urine pregnancy test, *Taiwan J. Obstet. Gynecol.* **47**, 296 (2008).
11. C.-M. Tîlmaciu and M. C. Morris, Carbon nanotube biosensors, *Front. Chem.* **3**, (2015).
12. S. Iijima, Helical microtubules of graphitic carbon, *Nature* **354**, 56 (1991).
13. K. Balasubramanian and M. Burghard, Biosensors based on carbon nanotubes, *Anal. Bioanal. Chem.* **385**, 452 (2006).
14. Y. Yun, Z. Dong, V. Shanov, W. R. Heineman, H. B. Halsall, A. Bhattacharya, L. Conforti, R. K. Narayan, W. S. Ball, and M. J. Schulz, Nanotube electrodes and biosensors, *Nano Today* **2**, 30 (2007).
15. M. Buch and J. Rishpon, An electrochemical immunosensor for C-reactive protein based on multi-walled carbon nanotube-modified electrodes, *Electroanalysis* **20**, 2592 (2008).
16. H. Su, R. Yuan, Y. Chai, Y. Zhuo, C. Hong, Z. Liu, and X. Yang, Multilayer structured amperometric immunosensor built by self-assembly of a redox multi-wall carbon nanotube composite, *Electrochim. Acta* **54**, 4149 (2009).
17. P. Mehrotra, Biosensors and their applications: A review, *J. Oral Biol. Craniofac. Res.* **6**, 153 (2016).
18. V. Naresh and N. Lee, A review on biosensors and recent development of nanostructured materials-enabled biosensors, *Sensors* **21**, 1109 (2021).
19. M. Shanehsaz, A. Mohsenifar, S. Hasannia, N. Pirooznia, Y. Samaei, and M. Shamsipur, Detection of *Helicobacter pylori* with a nanobiosensor

based on fluorescence resonance energy transfer using CdTe quantum dots, *Microchim. Acta* **180**, 195 (2013).

20. F. Ma, C. Li, and C. Zhang, Development of quantum dot-based biosensors: Principles and applications, *J. Mater. Chem. B* **6**, 6173 (2018).
21. P. W. K. Rothemund, Folding DNA to create nanoscale shapes and patterns, *Nature* **440**, 297 (2006).
22. M. Pfeiffer, K. Trofymchuk, S. Ranallo, F. Ricci, F. Steiner, F. Cole, V. Glembockyte, and P. Tinnefeld, Single antibody detection in a DNA origami nanoantenna, *IScience* **24**, 103072 (2021).
23. D. Deamer, M. Akeson, and D. Branton, Three decades of nanopore sequencing, *Nat. Biotechnol.* **34**, 518 (2016).
24. H. Wen and K. Park, *Oral Controlled Release Formulation Design and Drug Delivery: Theory to Practice* (John Wiley & Sons, 2011).
25. H. Jatzkewitz, Peptamin (glycyl-L-leucyl-mescaline) bound to blood plasma expander (polyvinylpyrrolidone) as a new depot form of a biologically active primary amine (mescaline), *Z. Naturforsch.* **10b**, 27 (1955).
26. G. Mathe, L. O. C. Tran Ba, and J. Bernard, Effect on mouse leukemia 1210 of a combination by diazo-reaction of amethopterin and gamma-globulins from hamsters inoculated with such leukemia by heterografts, *C. R. Hebd. Seances Acad. Sci.* **246**, 1626 (1958).
27. A. D. Bangham and R. W. Horne, Negative staining of phospholipids and their structural modification by surface-active agents as observed in the electron microscope, *J. Mol. Biol.* **8**, 660 (1964).
28. A. D. Bangham, M. W. Hill, and N. G. A. Miller, Preparation and use of liposomes as models of biological membranes, in *Methods in Membrane Biology* (Springer US, Boston, MA, 1974), pp. 1–68.
29. A. Varela-Moreira, Y. Shi, M. H. A. M. Fens, T. Lammers, W. E. Hennink, and R. M. Schiffelers, Clinical application of polymeric micelles for the treatment of cancer, *Mater. Chem. Front.* **1**, 1485 (2017).
30. B. Parshad, S. Prasad, S. Bhatia, A. Mittal, Y. Pan, P. K. Mishra, S. K. Sharma, and L. Fruk, Non-ionic small amphiphile based nanostructures for biomedical applications, *RSC Adv.* **10**, 42098 (2020).
31. B. Parshad, M. Kumari, V. Khatri, R. Rajeshwari, Y. Pan, A. K. Sharma, and I. Ahmed, Enzymatic synthesis of glycerol, azido-glycerol and azido-triglycerol based amphiphilic copolymers and their relevance as nanocarriers: A review, *Eur. Polym. J.* **158**, 110690 (2021).

32. S. Croy and G. Kwon, Polymeric micelles for drug delivery, *Curr. Pharm. Des.* **12**, 4669 (2006).

33. T. Nagano, M. Yasunaga, K. Goto, H. Kenmotsu, Y. Koga, J. Kuroda, Y. Nishimura, T. Sugino, Y. Nishiwaki, and Y. Matsumura, Antitumor activity of NK012 combined with cisplatin against small cell lung cancer and intestinal mucosal changes in tumor-bearing mouse after treatment, *Clin. Cancer Res.* **15**, 4348 (2009).

34. T. E. Nakajima, K. Yanagihara, M. Takigahira, M. Yasunaga, K. Kato, T. Hamaguchi, Y. Yamada, Y. Shimada, K. Mihara, T. Ochiya, and Y. Matsumura, Antitumor effect of SN-38–releasing polymeric micelles, NK012, on spontaneous peritoneal metastases from orthotopic gastric cancer in mice compared with irinotecan, *Cancer Res.* **68**, 9318 (2008).

35. M. Sumitomo, F. Koizumi, T. Asano, A. Horiguchi, K. Ito, T. Asano, T. Kakizoe, M. Hayakawa, and Y. Matsumura, Novel SN-38–incorporated polymeric micelle, NK012, strongly suppresses renal cancer progression, *Cancer Res.* **68**, 1631 (2008).

36. Y. Fujiwara, H. Mukai, T. Saeki, J. Ro, Y.-C. Lin, S. E. Nagai, K. S. Lee, J. Watanabe, S. Ohtani, S. B. Kim, K. Kuroi, K. Tsugawa, Y. Tokuda, H. Iwata, Y. H. Park, Y. Yang, and Y. Nambu, A multi-national, randomised, open-label, parallel, phase III non-inferiority study comparing NK105 and paclitaxel in metastatic or recurrent breast cancer patients, *Br. J. Cancer* **120**, 475 (2019).

37. C. L. Ventola, Progress in nanomedicine: Approved and investigational nanodrugs, *P. T.* **42**, 742 (2017).

38. S. Svenson, Clinical translation of nanomedicines, *Curr. Opin. Solid State Mater. Sci.* **16**, 287 (2012).

39. J. W. Valle, A. Armstrong, C. Newman, V. Alakhov, G. Pietrzynski, J. Brewer, S. Campbell, P. Corrie, E. K. Rowinsky, and M. Ranson, A phase 2 study of SP1049C, doxorubicin in P-glycoprotein-targeting pluronics, in patients with advanced adenocarcinoma of the esophagus and gastroesophageal junction, *Invest. New Drugs* **29**, 1029 (2011).

40. K. A. Autio, R. Dreicer, J. Anderson, J. A. Garcia, A. Alva, L. L. Hart, M. I. Milowsky, E. M. Posadas, C. J. Ryan, R. P. Graf, R. Dittamore, N. A. Schreiber, J. M. Summa, H. Youssoufian, M. J. Morris, and H. I. Scher, Safety and efficacy of BIND-014, a docetaxel nanoparticle targeting prostate-specific membrane antigen for patients with metastatic castration-resistant prostate cancer, *JAMA Oncol.* **4**, 1344 (2018).

41. S. Tomao, Albumin-bound formulation of paclitaxel (Abraxane&Reg; ABI-007) in the treatment of breast cancer, *Int. J. Nanomedicine* 99 (2009).

42. N. Desai, Nanoparticle albumin-bound paclitaxel (Abraxane®), in *Albumin in Medicine* (Springer Singapore, Singapore, 2016), pp. 101–119.

43. E. Karami, M. Behdani, and F. Kazemi-Lomedasht, Albumin nanoparticles as nanocarriers for drug delivery: Focusing on antibody and nanobody delivery and albumin-based drugs, *J. Drug Deliv. Sci. Technol.* **55**, 101471 (2020).

44. K. Maier-Hauff, F. Ulrich, D. Nestler, H. Niehoff, P. Wust, B. Thiesen, H. Orawa, V. Budach, and A. Jordan, Efficacy and safety of intratumoral thermotherapy using magnetic iron-oxide nanoparticles combined with external beam radiotherapy on patients with recurrent glioblastoma multiforme, *J. Neurooncol.* **103**, 317 (2011).

45. L. R. Hirsch, R. J. Stafford, J. A. Bankson, S. R. Sershen, B. Rivera, R. E. Price, J. D. Hazle, N. J. Halas, and J. L. West, Nanoshell-mediated near-infrared thermal therapy of tumors under magnetic resonance guidance, *Proc. Natl. Acad. Sci.* **100**, 13549 (2003).

46. S. K. Libutti, G. F. Paciotti, A. A. Byrnes, H. R. Alexander, W. E. Gannon, M. Walker, G. D. Seidel, N. Yuldasheva, and L. Tamarkin, Phase I and pharmacokinetic studies of CYT-6091, a novel PEGylated colloidal gold-RhTNF nanomedicine, *Clin. Cancer Res.* **16**, 6139 (2010).

47. M. L. Etheridge, S. A. Campbell, A. G. Erdman, C. L. Haynes, S. M. Wolf, and J. McCullough, The big picture on nanomedicine: The state of investigational and approved nanomedicine products, *Nanomedicine* **9**, 1 (2013).

48. R. Garg, K. J. H. Allen, W. Dawicki, E. M. Geoghegan, D. L. Ludwig, and E. Dadachova, 225Ac-labeled CD33-targeting antibody reverses resistance to Bcl-2 inhibitor venetoclax in acute myeloid leukemia models, *Cancer Med.* **10**, 1128 (2021).

49. S. Bonvalot, P. L. Rutkowski, J. Thariat, S. Carrère, A. Ducassou, M.-P. Sunyach, P. Agoston, A. Hong, A. Mervoyer, M. Rastrelli, V. Moreno, R. K. Li, B. Tiangco, A. C. Herraez, A. Gronchi, L. Mangel, T. Sy-Ortin, P. Hohenberger, T. de Baère, A. Le Cesne, S. Helfre, E. Saada-Bouzid, A. Borkowska, R. Anghel, A. Co, M. Gebhart, G. Kantor, A. Montero, H. H. Loong, R. Vergés, L. Lapeire, S. Dema, G. Kacso, L. Austen, L. Moureau-Zabotto, V. Servois, E. Wardelmann, P. Terrier, A. J. Lazar, J. V. M. G. Bovée, C. Le Péchoux, and Z. Papai, NBTXR3, a first-in-class radioenhancer hafnium oxide nanoparticle, plus radiotherapy versus radiotherapy alone in patients with locally advanced soft-tissue sarcoma (Act.In.Sarc): A multicentre, phase 2–3, randomised, controlled trial, *Lancet Oncol.* **20**, 1148 (2019).

50. E. Novo and M. Parola, Redox mechanisms in hepatic chronic wound healing and fibrogenesis, *Fibrog. Tissue Repair* **1**, 5 (2008).
51. H. Huang, W. Feng, Y. Chen, and J. Shi, Inorganic nanoparticles in clinical trials and translations, *Nano Today* **35**, 100972 (2020).
52. S. Vinogradov, E. Batrakova, and A. Kabanov, Poly(ethylene glycol)–polyethyleneimine NanoGel^{TM} particles: Novel drug delivery systems for antisense oligonucleotides, *Colloids Surf., B* **16**, 291 (1999).
53. K. Akiyoshi, S. Deguchi, N. Moriguchi, S. Yamaguchi, and J. Sunamoto, Self-aggregates of hydrophobized polysaccharides in water. Formation and characteristics of nanoparticles, *Macromolecules* **26**, 3062 (1993).
54. A. Sharma, T. Garg, A. Aman, K. Panchal, R. Sharma, S. Kumar, and T. Markandeywar, Nanogel—An advanced drug delivery tool: Current and future, *Artif. Cells Nanomed. Biotechnol.* **44**, 165 (2016).
55. M. Kumari, S. Prasad, L. Fruk, and B. Parshad, Polyglycerol-based hydrogels and nanogels: From synthesis to applications, *Future Med. Chem.* **13**, 419 (2021).
56. R. Sharma and P. Prajapati, Nanotechnology in medicine: Leads from ayurveda, *J. Pharm. Bioallied Sci.* **8**, 80 (2016).
57. Q. H. Tran, V. Q. Nguyen, and A.-T. Le, Corrigendum: Silver nanoparticles: Synthesis, properties, toxicology, applications and perspectives (Adv. Nat. Sci: Nanosci. Nanotechnol. **4** 033001), *Adv. Nat. Sci: Nanosci. Nanotechnol.* **9**, 049501 (2018).
58. M. Xing, L. Ge, M. Wang, Q. Li, X. Li, and J. Ouyang, Nanosilver particles in medical applications: Synthesis, performance, and toxicity, *Int. J. Nanomedicine* 2399 (2014).
59. A.-C. Burdușel, O. Gherasim, A. M. Grumezescu, L. Mogoantă, A. Ficai, and E. Andronescu, Biomedical applications of silver nanoparticles: An up-to-date overview, *Nanomaterials* **8**, 681 (2018).
60. V. B. P. Sudha, S. Ganesan, G. P. Pazhani, T. Ramamurthy, G. B. Nair, and P. Venkatasubramanian, Storing drinking-water in copper pots kills contaminating diarrhoeagenic bacteria, *J. Health Popul. Nutr.* **30**, 17 (2012).
61. O. Walusinski, The scientific illusion of Victor Burq (1822–1884), *Eur. Neurol.* **79**, 135 (2018).
62. S. Pal, Y. K. Tak, and J. M. Song, Does the antibacterial activity of silver nanoparticles depend on the shape of the nanoparticle? A study of the gram-negative bacterium *Escherichia coli*, *Appl. Environ. Microbiol.* **73**, 1712 (2007).

63. O. Choi, K. K. Deng, N.-J. Kim, L. Ross, R. Y. Surampalli, and Z. Hu, The inhibitory effects of silver nanoparticles, silver ions, and silver chloride colloids on microbial growth, *Water Res.* **42**, 3066 (2008).

64. X. Hong, J. Wen, X. Xiong, and Y. Hu, Shape effect on the antibacterial activity of silver nanoparticles synthesized via a microwave-assisted method, *Environ. Sci. Pollut. Res.* **23**, 4489 (2016).

65. V. K. Sharma, R. A. Yngard, and Y. Lin, Silver nanoparticles: Green synthesis and their antimicrobial activities, *Adv. Colloid Interface Sci.* **145**, 83 (2009).

66. R. E. Duval, J. Gouyau, and E. Lamouroux, Limitations of recent studies dealing with the antibacterial properties of silver nanoparticles: Fact and opinion, *Nanomaterials* **9**, 1775 (2019).

67. V. C. Karade, R. B. Patil, S. B. Parit, J. H. Kim, A. D. Chougale, and V. V. Dawkar, Insights into shape-based silver nanoparticles: A weapon to cope with pathogenic attacks, *ACS Sustain. Chem. Eng.* **9**, 12476 (2021).

68. R. Thomas, A. P. Nair, S. Kr, J. Mathew, and R. Ek, Antibacterial activity and synergistic effect of biosynthesized AgNPs with antibiotics against multidrug-resistant biofilm-forming coagulase-negative Staphylococci isolated from clinical samples, *Appl. Biochem. Biotechnol.* **173**, 449 (2014).

69. M. A. Wahab, L. Li, H. Li, and A. Abdala, Silver nanoparticle-based nanocomposites for combating infectious pathogens: Recent advances and future prospects, *Nanomaterials* **11**, 581 (2021).

70. W.-S. Kuo, C.-N. Chang, Y.-T. Chang, and C.-S. Yeh, Antimicrobial gold nanorods with dual-modality photodynamic inactivation and hyperthermia, *Chem. Commun.* 4853 (2009).

71. K. R. Raghupathi, R. T. Koodali, and A. C. Manna, Size-dependent bacterial growth inhibition and mechanism of antibacterial activity of zinc oxide nanoparticles, *Langmuir* **27**, 4020 (2011).

72. M. Banoee, S. Seif, Z. E. Nazari, P. Jafari-Fesharaki, H. R. Shahverdi, A. Moballegh, K. M. Moghaddam, and A. R. Shahverdi, ZnO nanoparticles enhanced antibacterial activity of ciprofloxacin against *Staphylococcus aureus* and *Escherichia coli*, *J. Biomed. Mater. Res. B Appl. Biomater.* **93B**, 557 (2010).

73. A. Cano, M. Ettcheto, M. Espina, A. López-Machado, Y. Cajal, F. Rabanal, E. Sánchez-López, A. Camins, M. L. García, and E. B. Souto, State-of-the-art polymeric nanoparticles as promising therapeutic tools against human bacterial infections, *J. Nanobiotechnology* **18**, 156 (2020).

74. A. Mittal, Krishna, Aarti, S. Prasad, P. K. Mishra, S. K. Sharma, and B. Parshad, Self-assembly of carbohydrate-based small amphiphiles and their applications in pathogen inhibition and drug delivery: A review, *Mater. Adv.* **2**, 3459 (2021).

75. N. F. Kamaruzzaman, L. P. Tan, R. H. Hamdan, S. S. Choong, W. K. Wong, A. J. Gibson, A. Chivu, and M. de F. Pina, Antimicrobial polymers: The potential replacement of existing antibiotics? *Int. J. Mol. Sci.* **20**, 2747 (2019).

76. I. Mela, P. P. Vallejo-Ramirez, S. Makarchuk, G. Christie, D. Bailey, R. M. Henderson, H. Sugiyama, M. Endo, and C. F. Kaminski, DNA nanostructures for targeted antimicrobial delivery, *Angew. Chem.* **132**, 12798 (2020).

77. Y. Zhang, W. Ma, Y. Zhu, S. Shi, Q. Li, C. Mao, D. Zhao, Y. Zhan, J. Shi, W. Li, L. Wang, C. Fan, and Y. Lin, Inhibiting methicillin-resistant *Staphylococcus aureus* by tetrahedral DNA nanostructure-enabled antisense peptide nucleic acid delivery, *Nano Lett.* **18**, 5652 (2018).

78. R. A. Siemieniuk, J. J. Bartoszko, D. Zeraatkar, E. Kum, A. Qasim, J. P. Díaz Martinez, A. Izcovich, F. Lamontagne, M. A. Han, A. Agarwal, T. Agoritsas, M. Azab, G. Bravo, D. K. Chu, R. Couban, T. Devji, Z. Escamilla, F. Foroutan, Y. Gao, L. Ge, M. Ghadimi, D. Heels-Ansdell, K. Honarmand, L. Hou, Q. Ibrahim, A. Khamis, B. Lam, C. Mansilla, M. Loeb, A. Miroshnychenko, M. Marcucci, S. L. McLeod, S. Motaghi, S. Murthy, R. A. Mustafa, H. Pardo-Hernandez, G. Rada, Y. Rizwan, P. Saadat, C. Switzer, L. Thabane, G. Tomlinson, P. O. Vandvik, R. W. Vernooij, A. Viteri-García, Y. Wang, L. Yao, Y. Zhao, G. H. Guyatt, and R. Brignardello-Petersen, Drug treatments for Covid-19: Living systematic review and network meta-analysis, *BMJ* m2980 (2020).

79. R. H. Green and D. W. Woolley, Inhibition by certain polysaccharides of hemagglutination and of multiplication of influenza virus, *J. Exp. Med.* **86**, 55 (1947).

80. B. N. Corder, B. L. Bullard, G. A. Poland, and E. A. Weaver, A decade in review: A systematic review of universal influenza vaccines in clinical trials during the 2010 decade, *Viruses* **12**, 1186 (2020).

81. P. A. Bovier, Epaxal®: A virosomal vaccine to prevent hepatitis A infection, *Expert Rev. Vaccines* **7**, 1141 (2008).

82. R. Rupp, S. L. Rosenthal, and L. R. Stanberry, VivaGel (SPL7013 Gel): A candidate dendrimer—Microbicide for the prevention of HIV and HSV infection, *Int. J. Nanomedicine* **2**, 561 (2007).

83. ClinicalTrials.gov, Study of ARB-001467 in subjects with chronic HBV infection receiving nucleos(t)Ide analogue therapy, https://clinicaltrials.gov/study/NCT02631096 (accessed in October 2023).

84. L. Lu, R. W.-Y. Sun, R. Chen, C.-K. Hui, C.-M. Ho, J. M. Luk, G. K. K. Lau, and C.-M. Che, Silver nanoparticles inhibit hepatitis B virus replication, *Antivir. Ther.* **13**, 253 (2008).

85. D. Baram-Pinto, S. Shukla, N. Perkas, A. Gedanken, and R. Sarid, Inhibition of herpes simplex virus type 1 infection by silver nanoparticles capped with mercaptoethane sulfonate, *Bioconjug. Chem.* **20**, 1497 (2009).

86. J. V. Rogers, C. V. Parkinson, Y. W. Choi, J. L. Speshock, and S. M. Hussain, A preliminary assessment of silver nanoparticle inhibition of monkeypox virus plaque formation, *Nanoscale Res. Lett.* **3**, 129 (2008).

87. I. Papp, C. Sieben, K. Ludwig, M. Roskamp, C. Böttcher, S. Schlecht, A. Herrmann, and R. Haag, Inhibition of influenza virus infection by multivalent sialic-acid-functionalized gold nanoparticles, *Small* **6**, 2900 (2010).

88. J. Kim, M. Yeom, T. Lee, H.-O. Kim, W. Na, A. Kang, J.-W. Lim, G. Park, C. Park, D. Song, and S. Haam, Porous gold nanoparticles for attenuating infectivity of influenza A virus, *J. Nanobiotechnol.* **18**, 54 (2020).

89. M. Sametband, S. Shukla, T. Meningher, S. Hirsh, E. Mendelson, R. Sarid, A. Gedanken, and M. Mandelboim, Effective multi-strain inhibition of influenza virus by anionic gold nanoparticles, *MedChemComm* **2**, 421 (2011).

90. D. Baram-Pinto, S. Shukla, A. Gedanken, and R. Sarid, Inhibition of HSV-1 attachment, entry, and cell-to-cell spread by functionalized multivalent gold nanoparticles, *Small* **6**, 1044 (2010).

91. D. Baram-Pinto, S. Shukla, N. Perkas, A. Gedanken, and R. Sarid, Inhibition of herpes simplex virus type 1 infection by silver nanoparticles capped with mercaptoethane sulfonate, *Bioconjug. Chem.* **20**, 1497 (2009).

92. J. L. Elechiguerra, J. L. Burt, J. R. Morones, A. Camacho-Bragado, X. Gao, H. H. Lara, and M. J. Yacaman, Interaction of silver nanoparticles with HIV-1, *J. Nanobiotechnol.* **3**, 6 (2005).

93. L. Lu, R. W.-Y. Sun, R. Chen, C.-K. Hui, C.-M. Ho, J. M. Luk, G. K. K. Lau, and C.-M. Che, Silver nanoparticles inhibit hepatitis B virus replication, *Antivir. Ther.* **13**, 253 (2008).

94. J. V. Rogers, C. V. Parkinson, Y. W. Choi, J. L. Speshock, and S. M. Hussain, A preliminary assessment of silver nanoparticle inhibition of monkeypox virus plaque formation, *Nanoscale Res. Lett.* **3**, 129 (2008).

95. X. X. Yang, C. M. Li, Y. F. Li, J. Wang, and C. Z. Huang, Synergistic antiviral effect of curcumin functionalized graphene oxide against respiratory syncytial virus infection, *Nanoscale* **9**, 16086 (2017).

96. B. Ziem, W. Azab, M. F. Gholami, J. P. Rabe, N. Osterrieder, and R. Haag, Size-dependent inhibition of herpesvirus cellular entry by polyvalent nanoarchitectures, *Nanoscale* **9**, 3774 (2017).

97. S. Bhatia, I. S. Donskyi, S. Block, C. Nie, A. Burdinski, D. Lauster, J. Radnik, A. Herrmann, R. Haag, K. Ludwig, and M. Adeli, Wrapping and blocking of influenza A viruses by sialylated 2D nanoplatforms, *Adv. Mater. Interfaces* **8**, 2100285 (2021).

98. C. Nie, M. Stadtmüller, B. Parshad, M. Wallert, V. Ahmadi, Y. Kerkhoff, S. Bhatia, S. Block, C. Cheng, T. Wolff, and R. Haag, Heteromultivalent topology-matched nanostructures as potent and broad-spectrum influenza A virus inhibitors, *Sci. Adv.* **7**, (2021).

99. C. Nie, B. Parshad, S. Bhatia, C. Cheng, M. Stadtmüller, A. Oehrl, Y. Kerkhoff, T. Wolff, and R. Haag, Topology-matching design of an influenza-neutralizing spiky nanoparticle-based inhibitor with a dual mode of action, *Angew. Chem.* **132**, 15662 (2020).

100. S. Bhatia, M. Hilsch, J. L. Cuellar-Camacho, K. Ludwig, C. Nie, B. Parshad, M. Wallert, S. Block, D. Lauster, C. Böttcher, A. Herrmann, and R. Haag, Adaptive flexible sialylated nanogels as highly potent influenza A virus inhibitors, *Angew. Chem., Int. Ed.* **59**, 12417 (2020).

101. P. Dey, T. Bergmann, J. L. Cuellar-Camacho, S. Ehrmann, M. S. Chowdhury, M. Zhang, I. Dahmani, R. Haag, and W. Azab, Multivalent flexible nanogels exhibit broad-spectrum antiviral activity by blocking virus entry, *ACS Nano* **12**, 6429 (2018).

102. M. N. Stadtmueller, S. Bhatia, P. Kiran, M. Hilsch, V. Reiter-Scherer, L. Adam, B. Parshad, M. Budt, S. Klenk, K. Sellrie, D. Lauster, P. H. Seeberger, C. P. R. Hackenberger, A. Herrmann, R. Haag, and T. Wolff, Evaluation of multivalent sialylated polyglycerols for resistance induction in and broad antiviral activity against influenza A viruses, *J. Med. Chem.* **64**, 12774 (2021).

103. S.-J. Kwon, D. H. Na, J. H. Kwak, M. Douaisi, F. Zhang, E. J. Park, J.-H. Park, H. Youn, C.-S. Song, R. S. Kane, J. S. Dordick, K. B. Lee, and R. J. Linhardt, Nanostructured glycan architecture is important in the inhibition of influenza A virus infection, *Nat. Nanotechnol.* **12**, 48 (2017).

104. M. Yamabe, K. Kaihatsu, and Y. Ebara, Sialyllactose-modified three-way junction DNA as binding inhibitor of influenza virus hemagglutinin, *Bioconjug. Chem.* **29**, 1490 (2018).

105. K.-J. Kim, W. S. Sung, B. K. Suh, S.-K. Moon, J.-S. Choi, J. G. Kim, and D. G. Lee, Antifungal activity and mode of action of silver nano-particles on *Candida albicans*, *BioMetals* **22**, 235 (2009).

106. Y. Zhang, H. Peng, W. Huang, Y. Zhou, and D. Yan, Facile preparation and characterization of highly antimicrobial colloid Ag or Au nanoparticles, *J. Colloid Interface Sci.* **325**, 371 (2008).

107. K.-J. Kim, W. S. Sung, S.-K. Moon, J.-S. Choi, J. G. Kim, and D. G. Lee, Antifungal effect of silver nanoparticles on dermatophytes, *J. Microbiol. Biotechnol.* **18**, 1482 (2008).

108. Y. Zhang, H. Peng, W. Huang, Y. Zhou, and D. Yan, Facile preparation and characterization of highly antimicrobial colloid Ag or Au nanoparticles, *J. Colloid Interface Sci.* **325**, 371 (2008).

109. K.-J. Kim, W. S. Sung, B. K. Suh, S.-K. Moon, J.-S. Choi, J. G. Kim, and D. G. Lee, Antifungal activity and mode of action of silver nano-particles on *Candida albicans*, *BioMetals* **22**, 235 (2009).

110. C. Jayaseelan, R. Ramkumar, A. A. Rahuman, and P. Perumal, Green synthesis of gold nanoparticles using seed aqueous extract of *Abelmoschus esculentus* and its antifungal activity, *Ind. Crops Prod.* **45**, 423 (2013).

111. N. Cioffi, L. Torsi, N. Ditaranto, G. Tantillo, L. Ghibelli, L. Sabbatini, T. Bleve-Zacheo, M. D'Alessio, P. G. Zambonin, and E. Traversa, Copper nanoparticle/polymer composites with antifungal and bacteriostatic properties, *Chem. Mater.* **17**, 5255 (2005).

112. M. Jalal, M. A. Ansari, A. K. Shukla, S. G. Ali, H. M. Khan, R. Pal, J. Alam, and S. S. Cameotra, Green synthesis and antifungal activity of Al_2O_3 NPs against fluconazole-resistant *Candida* spp. isolated from a tertiary care hospital, *RSC Adv.* **6**, 107577 (2016).

113. S. Nami, A. Aghebati-Maleki, and L. Aghebati-Maleki, Current applications and prospects of nanoparticles for antifungal drug delivery, *EXCLI J.* **20**, 562 (2021).

114. F. Sousa, D. Ferreira, S. Reis, and P. Costa, Current insights on antifungal therapy: Novel nanotechnology approaches for drug delivery systems and new drugs from natural sources, *Pharmaceuticals* **13**, (2020).

Chapter 2

Clinically Approved Nanotherapeutics

Susana I.C.J. Palma
(1) Associate Laboratory i4HB – Institute for Health and Bioeconomy, NOVA School of Science and Technology, Universidade NOVA de Lisboa, 2829-516 Caparica, Portugal
(2) UCIBIO – Applied Molecular Biosciences Unit, Department of Chemistry, NOVA School of Science and Technology, Universidade NOVA de Lisboa, 2829-516 Caparica, Portugal
s.palma@fct.unl.pt

Nanotherapeutics are medicines formulated in the form of nanoparticles. Their potential arises from the stability, safety, efficacy, and targeting benefits they can impart compared to conventional medicines. Several classes of nanotherapeutics are currently available in the market to treat and prevent diseases such as cancer, blood disorders, infectious diseases, or neurological conditions. Although the path from research to market faces challenges, the impact of nanotherapeutics in healthcare is undoubtful, reflected in improved clinical outcomes, better quality of life for patients, and even in public health solutions for an emergency health crisis as the COVID-19 pandemic. Opportunities to further expand the field are expected to arise from biotechnological advances in the next years, toward the widespread safe and effective use of nanotherapeutics.

Nanotherapeutics for Infectious Diseases
Edited by Íris Luz Batalha

ISBN 978-981-5129-36-6 (Hardcover), 978-1-003-56678-6 (eBook)
www.jennystanford.com

2.1 Introduction

The application of nanotechnology to the medical field is known as "nanomedicine," a multidisciplinary field of science that combines nanotechnology with biomedical and pharmaceutical sciences. Nanomedicine includes the study, development, and commercialization of products such as nanorobots, imaging agents for diagnostics, nanotherapeutics (also known as "nanopharmaceuticals" or "nanodrugs"), and theranostic agents (*i.e.*, combining therapeutic and diagnostic properties) [1]. Nanomedicine has brought critical improvements to healthcare, particularly in drug delivery, with many nanoparticle-based medicines and vaccines now available in the market. Nanotherapeutics are usually defined as nanoparticle formulations with at least one dimension in the 1–100 nm range, which can be engineered to address unmet clinical needs, such as delivering therapeutic molecules to target tissues, overcoming drug solubility issues, or controlling the site and temporal release of a drug.

Since the commercialization of the first nanotherapeutic in 1974 (INFeD®, to treat iron deficiency anemia) [2], extensive research has been conducted in the field. Still, it was only after the launch of the first anti-cancer nanotherapeutic (Doxil®, in 1995) [3] that the number of nanotherapeutics approvals by the U.S. Food and Drug Administration (FDA) and the European Medicines Agency (EMA) started to significantly increase, including products for a variety of applications beyond cancer, namely treatment and prevention of infections, blood deficiencies, and neurological disorders, bringing enormous health benefits and greatly improving a patient's quality of life.

The impact of nanomedicine in the society is leading to optimistic market projections [4] that suggest a continued growth in nanotechnology-based therapeutic products and an increasing rate of prescription of nanotherapeutics. This is possible because nanoparticle-based systems allow designing medicines with tuned properties and with the potential to outperform conventional drug delivery systems. To date, around 70 nanotherapeutics have been approved in Europe and the United States [5].

In this chapter, we highlight the properties and potential of nanotherapeutics, describe the different classes of clinically approved nanotherapeutics, explain the typical bench-to-market processes and what is limiting them, and discuss the opportunities envisaged for future nano-based medicines.

2.2 Properties of Nanotherapeutics

Many pharmaceutical agents present limitations when administered free or in conventional formulations (*e.g.,* in the form of tablets, capsules, or syrups). Such limitations are related to low water solubility, instability, hampered permeability through biological barriers, and thus low bioavailability, leading to poor pharmacokinetics. This results in limited delivery to the target site, which is a major inefficiency. To achieve clinical efficacy, most conventional drug formulations require high dosages and/or increased administration frequency, which may compromise the safety of non-targeted tissues, leading to toxic side effects. Nanotechnology has a significant impact on the mitigation of these limitations.

Nanotherapeutics are formulations of "medicines" in the form of nanoparticles. There are two types of nanotherapeutics:

- Those where the nanoparticle is a therapeutic agent itself, as is the case of protein–polymer conjugates, nanocrystals, and inorganic nanoparticles such as silver nanoparticles.
- Those where a therapeutic molecule is coupled to a nanoparticle carrier (*e.g.,* surface-functionalized, entrapped in a matrix, encapsulated in a vesicle), as is the case of liposomes, protein, and polymer-based nanotherapeutics.

The nanosized formulation presents benefits compared to the free therapeutic molecules or conventional drug delivery systems, namely:

- Protection against threats faced by the drugs after entering the human body (*e.g.,* chemical and enzymatic degradation, uptake by the reticuloendothelial system (RES) [6]). Nanoformulations improve the solubility and stability of the therapeutic molecule and, thus, increase its circulation time and bioavailability [7].
- Control over the drug release rate, allowing a constant release rate over a desired timescale [8].

- Passive targeting and accumulation at cancer tissues through the enhanced permeation and retention (EPR) effect [9].
- Active targeting and selective accumulation at target tissues based on molecular recognition interactions between ligands attached at the nanocarrier's surface and specific receptors expressed by the target cells [10].
- Improved efficacy and reduced toxicity [11,12].

As a result of these properties, the therapeutic molecules are delivered more precisely to the target tissues than in conventional formulations, promoting a more localized therapeutic effect and reducing the total dose required for efficacy.

2.3 Path from R&D to Market

The typical clinical translation path of a new nanotherapeutic starts with research, usually performed either by pharmaceutical and biotechnological companies or by research groups in the academia. When interesting research results are obtained and a certain technological maturity is reached (usually technology readiness level (TRL) 4–5) [13], the potential new nanotherapeutic development might be taken up by a company that conducts characterization assays under good laboratory practices (GLPs) and starts implementation of production under good manufacturing practices (GMPs). It must then progress through sequential clinical trial studies involving humans. Data collected during these studies are used to ask for market authorization from the competent regulatory entities: EMA in the European Union and the FDA in the United States. They assess the potential new nanotherapeutic based on the balance between its benefits and risks, making sure that it is only authorized if the benefits outweigh the risks. (Fig. 2.1).

Currently, there are no specific regulations for nanotherapeutics. The marketing authorization processes rely on the regulations available for conventional medicinal products. It is required that nanotherapeutics meet the same quality, safety, and efficacy criteria as conventional medicinal products (*e.g.,* small chemical drugs, biologics) in order to be translated to the market [14–17]. Given the unique properties that arise from the nanomaterial's complex structure, large surface/volume ratio, and polydispersity, additional considerations are usually needed to demonstrate quality, safety, and efficacy. Despite the lack of harmonized standards, "reflection papers" and "guidance for the industry" documents issued by the EMA and the FDA, respectively, complement the regulations and should be used as a reference for new nanomedicine-related products [18–20].

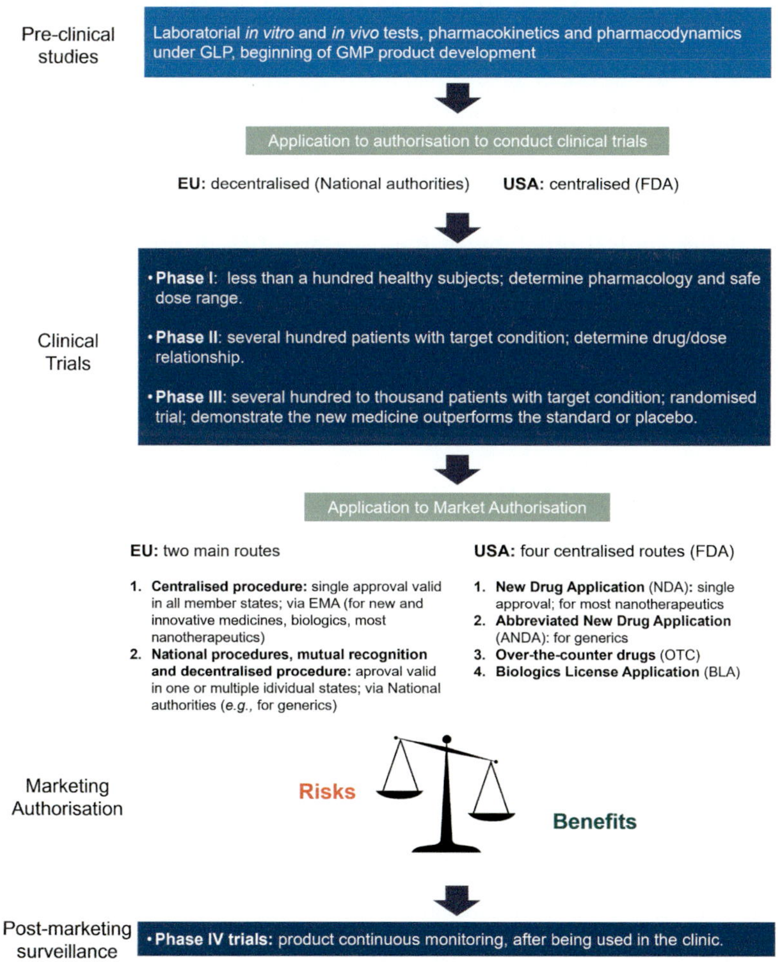

Figure 2.1 Steps from research and development to market authorization in the European Union (EU) and the United States. GLP: good laboratory practice; GMP: good manufacturing practice.

Developers of nanotherapeutics are encouraged to request guidance from the FDA and EMA since the early stages of development on the best study designs and methods to produce robust information, as well as on the type of information that needs to be included in the marketing authorization application.

Like conventional drugs, not all nanotherapeutic formulations are successfully translated to the market. Taking as an example nanotherapeutics for oncology applications, the success rate in Phase I was calculated as 94%, which is attributed to the good safety profile and improved pharmacokinetics of the drug. In Phase II, the success rate decreased to 48%, followed by 14% in Phase III [21]. Overall, the total success rate from Phase I to approval was estimated as 6%, thus outperforming conventional anti-cancer drugs, which have an estimated success rate of 3.4% [22]. Calculations for other clinical applications are not available, but it is recognized that the clinical outcome of nanotherapeutics in other areas has been poorer [15]. It is important to note that most of the approved oncologic nanotherapeutics are nanoformulations of previously approved drugs, which could have facilitated the approval process. On the other hand, completely new and innovative nanoformulations face a more challenging route from R&D to approval.

2.4 Historical Overview of Clinically Approved Nanotherapeutics

In 1974, the FDA approved the colloidal dispersion of ferric hydroxide stabilized with dextran, currently known as INFeD® (now marketed by Allergan Pharma), for the treatment of iron deficiency anemia in patients with chronic kidney disease. Later, in 1989, Diprivan®, a propofol emulsion based on lipids, was approved for anesthesia, and in the early 1990s, Adagen® (discontinued in 2018) and Oncaspar® were the first polymer–drug conjugates to enter the market in the United States, consisting of PEG-conjugated enzymes for enzyme replacement therapy and chemotherapy, respectively [23]. Shortly after that, the first liposomal formulations were commercialized in the United States for different indications, such as Doxil®, for refractory Kaposi's sarcoma in HIV patients, and Ambisome®, for fungal infections [24]. Since then, the nanotherapeutics potential has been increasingly explored, with an ever-growing number of new formulations entering the market.

In Europe, nanotherapeutics entered the market only in 1996, with Caelyx® (the European version of Doxil®). A significant delay has been observed between FDA and EMA approvals for some nanotherapeutics (*e.g.*, Diprivan®, USA: 1989, EU: 2001), but the number of approvals over the years follows a similar trend in the United States and the European Union, peaking between 2000 and 2005, followed by a fall and stabilization after 2006 (Fig. 2.2).

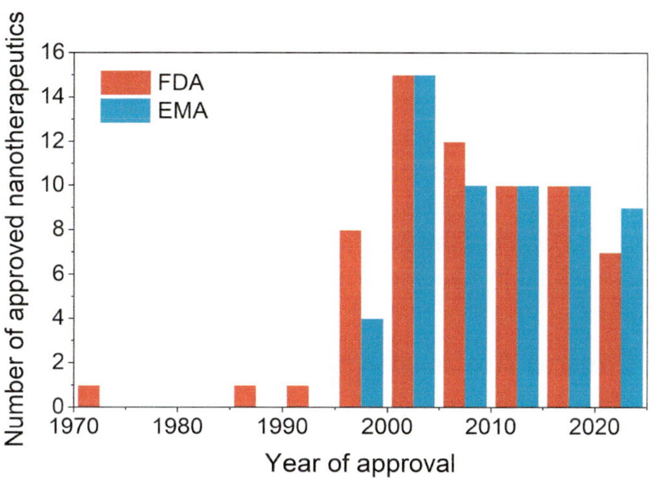

Figure 2.2 Evolution of the number of approved reference nanotherapeutics.

Table 2.1 (in the Appendix) provides a list of reference nanotherapeutics currently available for clinical use, with market authorization in the United States and Europe. The real number of commercialized nanotherapeutics is difficult to calculate due to the large number of generics and hybrid products that enter the market at a higher rate than the innovative formulations (which later become references for generics and hybrids) [25, 26]. As of today, there are around 77 approved reference formulations of nanotherapeutics. Of these, 19 have market approval only in the United States and 12 only in the European Union, while 46 are marketed in both regions.

Most of the marketed nanotherapeutics are nanoformulations of previously marketed drugs, engineered to improve pharmacokinetics and/or pharmacodynamics and hence drug safety and/or efficacy. For example, Doxil® was approved because it was as efficient as the free drug (doxorubicin) in the treatment of AIDS-related Kaposi's sarcoma but safer, as it caused significantly reduced cardiotoxicity [27]. Vyxeos®, a dual drug nanotherapeutic, improved the overall survival of acute myeloid leukemia patients (9.6 months) as compared to free cytarabine and daunorubicin combination therapy (standard of care, overall survival of 5.9 months) [28].

Engineered proteins, therapeutic oligonucleotides, and advanced engineered polymers have expanded the range of nanomaterials and therapeutic moieties incorporated in nanocarriers, resulting in new nanotherapeutics that are no longer just nano-formulated versions of existing drugs but completely new drugs (active therapeutic agents) themselves. Onpattro®, approved in 2018 for the treatment of neuropathy induced by hereditary transthyretin amyloidosis, is an excellent example of such next-generation nanotherapeutics, representing the first ever commercialized nucleic acid–based drug [29] and inspired the development of other nucleic acid–based therapies enabled by nanoparticle delivery.

2.5 Clinical Areas of Approved Nanotherapeutics

Out of the 77 clinically approved nanotherapeutics, approximately a fourth of them are indicated for cancer treatment. The remaining formulations are indicated for other therapeutical areas, spanning blood disorders, neurological diseases, chronic kidney diseases, and infections, among others (Fig. 2.3).

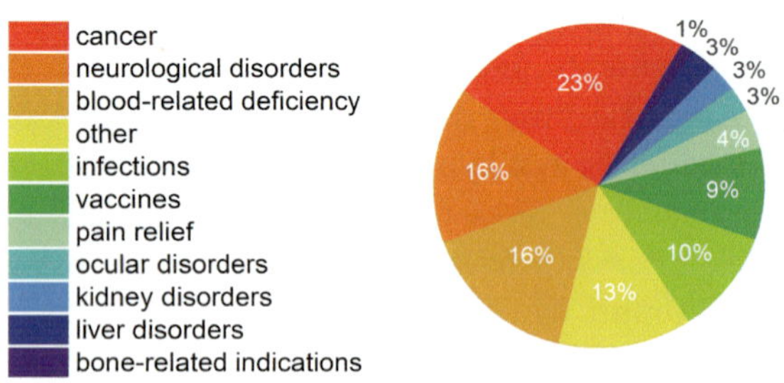

Figure 2.3 Approved reference nanotherapeutics per clinical area.

Infectious diseases are a case of particular interest due to their global impact and challenges faced in the prevention and treatment of diseases with high mortality rates, such as malaria, tuberculosis, and HIV [30]. Currently, there are eight distinct nanotherapeutics approved for the treatment (5) and prevention (3) of infections caused by fungi, bacteria, viruses, and parasites (Table 2.1 in the Appendix). The oldest in the market are the liposomal formulations of amphotericin B, for the treatment of systemic infections, namely Abelcet® (1995) and Ambisome® (1997), which improved the treatment efficacy compared to conventional amphotericin B formulations [31]. At the beginning of the 2000s, products for the management of infections by hepatitis viruses B and C and vaccines for the prevention of infections by hepatitis virus A, influenza, and zoster virus were marketed, but the most impactful and recent developments were in the fields of HIV, malaria, and COVID-19 (three major global health concerns).

Long-acting injectable nanoparticle-based antiretrovirals are breakthrough innovations for HIV management as they greatly reduce the administration frequency and promote patient adherence to therapy [32]. In 2021, Apretude® (cabotegravir) and Cabenuva Kit® (cabotegravir and rilpivirine) were approved by the FDA for pre-exposure prophylaxis and the treatment of HIV infections, respectively [33, 34]. These medicines consist of nanocrystals of the antiretroviral drugs coated with PEG and polysorbate surfactant [35], forming long-acting formulations. The extended half-life of the nanoformulations compared to the conventional formulation allows patients to substitute a daily pill with a bi-monthly intra-muscular injection with superior or similar efficacy [33, 36], which tremendously improves their quality of life.

Mosquirix™ is the only malaria vaccine that received a positive regulatory assessment. In 2015, the EMA approved the use of the vaccine outside of Europe as part of a cooperation with the World Health Organization (WHO) regarding medicines that are needed to prevent or treat diseases of major global public health importance [37]. Mosquirix™ is an injectable formulation where the antigen is formulated as a virus-like particle and combined with a liposomal adjuvant that helps protect children aged 6 weeks to 17 months against malaria caused by the parasite *Plasmodium falciparum* [38]. Although the efficiency is lower than 50%, a pilot implementation that started in 2019 in endemic countries demonstrated the potential of the vaccine as a public health tool in regions with moderate to high transmission risk, which led to a positive WHO recommendation, in 2021, for use in those regions [39].

Finally, another remarkable example of the outstanding impact of nanotherapeutics in the prevention of infectious diseases is the success of mRNA- and protein-based vaccines for SARS-CoV-2, where lipid nanoparticle–based delivery systems are employed to transport the vaccine into the body [40].

2.6 Classes of Clinically Approved Nanotherapeutics

Globally marketed nanotherapeutics have a variety of structures, depending on the nature of the active pharmaceutical moieties they carry and the target tissue.

Current nanotherapeutics can be grouped according to their structure into lipid-based nanotherapeutics (liposomes, lipid nanoparticles, and emulsions); polymer-based nanotherapeutics (polymer–protein conjugates, polymeric nanoparticles, dendrimers, and polymeric micelles); protein-based nanotherapeutics; nanocrystals; and inorganic nanoparticles. Nanocrystals and liposomes are currently the most commercialized formulations, representing around 50% of the available nanotherapeutics (Fig. 2.4).

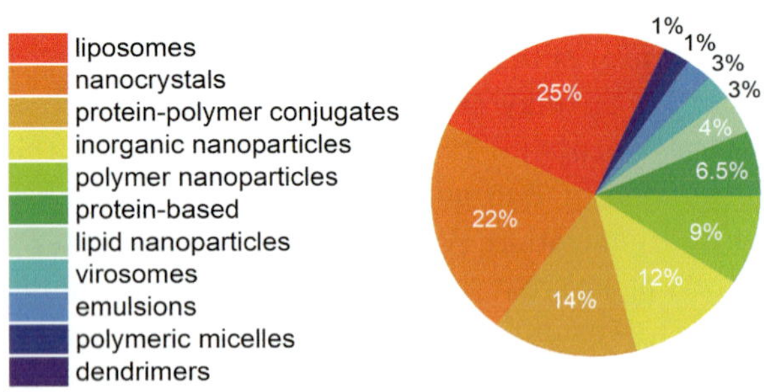

Figure 2.4 Proportion of currently commercialized products, distributed by subclasses.

2.6.1 Lipid-based Nanotherapeutics

Lipid-based nanotherapeutics include three main subclasses: liposomes, lipid nanoparticles, and emulsions (Fig. 2.5).

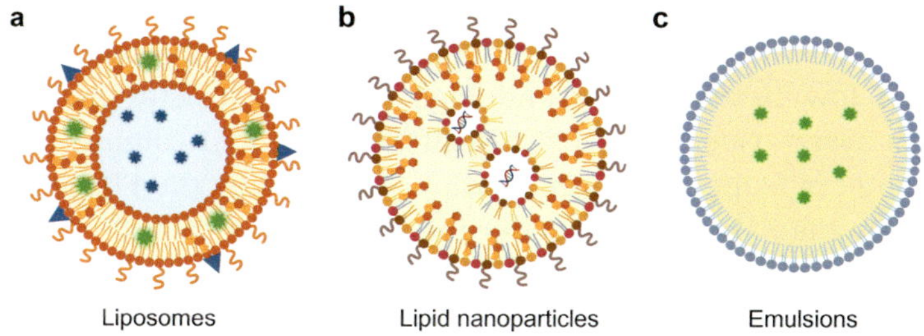

Figure 2.5 Lipid-based nanotherapeutics, with encapsulated hydrophilic and hydrophobic small drugs represented in blue and green, respectively, and targeting agents as blue triangles. **(a)** Liposomes and **(b)** lipid nanoparticles usually are coated with hydrophilic polymers, such as polyethylene glycol (PEG), to prolong the circulation time. **(c)** Oil-in-water emulsions encapsulating a hydrophobic drug, represented in green. Figure created with BioRender.com.

2.6.1.1 Liposomes

Liposomes are spherical self-assembled vesicles of phospholipid bilayers with hydrophilic cores, ranging between 20 nm and >1 µm [41]. The amphoteric structure of liposomes allows the transport of hydrophilic drug molecules in the aqueous core of the liposome, while hydrophobic drugs can be intercalated in the phospholipid bilayers (Fig. 2.5a). Also, more than one drug can be loaded in the same liposome within the two compartments (hydrophilic and hydrophobic), as in the case of Vyxeos® [42]. A common strategy to avoid fast clearance from circulation is liposome surface modification with hydrophilic polymers, such as polyethylene glycol (PEG) [43]. Cholesterol is also a frequent component of the lipidic bilayer, present in most commercial liposomal formulations as it improves properties such as fluidity, permeability, drug retention, phospholipid packing, and plasma stability [44].

The surface of liposomes can be decorated with specific functional biomolecules, as in the vaccine InflexalV®, which is functionalized with influenza virus glycoprotein antigens [45]. This type of liposomal formulation, where the liposome is decorated with viral moieties, is also known as a virosome.

Liposomes are the oldest nanotherapeutic class in clinical use and one of the most successful drug delivery systems to date. Liposomal drug nanocarriers were a revolution in medicine and pharmacology as they allowed to greatly improve the stability, biodistribution, and dose tolerability of existing drugs that suffered from low bioavailability and high systemic toxicity, associated with severe side effects [41]. Doxil® was the first commercialized nanotherapeutic for cancer therapy, approved by the FDA in 1995 and by the EMA in 1996, under the brand name Caelyx®. Doxil® is a liposome where doxorubicin molecules are encapsulated in the hydrophilic core (through a transmembrane pH gradient), and a PEG coating is used to increase the liposome circulation time. Doxil® is thought to operate by passive targeting, accumulating at the tumor sites by the EPR effect [46]. As a result of almost 30 years of research and development, liposomes are now well-established systems recognized as effective transporters for therapeutic agents. The clinical success of the liposomal drug delivery formulations motivated the issue, by both the EMA and the FDA, of specific guidelines for liposomes, discussing the technical aspects that should be included in an application for their market authorization [20, 47].

Nowadays, there are around 19 commercialized liposomal formulations, among anti-cancer [48], antibacterial [49], anti-fungal [31], pain-managing drugs [50], and vaccines [51] (Table 2.1).

Although liposomes are one of the most commercialized classes of nanotherapeutics, they still present limitations that prevent their wider utilization, namely the difficulty in modulating drug release *in vivo*, their limited drug-loading capacity, the possibility of oxidation of liposomal phospholipids, and short shelf-life [52].

2.6.1.2 Lipid nanoparticles

The development of lipid nanoparticles (LNPs) coincides with the development of nanocarriers for the delivery of oligonucleotides in gene therapy. LNPs were developed to overcome two major challenges of gene therapy: the need to deliver oligonucleotides intracellularly at the diseased tissue and the instability of free genetic material in a physiological medium. They, eventually, became the preferred genetic therapy carriers [53].

LNPs are multicomponent lipidic structures and differ from traditional liposomes mainly because they form micellar vesicles within the nanoparticle core (Fig. 2.5b). Nucleic acids can be transported in stable form within these micelles. LNPs are typically composed of phospholipids (that form the particle interface structure), cationic or ionizable lipids (that complex with negatively charged genetic material and aid endosomal escape), cholesterol (for stability and membrane fusion), and PEGylated lipids (to improve the stability of LNPs in body fluids and circulation time) [54]. Onpattro®, the first FDA- and EMA-approved LNP, in 2018, was also the first approved small interfering RNA (siRNA)-based drug and, therefore, a landmark in the history of both nanotherapeutics and genetic therapy. Onpattro® is a lipid nanoparticle–based siRNA drug with 60–100 nm used for the treatment of polyneuropathies induced by hereditary transthyretin amyloidosis. This nanodrug inhibits the synthesis of transthyretin protein by hepatocytes in the liver [29]. The success of Onpattro® paved the way for subsequent clinical studies and approval of other LNPs for gene delivery (Table 2.1), including the fast-paced development of Pfizer-BioNTech and Moderna mRNA COVID-19 vaccines [40], urgently needed to face the public health emergency caused by the pandemic that spread the world in late 2019. The vaccines, approved in late 2020 by the FDA and the EMA, reached the market through a remarkedly fast development process that allowed start using them worldwide in less than 2 years from the start of the pandemic, becoming the first mRNA nanotherapeutics to receive approval from the FDA and the EMA [24].

2.6.1.3 Emulsions

Emulsions consist of two immiscible liquids in which droplets of one are dispersed in the other, usually stabilized by an interface of lipidic surfactants. Only two of the 78 approved nanotherapeutics are emulsions (Fig. 2.5c). Diprivan®, an intravenous sedative-hypnotic agent used as an anesthetic, consists of an emulsion of the lipophilic anesthetic propofol utilizing soybean oil, purified egg lecithin, and glycerol [55]. Although it has been successfully used in the clinic, some drawbacks are the potential for hyperlipidemia caused by prolonged use, namely in intensive care units, and the thermodynamic instability of emulsions, which might lead to vesicle size increase during storage [56]. Norvir® is a self-emulsifying oil formulation for administration as an oral solution to treat HIV infections in conjunction with antiretrovirals. The emulsion formulation allows to carry the poorly soluble anti-HIV drug CSIC (5-chloro-3-phenylsulfonylindole-2-carboxamide) and improves its bioavailability. It is supplied as an anhydrous emulsion, but when dispersed in an aqueous medium, it transforms into a thermodynamically stable isotropic mixture of oil, surfactant, and co-surfactant, which self-emulsifies to yield micro- or nano-emulsions upon gentle agitation in an aqueous medium [57].

2.6.2 Protein-based Nanotherapeutics

Protein-based nanoparticles include conjugates where the protein acts simultaneously as a carrier and a targeting moiety for an attached therapeutic agent (Fig. 2.6a and Fig. 2.6b) and engineered fusion proteins combining targeting and therapeutic moieties (Fig. 2.6c).

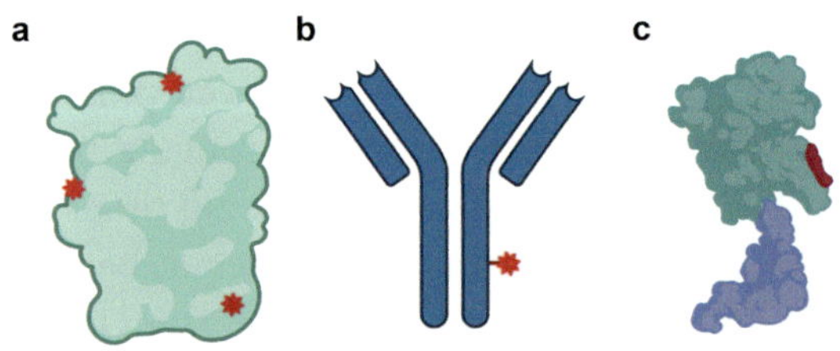

Figure 2.6 Examples of protein-based nanoparticles, with therapeutic moieties represented in red. **(a)** Albumin-bound drug nanoparticle. **(b)** Antibody–drug conjugate. **(c)** Therapeutic fusion protein. Figure created with BioRender.com.

Cancer therapy has been the main disease target for the currently approved protein nanotherapeutics (Table 2.1), but recently a protein-based nanotherapeutic, a vaccine in this case, has been approved for COVID-19 in the European Union in 2021 [58].

One of the key factors for the success of anti-cancer nanotherapeutics is the ability to be delivered specifically to tumor cells and not to healthy cells. Protein-based anti-cancer nanoparticles exploit tumor-targeting strategies. The human albumin protein, naturally present in blood serum, is considered a versatile carrier with specific characteristics that make it an ideal drug delivery agent, namely its availability, biodegradability, lack of toxicity, and lack of immunogenicity [5, 52]. In Abraxane®, human albumin is conjugated to the anti-cancer drug paclitaxel, forming nanoparticles that preferentially accumulate in solid tumors via albumin receptor–mediated transcytosis across the endothelium of tumor vessels [52, 59]. In addition, the extracellular matrix glycoprotein SPARC ("secreted protein, acidic and rich in cysteine"), overexpressed in many tumor types, binds albumin with high affinity, which further promotes the accumulation of albumin-bound drugs in the tumor interstitium [52, 59].

Another strategy for tumor targeting can be achieved by coupling the drug with a monoclonal antibody specific for characteristic surface biomarkers of target cells. These are known as antibody–drug conjugates (ADCs). Kadicla®, approved for clinical use in 2013, was the first ADC actively targeted to treat HER2-positive metastatic breast cancer. In Kadicla®, the cytotoxic drug DM1 is covalently coupled to the antibody trastuzumab (Herceptin®) [60], which specifically recognizes the biomarker HER-2, overexpressed in breast cancer cells. Thus, DM1 is delivered to HER2-positive breast cancer cells through trastuzumab recognition of the HER receptor. Internalized DM1 triggers apoptosis [61]. This is an example of a nanotherapeutic where the protein acts simultaneously as the transporter and the targeting agent. Ontak® is an example of a different strategy, where an engineered fusion protein was designed to combine a targeting moiety and a cytotoxic moiety in a single complex. The fusion protein is composed of the amino acid sequences for diphtheria toxin fragments A and B (the cytotoxic moiety) and the sequences for the receptor binding domain of human interleukin-2 (the targeting moiety), in order to selectively affect T cells that overexpress the interleukin-2 receptor, for the treatment of T-cell lymphoma through diphtheria toxin–mediated killing [62].

2.6.3 Nanocrystals

Encapsulating or entrapping a poorly soluble or unstable drug in a hydrophilic carrier helps to circumvent the solubility issue and facilitates delivery but can have limitations, namely, low drug-loading capacity, scale-up difficulties, or drug destabilization by the excipients [63]. Nanocrystallization is an alternative solubility-enhancement method that reduces the size of drug particles to the nanometer scale. Nanocrystals (Fig. 2.7a) have an increased surface area, which increases the dissolution speed and the saturation solubility compared to the bulk drug. Higher saturation solubility promotes the diffusion of particles across biological structures, such as the gastrointestinal tract walls [64].

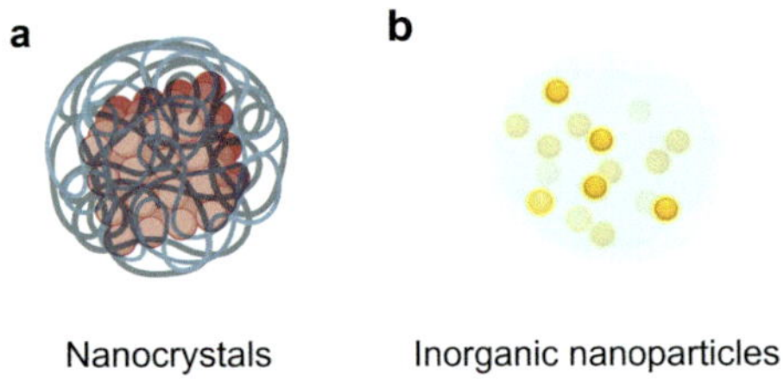

Figure 2.7 Nanocrystals and inorganic nanoparticle nanotherapeutics. **(a)** Nanocrystals are usually coated with hydrophilic polymers for improved circulation time. **(b)** Inorganic nanoparticles are usually formulated as aqueous colloidal suspensions stabilized by polymer coatings or matrices. Figure created with BioRender.com.

Nanocrystals are drug delivery systems that do not require a carrier. Instead, they are composed entirely of the drug compound and, in most cases, stabilizers (*e.g.,* PEG, PS80, xanthan gum) to prevent the aggregation of nanocrystals. Nanocrystalline formulations present fewer scalability problems and are easy to prepare using bottom-up or top-down approaches [63]. These properties make nanocrystals one of the preferred drug solubility enhancement methods for hydrophobic drugs, with promising results regarding drug bioavailability enhancement after oral administration [65]. Currently, 17 of the clinically approved nanotherapeutics are nanocrystal formulations, mostly commercialized for oral administration. They are mostly distributed as kits that include the nanocrystals in the form of dispersions, tablets/capsules, or powder, plus an aqueous solvent for the preparation of dispersions (nanosuspensions) just before administration (Table 2.1). Rapamune® was the first FDA- and EMA-approved organic nanocrystal, in 1999 and 2000, respectively. Its active ingredient is sirolimus, a bacteria-derived immunosuppressant to prevent kidney transplant rejection and is available as an oral solution or tablets.

2.6.4 Inorganic Nanoparticles

Inorganic nanoparticles (*e.g.*, gold, silver, silica, iron oxide, quantum dots, etc.) can have inherent therapeutic/diagnostic properties or be used as platforms for drug delivery (Fig. 2.7b) [66].

For example, iron oxide nanoparticles have been indicated for iron replacement therapy in chronic kidney disease patients suffering from anemia. All except one formulation (Triferic®) have stabilizing hydrophilic polymers forming coatings or entrapping matrices to increase the colloidal stability of the particles *in vivo*. These include INFed® (iron dextran), Ferrlecit® (sodium ferric gluconate complex in sucrose solution), Feraheme® (iron oxide nanoparticles coated with mannitol), Monoferric® (nanoparticle matrix structure composed of alternating layers of ferric hydroxide and the carbohydrate derisomaltose), among others (Table 2.1). Iron oxide nanoparticles have also been successfully used in the field of magnetic hyperthermia. NanoTherm™, a colloidal suspension of amino-silane-coated iron oxide nanoparticles of approximately 15 nm in size, has achieved European approval through CE marking, back in 2010 for the treatment of hard-to-treat cancers, including glioblastoma and prostate cancers [67, 68]. Magnetic hyperthermia is a technique that uses the heat generated by the magnetic nanoparticles when subjected to an alternating magnetic field to specifically kill cancer cells.

As mentioned previously, metal and metal oxide nanoparticles can have inherent therapeutic abilities, through different mechanisms, such as cellular membrane disruption, binding to intracellular components, and production of reactive oxygen species (ROS), which causes oxidative stress. For example, the antimicrobial properties of silver nanoparticles have been recognized for centuries, and these nanoparticles are already incorporated in hundreds of consumer products (*e.g.*, in food packaging, in textiles such as socks, or as coatings of surgical instruments). In the field of nanotherapeutics, so far this class of nanoparticles has been incorporated into antibacterial wound dressing gels (*e.g.*, SilvrSTAT®) [69].

Due to their optical, magnetic, radioactive, or plasmonic properties, inorganic nanoparticles have been extensively used in diagnostics and imaging (*e.g.*, gold nanoparticles are commonly used in rapid tests for pregnancy or COVID-19 and iron nanoparticles have been used to image live tumors). However, this is out of the scope of the current chapter and has been reviewed somewhere else [70].

The high number of ongoing clinical trials using inorganic nanoparticles (*e.g.*, AuroLase® therapy for the ablation of solid tumors) hints at a fast-changing scenario, with many new inorganic nanoparticle formulations expected to soon be seeing the light of day.

2.6.5 Polymer-based Nanotherapeutics

Polymer-based nanoparticles can be grouped into four subclasses: polymer–protein conjugates, polymeric nanoparticles, dendrimers, and polymeric micelles (Fig. 2.8).

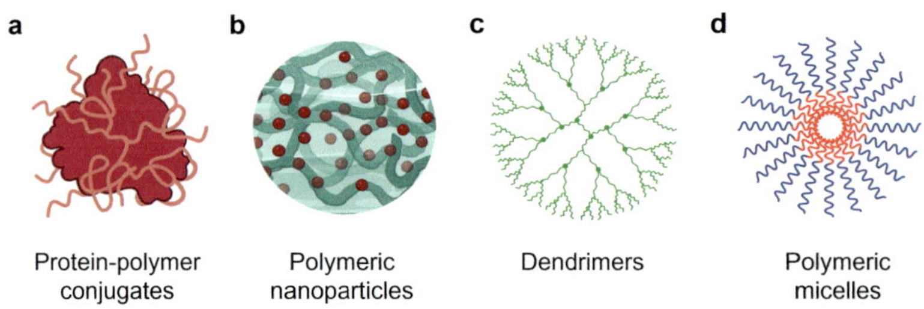

Figure 2.8 Polymer-based nanoparticles, where the active agent can be either a protein **(a)**; the polymer itself or a small drug molecule entrapped/encapsulated or conjugated to the polymer **(b)**; a dendrimer **(c)**; or a micelle **(d)**. Figure created with BioRender.com.

2.6.5.1 Polymer–protein conjugates

Recombinant proteins hold promise for the efficient and safe treatment of a range of diseases because they can be engineered to bind to specific cell surface receptors, replace natural proteins, or add new needed functionalities. To increase the stability and circulation time of therapeutic proteins, chains of a stabilizing polymer (mostly PEG) are conjugated covalently to proteins, resulting in polymer–protein conjugates (Fig. 2.8a). This subclass of nanotherapeutics represents around 15% of the total number of currently available formulations (Fig. 2.4).

Cimzia® is an example of a recombinant protein with specific binding properties and that adds new functions to the human body to treat a disease. It consists of a PEGylated recombinant humanized antibody Fab fragment that specifically recognizes and binds to tumor necrosis factor-alpha (TNF-α), blocking its activity. It has been clinically available since 2008 (the United States) and 2009 (European Union) to treat patients with autoimmune diseases such as rheumatoid arthritis, since TNF-α is a pro-inflammatory cytokine involved in promoting autoimmune attacks [71]. Mircera® is an injectable long-acting erythropoiesis-stimulating agent used to treat anemia associated with chronic kidney disease. It was approved by the FDA and the EMA in 2007 as a weekly injection and consists of PEGylated recombinant erythropoietin (epoetin-beta), used as a replacement for the natural erythropoietin, which is lacking in chronic kidney disease patients [72]. The most recently approved polymer–protein conjugate, in 2015 (UE) and 2018 (the United States), was Adynovate®, another example of a long-acting formulation. In this case, the therapeutic protein is a recombinant blood coagulation factor VIII to treat hemophilia A in patients suffering from frequent bleeding events [73].

2.6.5.2 Polymeric nanoparticles

Most polymeric nanoparticles currently available use functional synthetic polymers other than PEG, working either as the active therapeutic agent itself (*e.g.*, Copaxone®, Renagel®, Renvela®), or as an entrapping matrix for the therapeutic agent (*e.g.*, Eligard®, Restasis®, Zilretta®) (Fig. 2.8b). Different levels of structure complexity are found between these nanotherapeutics, ranging from a simple peptide mixture solution (Copaxone®) to engineered microspheres for prolonged release of the entrapped therapeutic substance (Zirletta®). Overall, the main advantage of polymeric nanoparticles is the versatility and variability of available polymers allowing to tune properties that are important for proper performance, namely, biodegradability, water solubility, or storage stability [54]. For example, Renagel® consists of tablets for oral administration made of sevelamer hydrochloride, a hydrogel of poly(allylamine hydrochloride) crosslinked with epichlorohydrin. Sevelamer binds to ingested phosphate, preventing its absorption into the blood. It is indicated to control the levels of serum phosphorus in patients with chronic kidney disease undergoing dialysis [74]. Renvela® is a second generation of Renagel®, resulting from tuning the counterion in sevelamer from hydrochloride to carbonate to mitigate the risk of metabolic acidosis associated with sevelamer hydrochloride [75]. Zilretta® is the first long-acting medicine approved by the FDA in 2017 for the treatment of osteoarthritis knee pain [64]. It represents an example of tuned biodegradability through the design of the nanoparticle and the choice of the polymer. The release of triamcinolone acetonide is controlled by the slow dissolution of a poly(lactic-co-glycolic acid) (PLGA) microsphere, opening nanochannels that allow a gradual drug release for a period of 3 months [76].

In polymer–small molecule drug conjugates, the therapeutic moiety is a small molecule chemically linked to the chains of the polymer (PEG or others). The choice of the linker chemistry allows to control the release and subsequent activity of the therapeutic molecule, while the polymer contributes to increasing the nanoparticle circulation time. A high number of clinical trials are ongoing (*e.g.*, Onzelad® therapy for breast cancer), suggesting that many polymer–small molecule drug conjugates will soon be available [77], but Movantik® (Moventig® in the European Union) is only currently available in the market, after receiving FDA and EMA approval (2014) [78, 79] for the treatment of opioid-induced constipation in patients with chronic pain. In Movantik®, the conjugation of the opioid antagonist nolaxone to PEG increases the circulation time and reduces the system penetration into the central nervous system compared to the free nolaxone. Thus, it treats the opioid-related constipation but prevents nolaxone from counter-acting opioid-induced analgesia [80].

2.6.5.3 Dendrimers and polymeric micelles

Dendrimers and polymeric micelles are nanoparticles formed by synthetic materials assembled as a repetitive pattern of basic units.

Dendrimers are globular nanoparticles composed of iterative monomers arranged as repetitive branches starting from an initiator core and finishing in terminal surface groups (Fig. 2.8c) [81]. They possess a predictable and ordered structure, and their production permits precise control over properties such as shape, size, charge, or surface properties [64, 81], a major difference when compared to other vesicle-like nanotherapeutics (*e.g.*, liposomes, LNPs) and a promising feature for drug delivery products. The architecture of dendrimers allows to incorporate pharmaceutically active moieties within the interior cavity or at the surface, conjugated to the terminal groups. VivaGel® (EMA approved in 2015) is the first and single commercial dendrimer-based nanotherapeutic, formulated as a topical gel, indicated to prevent and treat bacterial vaginosis infections. It is composed of a poly-L-lysine (PLL) dendrimer with anionic naphthalene disulfonate groups at the surface, which bind to the virus or bacterial membranes, preventing host infection [81].

The production of polymeric micelles (Fig. 2.8d) is not as precisely controlled as the synthesis of dendrimers because micelles form via a self-assembling mechanism. They consist of amphiphile polymers forming nanoparticles with a hydrophobic core, which can be loaded with poorly aqueous soluble drugs, and a hydrophilic surface, to allow stable dispersion in aqueous media [64]. In Apealea®, which was approved by the EMA in 2018 for use in combination with carboplatin, the anti-cancer drug paclitaxel is encapsulated in micelles, supplied as a powder for reconstitution as an intravenous infusion to treat ovarian, fallopian tube, and peritoneal cancers [82].

2.7 Challenges for Nanotherapeutics Clinical Approval

There are still challenges in the clinical translation of nanotherapeutics [5, 15, 83], and thus, the path from the laboratory bench to the market is usually long (up to 20 years [71]) and costly (average US$1.3 billion [84]). There is an ongoing discussion on the real success of nanotherapeutics [15, 85], despite the variety and number of products available. The challenges faced in clinical translatability have been thoroughly reviewed in the last years [14, 83] and are mainly related to:

- Difficulties in performing proper preclinical characterization and safety assessment of nanotherapeutics from early stages due to the lack of harmonized standard protocols and analytical tools.

  International standardization entities have agreed that additional physicochemical properties must be characterized for nanotherapeutics, namely, size distribution, surface charge, solubility, stability, drug loading, and drug release profile [86], in addition to the measurement of the typical parameters required for chemical and biological drugs (*e.g.*, identity, strength, potency, stability and impurities, bacterial endotoxins, and bioburden of different chemical ingredients) [15]. However, harmonized protocols for characterization have not yet been defined.
- Difficulties in scaling up production and absence of GMP standards, due to the complexity and variety of nanotherapeutics preparation methods.

  The properties of the small lab batch should be kept along the scale-up to large industrial batch volumes to ensure both the *in vitro* and *in vivo* performance of the nanotherapeutics. Challenges such as batch-to-batch variability, limited shelf-lives of the formulations, and nanoparticle damage upon sterilization, among others, are faced many times when trying to scale up production. GMPs are needed to ensure that products are consistently produced with controlled quality.
- The fragmented and unclear regulatory framework, lacking specific guidance and processes for nanotherapeutics.

  The guidelines on quality requirements for nanoparticle-based medicines issued by the EMA and the FDA are not legally binding and are limited to certain classes of nanoparticles (*e.g.*, liposomes, iron nanoparticles, copolymer micelles). Specific regulations are needed for the different classes of nanotherapeutics as well as for those with features that bridge the definitions of "medicine" and "medical devices."
- Costly development and manufacture along with lack of cost-effectiveness studies.

  Nanotherapeutics manufacture is considerably more expensive than conventional medicines, namely, owing to the small-scale production, which leads to higher selling prices or acquisition costs for hospitals, limiting their daily availability and success [87]. Improved cost-effectiveness of the nanotherapeutic over the conventional drug counterpart can justify the high acquisition cost, as demonstrated for PEGylated liposomal doxorubicin [88].

2.8 Opportunities and Future Trends for Clinical Use of Nanotherapeutics

The progress of nanomedicine in the last two decades is predicted to deliver positive results in the future. The first generation of nanotherapeutics (liposomes and nanocrystals) is proving safe in the long term. At the same time, patent exclusivity periods are ending. Thus, generic nanotherapeutics are expected to become even more frequent in the market, possibly lowering the acquisition costs for hospitals and increasing availability to patients. The knowledge acquired regarding the characterization and safety evaluation of the first-generation nanotherapeutics can be used to accelerate the establishment of specific and harmonized characterization protocols and regulations toward simpler and faster market approval processes.

Both Europe and the United States have launched nanomedicine initiatives to support the field, including the European Technology Platform on Nanomedicine (EPTN) and the US National Nanotechnology Initiative (NNI). Through these initiatives, nanotherapeutics developers can get mentoring on preclinical characterization, risk assessment, and regulatory frameworks from the beginning of development. For example, the European Nanomedicine Characterisation Laboratory (EUNCL), within the EPTN, provides a set of free-of-charge preclinical characterization services and GMP for nanomedicine formulations to help advance the translation to the clinic [15]. Similarly, in the United States, researchers can send their materials for characterization to the Nanotechnology Characterization Laboratory (NCL), according to a series of emerging protocols [64].

Trends in research publications and clinical trials suggest a direction toward nucleic acid–based therapeutics, multicomponent nanosystems encapsulating more than one therapeutic agent, or offering both therapeutic and imaging functions (theranostics) [89], and engineered nanoplatforms to overcome heterogeneous barriers to delivery, for more efficient, safe, and personalized new medicines in the future [54]. The attraction of private investment to boost market translation of such promising innovations is required. The conjugation of (1) a clearer understanding of the safety and cost-effectiveness of nanotherapeutics, (2) harmonized characterization GMP standards, and (3) specific regulations for nanotherapeutics is expected to promote larger interest and investment of big pharma industries in the nanomedicine field, thus contributing to faster widespread use of nanotherapeutics.

2.9 Conclusion

Current clinically approved nanotherapeutics include different classes of formulations with multicomponent structures. Nanotherapeutics have the advantage of mitigating solubility problems, stabilizing and delivering pharmaceutically active moieties to the site of action more efficiently and in a sustained manner compared to conventional medicines.

The potential of nanotherapeutics is enormous. However, the innovative and complex structures of nanotherapeutics compared to conventional nanomedicines bring challenges to clinical translation. There is an ongoing debate between academia, industry, and regulatory bodies regarding the definition of harmonized product characterization, GMP, and specific regulations for marketing authorization of nanotherapeutics.

A nanomedicine-friendly regulatory framework is expected to accelerate the commercialization of more and more products in varied clinical areas. Nanotherapeutics represent an important pathway for complementing conventional medicines. The know-how acquired in the last two decades offers unprecedented opportunities to evolve the nanotherapeutics field toward the promised societal impact of nanotechnology, offering solutions for improved drug delivery, safer therapies, and cost-effective medicines available to patients in their daily routines, overall promoting an improved quality of life.

Appendix

Table 2.1 Representative list of the clinically approved nanotherapeutics currently in the market, grouped by class. Generics are not included.

Commercial Name	Active Ingredients	Marketing Authorization Holder	Formulation	Dosage Form/Route	Indication	First FDA Approval	First EMA Approval
NANOCRYSTALS							
Rapamune®	Sirolimus (rapamycin)	Pfizer	Rapamycin plus a stabilizing excipient mixture, including PS80, PC, PEG	Solution or tablets for oral administration	Prevents kidney transplant rejection	1999	2001
Ritalin LA®	Methylphenidate hydrochloride	Novartis	Uses the Novartis proprietary SODAS® technology (Spheroidal Oral Drug Absorption System) technology); ritalin filled in uncoated and coated beads for a 2-stage (immediate plus delayed) drug release	Oral administration; capsule	Treats attention deficit hyperactivity disorder in children	2001	2002*
Concerta®	Methylphenidate hydrochloride	Jansen Pharmaceuticals (USA); Janssen-Cilag (EU)	Uses OROS® technology. Osmotically active trilayer drug core (for prolonged release), surrounded by a semipermeable membrane with an immediate-release drug overcoat	Oral administration; capsule	Treats attention deficit hyperactivity disorder in children	2000	2002*

(*Continued*)

Table 2.1 *(Continued)*

Commercial Name	Active Ingredients	Marketing Authorization Holder	Formulation	Dosage Form/Route	Indication	First FDA Approval	First EMA Approval
Zanaflex®	Tizanidine hydrochloride	Acorda	Capsules and tablets	Capsules and tablets; oral administration	Muscle relaxant; treats muscle stiffness associated with cerebral palsy or other neurological disorders	2002	n.a.
Emend®	Aprepitant	Merck Sharp & Dohme	Suspension and capsules	Suspension; capsules; oral administration	Antiemetic (prevents nausea and vomiting associated with chemotherapy)	2003	2003
TriCor®	Fenofibrate	Abbott Laboratories	Fenofibrate with several stabilizer excipients (*e.g.*, sodium lauryl sulfate, soybean lecithin, xanthan gum)	Tablet; oral administration	Treats hyperlipidemia	2004	Unknown*
Focalin XR®	Dexmethylphenidate HCl	Novartis	Uses the Novartis proprietary SODAS® for a 2-stage drug release	Oral administration; capsule	Psychostimulant; treats attention deficit hyperactivity disorder in children	2005	Unknown*

Commercial Name	Active Ingredients	Marketing Authorization Holder	Formulation	Dosage Form/Route	Indication	First FDA Approval	First EMA Approval
Megace ES®	Magestrol acetate	Par Pharmaceutical, Inc.	Uses nanocrystal technology to enhance solubility	Suspension; oral	Treats loss of appetite and wasting syndrome in people with acquired immunodeficiency syndrome (AIDS)	2005	Unknown*
Invega®	Paliperidone	Jansen Pharmaceuticals (USA); Jansen-Cilag International N.V. (EU)	Uses OROS® osmotic drug release technology	Oral administration; tablets	Antipsychotic; treats schizophrenia or schizoaffective disorder in adults	2006	2007
Emend injection (USA); IVEMEND (EU)	Fosaprepitant glutamine	Merck Sharp & Dohme	Nanoparticles of fosaprepitant dimeglumine stabilized by PS80	Powder; injectable; intravenous; infusion	Antiemetic	2008	2008
Zyprexa Relprevv (USA)/Zypadhera® (EU)	Olanzapine	Eli Lilly Nederland B.V.(EU)	Olanzapine plus, a stabilizing excipient mixture, including Mannitol and PS80	Powder + solvent; injection; intramuscular	Maintenance treatment of adult patients with schizophrenia	2009	2008

(*Continued*)

Table 2.1 (*Continued*)

Commercial Name	Active Ingredients	Marketing Authorization Holder	Formulation	Dosage Form/Route	Indication	First FDA Approval	First EMA Approval
Invega Sustenna®; Invega Trinza; Invega Hafyera (USA) / Xeplion®; Trevicta; Byannli (EU)	Paliperidone Palmitate	Jansen Pharmaceuticals (USA); Jansen-Cilag International N.V. (EU)	Aqueous suspension that uses Alkermes' nanoparticle technology. Attached to a fatty acid to allow a slow release.	Suspension; extended release; injection; intramuscular	Antipsychotic; treats schizophrenia or schizoaffective disorder in adults.	2009; 2015; 2021	2011; 2014; 2020
Ryanodex®	Dantrolene sodium	Eagle Pharmaceuticals	Powder; reconstituted into a nanosuspension in sterile water	Suspension; injectable	Skeletal muscle relaxant to treat malignant hyperthermia	2014	2021◊
Apretude®	Cabotegravir	ViiV Healthcare	Surfactant-stabilized drug nanocrystals (crystalline-free acid form of cabotegravir, stabilized by a PEG and a polysorbate)	Suspension, extended release; intramuscular	Prevents HIV (pre-exposure prophylaxis – reduces the risk of sexually acquired HIV-1 infection)	2021	n.a.
Cabenuva Kit® (USA); Rekambys® and Vocabria® (EU)	Cabotegravir and ripilvirine (co-packaged as independent injections)	ViiV Healthcare	Surfactant-stabilized drug nanocrystals	Suspension, extended release; intramuscular	Treats HIV	2021	2021

Commercial Name	Active Ingredients	Marketing Authorization Holder	Formulation	Dosage Form/Route	Indication	First FDA Approval	First EMA Approval
LIPID-BASED NANOPARTICLES							
Liposomes							
Doxil® USA); Caelyx® (EU)	Doxorubicin	Johnson & Johnson (USA); Baxter Holding B.V (EU)	PEGylated liposomal doxorubicin hydrochloride; HSPC:cholesterol:DSPE-PEG	Injectable; intravenous	Treats breast neoplasm, ovarian neoplasm, multiple myeloma, Kaposi's sarcoma	1995	1996
Abelcet®	Amphotericin B	Lediant Biosciences	Amphoterecin B complexed with 2 phospholipids; DMPC:DMPG; ribbon structure	Injectable; intravenous	Treats life-threatening fungal infections	1995	n.a.
DaunoXome®	Daunorubinin	Galen Ltd.; Gilead Sciences	Liposome; DSPC:cholesterol	Aqueous dispersion; injectable	Treats cancer, advanced HIV-associated Kaposi's sarcoma, acute myeloid leukemia	n.a.	1996*
AmBisome®	Amphotericin B	Astellas Pharma US, Inc.	Liposome with drug intercalated with the membrane; HSPC:DSPG, cholesterol	Powder; injectable; intravenous	Treats systemic fungal infection, visceral leishmaniasis, cryptococcal meningitis in HIV patients	1997	n.a.

(Continued)

Table 2.1 (*Continued*)

Commercial Name	Active Ingredients	Marketing Authorization Holder	Formulation	Dosage Form/Route	Indication	First FDA Approval	First EMA Approval
Curosurf®	Poractant alfa	Chiesi Farmaceutici, S.p.A.	Natural surfactant from porcine lungs (mainly phosphatidylcholine)	Suspension; injectable; intratracheal suspension	Rescues treatment of respiratory distress syndrome in newborn babies	1999	n.a.
Myocet™ liposomal	Doxorubicin	Teva B. V.	Non-pegylated liposome; EPC:cholesterol	Powder; injectable; intravenous	Treats metastatic breast neoplasm	n.a.	2000
Visudyne®	Verteporfin	Novartis Pharma	EGP:DMPC	Powder; intravenous perfusion to be used in photodynamic therapy	Photosensitizer to treat choroidal neovascularization	2000	2000
Lipoplatin™ Nanoplatin™	Cisplatin	Regulon AE	DPPG:soy PC:MPEG-DSPE:Cholesterol	Suspension; injection; intravenous	Treats pancreatic adenocarcinoma (lipoplatin), non-small lung cancer carcinoma (nanoplatin)	n.a.	2007◊
Mepact®	Mifamurtide	Takeda France SAS	DOPC:DOPS	Powder; injectable; intravenous	Treats osteosarcoma	n.a.	2009
Exparel®	Bupivacaine	Pacira Biosciences, Inc.	DEPC:DPPG:Cholesterol:Tricaprylin	Dispersion; injectable; infiltration	Post-surgery pain relief (anesthetic)	2011	2020

Commercial Name	Active Ingredients	Marketing Authorization Holder	Formulation	Dosage Form/Route	Indication	First FDA Approval	First EMA Approval
Marqibo®	Vincristine	Acrotech Biopharma LLC	SPH:Cholesterol	Suspension; injectable; intravenous	Treats pancreatic adenocarcinoma, Philadelphia chromosome-negative acute lymphoblastic leukemia	2012	n.a.
Lipodox®	Doxorubicin	Sun Pharmaceutical Industries Ltd.	PEGylated liposome; HSPC:Cholesterol:DSPE-PEG	Aqueous dispersion; injection; intravenous	Treats breast neoplasm, Kaposi sarcoma (AIDS patients)	2013	n.a.
Onivyde®	Iriontecan hydrochloride trihydrate	Ipsen Biopharm Ltd. (USA); Les Laboratoires Servier (EU)	DSPC:Cholesterol:DSPE-PEG	Suspension; injection; intravenous	Treats metastatic pancreatic adenocarcinoma	2015	2016 ◊
Mosquirix™	RTS,S antigen based vaccine	GlaxoSmithKline	DOPC:Cholesterol	Powder; injection; intramuscular	Prevents malaria in young children	n.a.	2015
Vyxeos®	Daunorubicin and cytrabine	Jazz Pharmaceuticals, Inc.	DSPC:DSPG:Cholesterol	Injection; intravenous	Treats acute myeloid leukemia	2017	2018
Shingrix®	Varicella Zoster Virus glycoprotein E	GlaxoSmithKline Biologicals S.A.	DOPC:Cholesterol	Powder and suspension; injection; intramuscular	Prevents and treats Herpes Zoster in adults >50 years old	2017	2018

(Continued)

Table 2.1 *(Continued)*

Commercial Name	Active Ingredients	Marketing Authorization Holder	Formulation	Dosage Form/Route	Indication	First FDA Approval	First EMA Approval
Arikayce®	Amikacin (aminoglycoside antibacterial)	Insmed Inc. (USA); Insmed Netherlands B.V. (EU)	DPPC:Cholesterol	Suspension; oral inhalation	Treats *Mycobacterium avium complex (MAC)* lung disease	2018	2020 ◇
Zolsketil™ pegylated liposomal	Doxorubicin	Accord Healthcare S.L.U.	PEGylated liposome	Aqueous dispersion; injection; intravenous	Treats metastatic breast cancer, advanced ovarian cancer, progressive multiple myeloma, and AIDS-related Kaposi's sarcoma	n.a.	2022
Lipid Nanoparticles							
Onpattro®	Patisiran sodium (a siRNA)	Alnylam Pharmaceuticals, Inc. (USA); Alnylam Netherlands B.V. (EU)	siRNA:DLin-MC3-DMA:DSPC:Cholesterol:PEG2000-C-DMG	Injection; intravenous	Treats polyneuropathy of hereditary transthyretin-mediated amyloidosis	2018	2018
COMIRNATY (previously Pfizer-BioNTech Vaccine)	mRNA encoding the viral spike glycoprotein of SARS-CoV-2	Pfizer (USA); BioNTech Manufacturing GmbH (EU)	AL0315:DSPC:Cholesterol:DMA-PEG	Suspension; injection; intramuscular	Prevents COVID-19 infection	2020	2020

Commercial Name	Active Ingredients	Marketing Authorization Holder	Formulation	Dosage Form/Route	Indication	First FDA Approval	First EMA Approval
Spikevax (previously Moderna COVID-19 Vaccine)	mRNA encoding the viral spike glycoprotein of SARS-CoV-2	Moderna TX Inc. (USA); Moderna Biotech Spain S.L. (EU)	SM-102: DSPC:Cholesterol:DMG-PEG	Suspension; injection; intramuscular	Prevents COVID-19 infection	2020	2021
Emulsions							
Diprivan®	Propofol	Fresenius Kabi USA, LLC (USA); Aspen Pharma Trading Ltd. (EU)	Oil-in-water emulsion; soybean oil: glycerol: egg lecithin	Emulsion; injectable; parenteral infusion	Anesthetic and sedative	1989	2001
Norvir®	Ritonavir, a peptidomimetic inhibitor of HIV-1 and HIV-2 proteases	Abbot Laboratories (USA); AbbVie Deutschland GmbH Co. KG (EU)	Self-emulsifying oil formulation; mixture of oil, surfactant, and co-surfactant, which on gentle agitation in aqueous medium self-emulsifies to yield micro- or nano-emulsion	Oral solution; powder for oral solution; capsules and tablets	Treats HIV infections	1996	1996
VIROSOMES							
Inflexal V	Inactivated influenza virus components	Crucell Berna Biotech	Mixture of three monovalent unilamellar virosomes (liposomes with surface decorated with viral antigens HA and NA); PC:HA:NAPC:cephalin:DOPC:DOPE	Suspension; injectable; intramuscular; sub-cutaneous	Prevents influenza infection	n.a.	1997

(*Continued*)

Table 2.1 *(Continued)*

Commercial Name	Active Ingredients	Marketing Authorization Holder	Formulation	Dosage Form/Route	Indication	First FDA Approval	First EMA Approval
Epaxal®	Inactive hepatitis A virus	Crucell (Johnson & Johnson)	Inactivated HAV bound to influenza virosomes (liposomes composed of PC and PCA with influenza glycoproteins HA and NA intercalated in the bilayer)	Injection; intramuscular	Prevents hepatitis A virus	n.a.	2000
PROTEIN-BASED NANOPARTICLES							
Ontak®	Diphtheria toxin	Eisai (USA); Clinipace GmbH (EU)	Fusion protein composed of the amino acid sequences for diphtheria toxin fragments A and B and the sequences for the receptor binding domain of human interleukin-2	Solution; injection; intravenous	Treats leukemia, T-cell lymphoma	2008	2001 ◊
Zevalin®	^{90}Y-ibritumomab tiuxetan	CIS bio international	Monoclonal antibody ibritumomab conjugated with tiuexan. Mixed with a solution of radioactive lithium chloride (^{90}Y) immediately before use	Suspension, kit for radiopharmaceutical preparation; injection; intravenous	Treats B cell non-Hodgkin's lymphoma	2022	2004
Abraxane®	Paclitaxel	Celgene Distribution B.V.	Albumin-bound paclitaxel nanoparticles	Powder for dispersion; perfusion; intravenous	Treats breast neoplasm, pancreatic neoplasm, non-small-cell lung carcinoma	2005	2008

Commercial Name	Active Ingredients	Marketing Authorization Holder	Formulation	Dosage Form/Route	Indication	First FDA Approval	First EMA Approval
Nuvaxovid®	Recombinant SARS-CoV-2 spike protein	Novavax	PS80 micelle nanoparticle decorated with recombinant SARS-CoV-2 spike protein; co-formulated with adjuvant matrix M	Dispersion; injection; intramuscular	Prevents COVID-19 infection	n.a.	2021
Kadicla®	DM1 (cytotoxic maytansinoid)	Genentech, Inc. (USA); Roche Registration GmbH (EU)	Lysine-linked monoclonal antibody drug conjugate; trastuzumab coupled to DM1 (cytotoxic microtubule-inhibiting drug)	Power; injection; intravenous	Treats HER2+ metastatic breast cancer	2013	2013
POLYMER-BASED NANOTHERAPEUTICS							
Polymer-Protein Conjugates							
Oncaspar®	L-asparaginase	Enzon Pharmaceuticals (USA); Les Laboratoires Servier (EU)	PEGylated L-asparaginase	Solution; intramuscular injection or intravenous infusion	Treats acute lymphoblastic leukemia (ALL)	1994	2016
PegIntron®	Recombinant interferon alpha 2b	SCHERING	PEGylated recombinant interferon alpha 2b	Powder; injection; subcutaneous	Treats hepatitis C infection	2001	n.a.
Pegasys®	Recombinant interferon alpha 2a	Zr Pharma & GmbH (Eu)	PEGylated recombinant interferon alpha 2a	Injection; subcutaneous	Treats chronic hepatitis B and chronic hepatitis C	2002	2002

(*Continued*)

Table 2.1 *(Continued)*

Commercial Name	Active Ingredients	Marketing Authorization Holder	Formulation	Dosage Form/Route	Indication	First FDA Approval	First EMA Approval
Neulasta®	Recombinant methionyl human G-CSF	Amgen (USA); Amgen Europe B.V.	PEGylated recombinant methionyl human G-CSF (filgrastim)	Solution; injection; subcutaneous	Reduces the duration of neutropenia and prevents febrile neutropenia in cancer patients undergoing chemotherapy	2002	2002
Somavert®	Recombinant pegvisomant	Pfizer Pharmaceuticals (USA); Pfizer Europe MA EEIG (EU)	PEGylated recombinant pegvisomant (analogue of human growth hormone (GH) that has been structurally altered to act as a GH receptor antagonist)	Powder; injection; subcutaneous	Treats acromegaly (growth hormone-related disorder)	2003◊	2002
Mircera®	Epoetin beta	Hoffman la Roche (UEA); Roche Registration GmbH (EU)	Recombinant erythropoietin (epoetin-beta) covalently linked with methoxypolyethylene glycol	Solution; injection	Treats anemia associated with chronic kidney disease	2007	2007
Cimzia®	Recombinant anti-TNF-α IgG Fab's fragment	UCB Inc (USA); UCB Pharma SA (EU)	PEGylated recombinant humanized antibody Fab fragment against tumor necrosis factor alpha (TNF-α) (Certolizumab pegol)	Powder; injection; subcutaneous	Rheumatoid arthritis, Crohn's disease, psoriatic arthritis, ankylosing spondylitis	2008	2009

Commercial Name	Active Ingredients	Marketing Authorization Holder	Formulation	Dosage Form/Route	Indication	First FDA Approval	First EMA Approval
Krystexxa®	Pegloticase	Horizon Pharma (USA)	PEGylated recombinant porcine-like uricase (pegloticase)	Concentrate; injection; intravenous	Treats gout disease	2010	n.a.
Plegridy®	Recombinant interferon beta-1a	Biogen Idec. (USA); Biogen Netherlands B.V.(EU)	PEGylated interferon beta 1-a	Solution in prefilled pen or syringe; injection; subcutaneous	Treats relapsing forms of multiple sclerosis	2014	2014
Rebinyn®(USA); Refixia® (EU)	Recombinant coagulation factor IX	Novo Nordisk Inc. (USA); Novo Nordisk A/S (EU)	Recombinant Coagulation Factor IX, GlycoPEGylated	Powder and solvent solution; injection; intravenous infusion	Treats and controls bleeding episodes or perioperative management of bleeding in patients with hemophilia B	2017	2017
Adynovate® (USA); Adynovi (EU)	Recombinant coagulation factor VIII	Baxalta Incorporated (USA); Baxalta Innovations GmbH (EU)	PEGylated coagulation factor VIII	Powder and solvent; solution; injection; intravenous	Treats and prevents bleeding in patients with hemophilia A	2018	2015
Polymeric Nanoparticles							
Copaxone®	Glatiramer acetate	Teva Neuroscience, Inc	Mixture of random polymers (peptides) composed of the four amino acids found in myelin basic protein (glutamic acid, lysine, alanine, and tyrosine)	Solution; injection; subcutaneous	Treats relapsing forms of multiple sclerosis	1996	Unknown*

(Continued)

Table 2.1 (*Continued*)

Commercial Name	Active Ingredients	Marketing Authorization Holder	Formulation	Dosage Form/Route	Indication	First FDA Approval	First EMA Approval
Renagel®	Sevelamer hydrochloride	Genzyme Europe B.V.	Film-coated tablets of sevelamer hydrochloride, a hydrogel of poly(allylamine hydrochloride) crosslinked with epichlorohydrin	Oral	Controls serum phosphorus in patients with chronic kidney disease under dialysis	2000	2000
Eligard®	Leuprolide acetate	Tolmar Pharmaceuticals Inc. (USA)	Leuprolide acetate (nonapeptide analogue of the gonadotropin-releasing hormone receptor) entrapped in a polymeric matrix of poly (DL-lactide-co-glycolide) (PLGH or PLG) dissolved in N-methyl-2-pyrrolidone (NMP)	Two separate prefilled syringes to mix right before injection; subcutaneous	Treats prostate cancer	2002	2003*
Restasis®	Cyclosporine	Allergan	Emulsion; glycerin:castor oil:polysorbate 80:carbomer copolymer type A	Ophthalmic	Chronic dry eye	2003	n.a.
Renvela®	Sevelamer carbonate	Sanofi (USA); Genzyme Europe B.V (EU)	Powder or film-coated tablets of sevelamer carbonate (poly (allylamine-co-N,N'-diallyl1,3-diamino-2-hydroxypropane) carbonate)	Oral	Controls serum phosphorus in patients with chronic kidney disease under dialysis	2009	2007

Commercial Name	Active Ingredients	Marketing Authorization Holder	Formulation	Dosage Form/Route	Indication	First FDA Approval	First EMA Approval
Movantik® (USA) Moventig® (EU)	Nolaxone	AstraZeneca Pharmaceuticals LP (USA) Kyowa Kirin (EU)	Film-coated tablets of PEGylated nolaxone	Oral	Treats opioid-induced constipation	2014	2014
Zilretta®	Triamcinolone acetonide	Flexion Therapeutics	Microspheres composed of triamcinolone acetonide embedded in a biodegradable PLGA matrix	Powder; extended-release injection; intra-articular	Manages knee osteoarthritis	2017	n.a.
Dendrimers							
VivaGel® BV	SPL7013 (polyanionic dendrimer with a benzylhydramine-amide-lysine core and 32 naphthalene disulfonate surface functional groups)	Mundipharma	Mucoadhesive gel with 1% SPL7013	Topical	Prevents and treats bacterial vaginosis infections	n.a.	2015
Polymeric Micelles							
Apealea®	Paclitaxel	Inceptua AB	Paclitaxel encapsulated in the proprietary XR17 micelle platform technology	Powder; infusion; intravenous	Treats ovarian cancer, fallopian tube cancer, peritoneal cancer	n.a.	2018

(Continued)

Table 2.1 *(Continued)*

Commercial Name	Active Ingredients	Marketing Authorization Holder	Formulation	Dosage Form/Route	Indication	First FDA Approval	First EMA Approval
INORGANIC NANOPARTICLES							
INFeD® (USA); Cosmofer (EU)	Ferric oxyhydroxide	Allergan USA, Inc. (USA)	A complex of ferric hydroxide and dextran in aqueous solution	Solution; injection or infusion; intravenous	Treats iron deficiency anemia in chronic kidney disease	1974	2013*
Ferrlecit®	Ferric oxyhydroxide	Sanofi Aventis, US (USA)	Sodium ferric gluconate complex in sucrose aqueous solution	Solution; injection; intravenous	Treats iron deficiency anemia in chronic kidney disease	1999	Unknown*
Venofer®(USA)	Ferric oxyhydroxide	American Reagent, Inc. (USA)	A complex of polynuclear iron(III)-hydroxide in sucrose in aqueous solution	Solution; injection; intramuscular or intravenous	Treats iron deficiency anemia in chronic kidney disease	2000	2013*
Feraheme® (USA); Rienso® (EU)	Ferumoxytol (non-stoichiometric magnetite coated with polyglucose sorbitol carboxymethylether)	Covis® (USA) / Takeda Pharma A/S (EU)	Aqueous colloidal formulation of iron oxide nanoparticles coated with mannitol	Solution; injection; intravenous	Treats iron deficiency anemia in chronic kidney disease	2009	n.a.
Injectafer®(USA); Ferinject® (EU)	Ferric carboxymaltose	American Reagent (USA); VIFOR France SA (EU)	Ferric carboxymaltose aqueous solution	Solution; injection; intravenous	Treats iron deficiency anemia in chronic kidney disease	2013	2013*

Commercial Name	Active Ingredients	Marketing Authorization Holder	Formulation	Dosage Form/Route	Indication	First FDA Approval	First EMA Approval
Monoferric® (USA); Monofer (EU)	Ferric derisomaltose	Pharmacosmos A/S	Colloidal particle solution, nanoparticle matrix composed of alternating layers of ferric hydroxide and the carbohydrate derisomaltose	Solution; injection, intravenous	Treats iron deficiency anemia in chronic kidney disease	2020	2009*
Triferic®	Ferric pyrophosphate citrate	Rockwell medical Inc.	Carbohydrate-free, water-soluble, complex iron salt added to the dialysate of patients	Solution added to the dialysate; injection; intravenous	Treats iron deficiency anemia in chronic kidney disease	2015	n.a.
Nanotherm™	Iron oxide nanoparticles coated with aminosilane	MagForce NT GmbH	Aqueous dispersion of aminosilane-coated iron oxide nanoparticles	Water dispersion; intratumoral injection	Treats glioblastoma and prostate cancer through hyperthermia	n.a.	2010.
SilvrSTAT™	Silver nanoparticles	ABL Medical, LCC	Hydrogel impregnated with silver nanoparticles coated with tetrasilver tetroxide	Water-based gel; topical application	Treats and prevents bacterial infection in wounds	2014	n.a.

*Authorized via National Procedures in the EU, available under different names; ◊Orphan designation (simpler approval process in the United States and in the European Union for medicines targeting rare diseases); n.a. not available in the market currently

Abbreviations:

DMPC: l-α-dimyristoylphosphatidylcholine; **DMPG**: l-α-dimyristoylphosphatidylglycerol; **DOPC**: dioleylphosphatidylcholine; **DPPG**: 1,2-dipalmitoyl-sn-glycero-3-phosphorylglycerol; **DPPG**: 1,2-Dipalmitoyl-sn-glycero-3-phosphorylglycerol; **DSPE-PEG 2000**: 1,2-distearoyl-sn-glycero-3-phosphoethanol-amine-N-[poly-(ethyleneglycol)-2000]; **G-CSF:** granulocyte colony stimulating factor; **HA**: hemagglutinin (glycoprotein from influenza virus); **HAV**: hepatitis A virus; **HER-2**: human epidermal growth factor 2; **MPEG-DSPE**: monomethoxy-polyethylene-glycol-1,2-distearoyl-sn-glycero-3-phosphoethanolamine; **NA**: neuraminidase (glycoprotein from influenza virus); **PC**: phosphatidylcholine (lecithin); **PCA**: phosphatidylethanolamine (cephalin); **PEG 4000**: polyethylene glycol 4000; **POPC**: palmitoyl-oleoyl-phosphatidylcholine; **PS20**: polysorbate 20; **PS80**: polysorbate 80

References

1. S. Bayda, M. Adeel, T. Tuccinardi, M. Cordani, and F. Rizzolio, The history of nanoscience and nanotechnology: From chemical–physical applications to nanomedicine, *Molecules* **25**, 112 (2020).
2. D. K. Wysowski, L. Swartz, B. V. Borders-Hemphill, M. R. Goulding, and C. Dormitzer, Use of parenteral iron products and serious anaphylactic-type reactions, *Am. J. Hematol.* **85**, 650 (2010).
3. Y. Barenholz, Doxil®: The first FDA-approved nano-drug: Lessons learned, *J. Control. Release* **160**, 117 (2012).
4. Coherent Market Insights, Nanopharmaceuticals market projection 2017–2026, https://www.coherentmarketinsights.com/press-release/nanopharmaceuticals-market-to-surpass-us-1689-billion-by-2026-1287 (accessed in October 2023).
5. A. A. Halwani, Development of pharmaceutical nanomedicines: From the bench to the market, *Pharmaceutics* **14**, 106 (2022).
6. M. J. Ernsting, M. Murakami, A. Roy, and S. D. Li, Factors controlling the pharmacokinetics, biodistribution and intratumoral penetration of nanoparticles, *J. Control. Release* **172**, 782 (2013).
7. L. M. Tatham, S. P. Rannard, and A. Owen, Nanoformulation strategies for the enhanced oral bioavailability of antiretroviral therapeutics, *Ther. Deliv.* **6**, 469 (2015).
8. N. J. Carter, Extended-release intramuscular paliperidone palmitate, *Drugs* **72** (8), 1137 (2012).
9. M. A. Subhan, S. S. K. Yalamarty, N. Filipczak, F. Parveen, and V. P. Torchilin, Recent advances in tumor targeting via EPR effect for cancer treatment, *J. Pers. Med.* **11**, 571 (2021).
10. R. Li, K. Zheng, C. Yuan, Z. Chen, and M. Huang, Be active or not: The relative contribution of active and passive tumor targeting of nanomaterials, *Nanotheranostics* **1**, 346 (2017).
11. M. E. R. O'Brien et al., Reduced cardiotoxicity and comparable efficacy in a phase III trial of pegylated liposomal doxorubicin HCl (CAELYX/Doxil) versus conventional doxorubicin for first-line treatment of metastatic breast cancer, *Ann. Oncol. Off. J. Eur. Soc. Med. Oncol.* **15**, 440 (2004).
12. M. M. Jain et al., The efficacy and safety of paclitaxel injection concentrate for nano-dispersion (PICN) at two different doses versus paclitaxel albumin-stabilized nanoparticle formulation in subjects with metastatic breast cancer (MBC), *J. Clin. Oncol.* **32**, 1069 (2014).

13. E. B. Souto, G. F. Silva, J. Dias-ferreira, A. Zielinska, F. Ventura, A. Durazzo, M. Lucarini, E. Novellino, and A. Santini, Nanopharmaceutics: Part II—production scales and clinically compliant production methods, *Nanomaterials* **10**, 455 (2020).

14. S. Đorđević, M. M. Gonzalez, I. Conejos-Sánchez, B. Carreira, S. Pozzi, R. C. Acúrcio, R. Satchi-Fainaro, H. F. Florindo, and M. J. Vicent, Current hurdles to the translation of nanomedicines from bench to the clinic, *Drug Deliv. Transl. Res.* **12** (3), 500 (2022).

15. M. Germain, F. Caputo, S. Metcalfe, G. Tosi, K. Spring, A. K. O. Åslund, A. Pottier, R. Schiffelers, A. Ceccaldi, and R. Schmid, Delivering the power of nanomedicine to patients today, *J. Control. Release* **326**, 164 (2020).

16. K. M. Tyner, P. Zou, X. Yang, H. Zhang, C. N. Cruz, and S. L. Lee, Product quality for nanomaterials: Current U.S. experience and perspective, *Wiley Interdiscip. Rev. Nanomed. Nanobiotechnol.* **7** (5), 640 (2015).

17. Y. Min, J. M. Caster, M. J. Eblan, and A. Z. Wang, Clinical translation of nanomedicine, *Chem. Rev.* **115**, 11147 (2015).

18. FDA, Drug products, including biological products, that contain nanomaterials: Guidance for industry, https://www.fda.gov/regulatory-information/search-fda-guidance-documents/drug-products-including-biological-products-contain-nanomaterials-guidance-industry (accessed in October 2023).

19. EMA, Multidisciplinary: Nanomedicines | guidelines on nanomedicines, https://www.ema.europa.eu/en/human-regulatory/research-development/scientific-guidelines/multidisciplinary/multidisciplinary-nanomedicines (accessed in October 2023).

20. FDA, Liposome drug products: Chemistry, manufacturing, and controls; human pharmacokinetics and bioavailability; and labeling documentation, https://www.fda.gov/regulatory-information/search-fda-guidance-documents/liposome-drug-products-chemistry-manufacturing-and-controls-human-pharmacokinetics-and (accessed in October 2023).

21. H. He, L. Liu, E. E. Morin, M. Liu, and A. Schwendeman, Survey of clinical translation of cancer nanomedicines: Lessons learned from successes and failures, *Acc. Chem. Res.* **52**, 2673 (2019).

22. C. H. Wong, K. W. Siah, and A. W. Lo, Estimation of clinical trial success rates and related parameters, *Biostatistics* **20**, 273 (2019).

23. S. Bisso and J. C. Leroux, Nanopharmaceuticals: A focus on their clinical translatability, *Int. J. Pharm.* **578**, 119098 (2020).

24. T. T. H. Thi, E. J. A. Suys, J. S. Lee, D. H. Nguyen, K. D. Park, and N. P. Truong, Lipid-based nanoparticles in the clinic and clinical trials: From cancer nanomedicine to Covid-19 vaccines, *Vaccines* **9**, 1 (2021).

25. EMA, Generic and hybrid medicines, https://www.ema.europa.eu/en/human-regulatory/marketing-authorisation/generic-hybrid-medicines (accessed in October 2023).

26. FDA, What is the approval process for generic drugs? https://www.fda.gov/drugs/generic-drugs/what-approval-process-generic-drugs (accessed in October 2023).

27. D. S. Alberts and D. J. Garcia, Safety aspects of pegylated liposomal doxorubicin in patients with cancer, *Drugs* **54** (4), 30 (1997).

28. A. C. Krauss et al., FDA approval summary: (Daunorubicin and cytarabine) liposome for injection for the treatment of adults with high-risk acute myeloid leukemia, *Clin. Cancer Res.* **25**, 2685 (2019).

29. A. Akinc et al., The Onpattro story and the clinical translation of nanomedicines containing nucleic acid-based drugs, *Nat. Nanotechnol.* **14** (12), 1084 (2019).

30. G. A. Roth et al., Global, regional, and national age-sex-specific mortality for 282 causes of death in 195 countries and territories, 1980–2017: A systematic analysis for the global burden of disease study 2017, *Lancet* (London, England) **392**, 1736 (2018).

31. J. P. Adler-moore and R. T. Proffitt, Amphotericin B lipid preparations: What are the differences? *Clin. Microbiol. Infect.* **14**, 25 (2008).

32. D. H. Surve and A. B. Jindal, Recent advances in long-acting nanoformulations for delivery of antiretroviral drugs, *J. Control. Release* **324**, 379 (2020).

33. FDA, FDA approves first injectable treatment for HIV pre-exposure prevention, https://www.fda.gov/news-events/press-announcements/fda-approves-first-injectable-treatment-hiv-pre-exposure-prevention (accessed in October 2023).

34. FDA, FDA approves first extended-release, injectable drug regimen for adults living with HIV, https://www.fda.gov/news-events/press-announcements/fda-approves-first-extended-release-injectable-drug-regimen-adults-living-hiv (accessed in October 2023).

35. A. R. Kirtane, M. Verma, P. Karandikar, J. Furin, R. Langer, and G. Traverso, Nanotechnology approaches for global infectious diseases, *Nat. Nanotechnol.* **16**, 369 (2021).

36. E. T. Overton et al., Long-acting cabotegravir and rilpivirine dosed every 2 months in adults with HIV-1 infection (ATLAS-2M), 48-

week results: A randomised, multicentre, open-label, phase 3b, non-inferiority study, *Lancet* **396**, 1994 (2020).

37. C. Morrison, Landmark green light for Mosquirix malaria vaccine, *Nat. Biotechnol.* **33**, 1015 (2015).

38. M. B. Laurens, RTS,S/AS01 vaccine (Mosquirix^{TM}): An overview, *Hum. Vaccin. Immunother.* **16**, 480 (2020).

39. WHO, WHO recommends groundbreaking malaria vaccine for children at risk, https://www.who.int/news/item/06-10-2021-who-recommends-groundbreaking-malaria-vaccine-for-children-at-risk (accessed in October 2023).

40. R. Verbeke, I. Lentacker, S. C. De Smedt, and H. Dewitte, The dawn of MRNA vaccines: The COVID-19 case, *J. Control. Release* **333**, 511 (2021).

41. P. Liu, G. Chen, and J. Zhang, A review of liposomes as a drug delivery system: Current status of approved products, regulatory environments, and future perspectives, *Molecules* **27**, 1372 (2022).

42. K. Tzogani et al., EMA review of daunorubicin and cytarabine encapsulated in liposomes (Vyxeos, CPX-351) for the treatment of adults with newly diagnosed, therapy-related acute myeloid leukemia or acute myeloid leukemia with myelodysplasia-related changes, *Oncologist* **25**, e1414 (2020).

43. K. Maruyama, T. Yuda, A. Okamoto, S. Kojima, A. Suginaka, and M. Iwatsuru, Prolonged circulation time *in vivo* of large unilamellar liposomes composed of distearoyl phosphatidylcholine and cholesterol containing amphipathic poly(ethylene glycol), *Biochim. Biophys. Acta* **1128**, 44 (1992).

44. P. Nakhaei, R. Margiana, D. O. Bokov, W. K. Abdelbasset, M. A. Jadidi Kouhbanani, R. S. Varma, F. Marofi, M. Jarahian, and N. Beheshtkhoo, Liposomes: Structure, biomedical applications, and stability parameters with emphasis on cholesterol, *Front. Bioeng. Biotechnol.* **9**, 705886 (2021).

45. R. Mischler and I. C. Metcalfe, Inflexal®V: A trivalent virosome subunit influenza vaccine: Production, *Vaccine* **20**, B17 (2002).

46. A. A. Gabizon, H. Shmeeda, and S. Zalipsky, Pros and cons of the liposome platform in cancer drug targeting, *J. Liposome Res.* **16**, 175 (2006).

47. EMA, Data requirements for intravenous liposomal products developed with reference to an innovator liposomal product, https://www.ema.europa.eu/en/data-requirements-intravenous-liposomal-products-

developed-reference-innovator-liposomal-product-0 (accessed in October 2023).

48. L. D. Mayer, P. Tardi, and A. C. Louie, CPX-351: A nanoscale liposomal co-formulation of daunorubicin and cytarabine with unique biodistribution and tumor cell uptake properties, *Int. J. Nanomedicine* **14**, 3819 (2019).

49. O. Khan and N. Chaudary, The use of amikacin liposome inhalation suspension (Arikayce) in the treatment of refractory nontuberculous mycobacterial lung disease in adults, *Drug Des. Devel. Ther.* **14**, 2287 (2020).

50. P. Discepola, M. Bouhara, M. Kwon, B. A. Siddiqui, T. A. Whitwell, S. Y. Sanghvi, K. D. Cook, R. E. Moore, A. Korban, and J. D. Eloy, EXPAREL® (long-acting liposomal bupivacaine) use for popliteal nerve block in postoperative pain control after ankle fracture fixation, *Pain Res. Manag.* **2020**, 5982567 (2020).

51. L. Schoenmaker, D. Witzigmann, J. A. Kulkarni, R. Verbeke, G. Kersten, W. Jiskoot, and D. J. A. Crommelin, MRNA-lipid nanoparticle COVID-19 vaccines: Structure and stability, *Int. J. Pharm.* **601**, 120586 (2021).

52. A. Loureiro, N. G. Azoia, A. C. Gomes, and A. Cavaco-Paulo, Albumin-based nanodevices as drug carriers, *Curr. Pharm. Des.* **22**, 1371 (2016).

53. X. Hou, T. Zaks, R. Langer, and Y. Dong, Lipid nanoparticles for MRNA delivery, *Nat. Rev. Mater.* **6** (12), 1078 (2021).

54. M. J. Mitchell, M. M. Billingsley, R. M. Haley, M. E. Wechsler, N. A. Peppas, and R. Langer, Engineering precision nanoparticles for drug delivery, *Nat. Rev. Drug Discov.* **20**, 101 (2021).

55. K. A. Thompson and D. B. Goodale, The recent development of propofol (DIPRIVAN®), *Intensive Care Med.* **26**, S400 (2000).

56. D. C. Warltier, M. T. Baker, M. Naguib, and B. Ch, Propofol: The challenges of formulation, *Anesthesiology* **103**, 860 (2005).

57. N. C. Obitte, L. C. Rohan, C. M. Adeyeye, M. A. Parniak, and C. O. Esimone, The utility of self-emulsifying oil formulation to improve the poor solubility of the anti HIV drug CSIC, *AIDS Res. Ther.* **10**, 1 (2013).

58. J. H. Tian et al., SARS-CoV-2 spike glycoprotein vaccine candidate NVX-CoV2373 immunogenicity in baboons and protection in mice, *Nat. Commun.* **12** (1), 372 (2021).

59. N. Desai et al., Increased antitumor activity, intratumor paclitaxel concentrations, and endothelial cell transport of cremophor-free, albumin-bound paclitaxel, ABI-007, compared with cremophor-based paclitaxel, Clin. Cancer Res. **12**, 1317 (2006).

60. M. T. Kim, Y. Chen, J. Marhoul, and F. Jacobson, Statistical modeling of the drug load distribution on trastuzumab emtansine (Kadcyla): A lysine-linked antibody drug conjugate, *Bioconjug. Chem.* **25**, 1223 (2014).

61. G. von Minckwitz et al., Trastuzumab emtansine for residual invasive HER2-positive breast cancer, *N. Engl. J. Med.* **380**, 617 (2019).

62. D. Kaminetzky and K. B. Hymes, Denileukin diftitox for the treatment of cutaneous T-cell lymphoma, *Biol. Targets Ther.* **2**, 717 (2008).

63. K. Joshi, A. Chandra, K. Jain, and S. Talegaonkar, Nanocrystalization: An emerging technology to enhance the bioavailability of poorly soluble drugs, *Pharm. Nanotechnol.* **7**, 259 (2019).

64. C. L. Ventola, Progress in nanomedicine: Approved and investigational nanodrugs, *Pharm. Ther.* **42**, 742 (2017).

65. M. R. Gigliobianco, C. Casadidio, R. Censi, and P. Di Martino, Nanocrystals of poorly soluble drugs: Drug bioavailability and physicochemical stability, *Pharmaceutics* **10**, 134 (2018).

66. H. Huang, W. Feng, Y. Chen, and J. Shi, Inorganic nanoparticles in clinical trials and translations, *Nano Today* **35**, 100972 (2020).

67. E. Union, Nanotherm CE certificate, https://ec.europa.eu/tools/eudamed/#/screen/search-device/2d0eb76d-1dec-4a2b-b98f-36fba5433988 (accessed in October 2023).

68. X. Liu et al., Comprehensive understanding of magnetic hyperthermia for improving antitumor therapeutic efficacy, *Theranostics* **10**, 3793 (2020).

69. M. E. Zayed and S. G. Ibrahim, Comparative study between silver nanoparticles dressing (SilvrSTAT gel) and conventional dressing in diabetic foot ulcer healing: A prospective randomized study, *Int. J. Low. Extrem. Wounds* **22**, 48 (2023).

70. X. Han, K. Xu, O. Taratula, and K. Farsad, Applications of nanoparticles in biomedical imaging, *Nanoscale* **11**, 799 (2019).

71. F. Farjadian, A. Ghasemi, O. Gohari, A. Roointan, M. Karimi, and M. R. Hamblin, Nanopharmaceuticals and nanomedicines currently on the market: Challenges and opportunities, *Nanomedicine* **14**, (2019).

72. N. W. Levin, S. Fishbane, F. V. Cañedo, S. Zeig, G. M. Nassar, J. E. Moran, G. Villa, U. Beyer, and D. Oguey, Intravenous methoxy polyethylene glycol-epoetin beta for haemoglobin control in patients with chronic kidney disease who are on dialysis: A randomised non-inferiority trial (MAXIMA), *Lancet* **370**, 1415 (2007).

73. P. L. Turecek, S. Romeder-Finger, C. Apostol, A. Bauer, A. Crocker-Buqué, D. A. Burger, R. Schall, and H. Gritsch, A world-wide survey and field study in clinical haemostasis laboratories to evaluate FVIII:C activity assay variability of ADYNOVATE and OBIZUR in comparison with ADVATE, *Haemophilia* **22**, 957 (2016).

74. E. A. Slatopolsky et al., RenaGel®: A nonabsorbed calcium- and aluminum-free phosphate binder, lowers serum phosphorus and parathyroid hormone, *Kidney Int.* **55**, 299 (1999).

75. A. B. Pai and B. M. Shepler, Comparison of sevelamer hydrochloride and sevelamer carbonate: Risk of metabolic acidosis and clinical implications, *Pharmacotherapy* **29**, 554 (2009).

76. Knee injection for osteoarthritis pain | ZILRETTA® (triamcinolone acetonide extended-release injectable suspension), https://zilretta.com/about-zilretta/the-microspheres/ (accessed in October 2023).

77. I. Ekladious, Y. L. Colson, and M. W. Grinstaff, Polymer–drug conjugate therapeutics: Advances, insights and prospects, *Nat. Rev. Drug Discov.* **18**, 273 (2019).

78. Moventig | European Medicines Agency, https://www.ema.europa.eu/en/medicines/human/EPAR/moventig (accessed in October 2023).

79. Movantik | Drugs@FDA, https://www.accessdata.fda.gov/drugsatfda_docs/nda/2014/204760Orig1s000TOC.cfm (accessed in October 2023).

80. T. Anantharamu, S. Sharma, A. K. Gupta, N. Dahiya, D. B. S. Singh Brashier, and A. K. Sharma, Naloxegol: First oral peripherally acting Mu opioid receptor antagonists for opioid-induced constipation, *J. Pharmacol. Pharmacother.* **6**, 188 (2015).

81. A. Santos, F. Veiga, and A. Figueiras, Dendrimers as pharmaceutical excipients: Synthesis, properties, toxicity and biomedical applications, *Materials* (*Basel*). **13**, 65 (2020).

82. Apealea®: Elevar Therapeutics, https://elevartherapeutics.com/apealea-paclitaxel-micellar-elevar/ (accessed in October 2023).

83. V. Gadekar, Y. Borade, S. Kannaujia, K. Rajpoot, N. Anup, V. Tambe, K. Kalia, and R. K. Tekade, Nanomedicines accessible in the market for clinical interventions, *J. Control. Release* **330**, 372 (2021).

84. O. J. Wouters, M. McKee, and J. Luyten, Estimated research and development investment needed to bring a new medicine to market, 2009-2018, *JAMA* **323**, 844 (2020).

85. K. Greish, A. Mathur, M. Bakhiet, and S. Taurin, Nanomedicine: Is it lost in translation? *Ther. Deliv.* **9**, 269 (2018).

86. ISO/IEC/NIST/OECD, ISO, IEC, NIST and OECD International Workshop on Documentary Standards for Measurement and Characterization for Nanotechnologies: Final Report (2008).

87. R. Bosetti and S. L. Jones, Cost-effectiveness of nanomedicine: Estimating the real size of nano-costs, *Nanomedicine* **14**, 1367 (2019).

88. R. Bosetti, G. Ferrandina, W. Marneffe, G. Scambia, and L. Vereeck, Cost-effectiveness of gemcitabine versus PEGylated liposomal doxorubicin for recurrent or progressive ovarian cancer: Comparing chemotherapy with nanotherapy, *Nanomedicine* (Lond). **9**, 2175 (2014).

89. M. B. Shete, T. S. Patil, A. S. Deshpande, G. Saraogi, N. Vasdev, M. Deshpande, K. Rajpoot, and R. K. Tekade, Current trends in theranostic nanomedicines, *J. Drug Deliv. Sci. Technol.* **71**, 103280 (2022).

Chapter 3

High-Priority Pathogens: Where Do We Stand?

Raquel Portela,[a*] Klementina Borovnik,[a*] Constança M.P. Costa,[a*] Inês R. Grilo,[b] and Pedro M. Pereira[a]
[a]*Mostmicro, ITQB NOVA, Av. da República, EAN. Oeiras, 2780-157, Portugal*
[b]*Lysando Innovations Lab GmbH, Am BioPark 13, 93053 Regensburg, Germany*
pmatos@itqb.unl.pt

*These authors contributed equally.

Infectious diseases (also known as communicable diseases or transmissible diseases) can be defined as an illness caused by infection/presence/growth of a pathogenic biological agent in a host (human or otherwise). These pathogenic biological agents can be separated into four groups: parasites (protozoa and helminths—worms), fungi, viruses, and bacteria.

3.1 Introduction

Infectious diseases are currently a chief cause of morbidity and mortality across the globe and are a clear indicator of social justice as these have an increased burden in developing countries. In fact,

Nanotherapeutics for Infectious Diseases
Edited by Íris Luz Batalha

ISBN 978-981-5129-36-6 (Hardcover), 978-1-003-56678-6 (eBook)
www.jennystanford.com

people living in low-income countries are far more likely to die from an infectious disease, and even when death is not the outcome, they usually suffer a higher socioeconomic burden of the disease [1]. However, and evidently, the burden of infectious diseases is not exclusive to developing countries. It is currently a considerable civilizational threat as we live in a time where pandemics are more likely to occur due to increased global travel, urbanization, and greater exploitation of the natural environment [2, 3], and where there is a real risk of antibiotic-resistant bacteria taking us back to pre-penicillin times [4, 5]. In developing and developed countries alike, infectious diseases, more than other human death causes (such as heart disease, the leading cause of death worldwide in 2019 [6]), elicit a great deal of attention due to the human suffering they cause and the global scale they have or have the potential to achieve (Table 3.1).

HIV/AIDS, malaria, and tuberculosis, the "Big Three" infectious diseases and historically the world's deadliest infections, have sparked enormous attention due to the difficulty of finding a vaccine (more on that in this chapter) and their prevalence in developing countries [7–9]. However, these are not the only priority pathogens, nor are priority pathogens "restricted to" pandemic-causing agents or historically important pathogens. In this chapter, we will cover the latter, but we will also give emphasis to other pathogens that can be and, if we don't do something, will most definitely be threats to our society as it is currently organized. This is especially important at a time when the focus has been, understandably, diverted to COVID-19. The COVID-19 pandemic and its concomitant response will have long-term consequences on the burden caused by other infectious diseases. It is projected that the 2.7 million deaths caused by the "Big Three" could double due to the disruption in healthcare procedures [10–12]. This illustrates the threat infectious diseases pose to global healthcare, with an additive effect caused by the high number of people suffering from the disease and the lack of care provided for other diseases. Here, we aim to outline the current and future priority pathogens. These are deemed so not necessarily, or solely, due to their immediate pandemic potential. Rather, they were selected due to their current heavy social and/or economic burden and due to the limited treatment options available.

Table 3.1 Examples of infectious diseases through history

Pathogen	Year(s)	Spread	Deaths (M)	Ref.
Bubonic plague (*Yersinia pestis*)	541–543 1347–1351 1885–1940	Fleas	≤200	[13–15]
Smallpox (Variola virus)	1347–1979	Fleas	≤500	[14]
Cholera pandemic (*Vibrio cholerae*)	1817–1886 1889–1923 1961–	Water	≤8	[16, 17]
Avian influenza A Russian flu/H3N8 Spanish flu/H1N1 Asian flu/ H2N2	1889–1894 1918–1919 1957–1959	Birds	≤110	[18–20]
AIDS HIV-1	1981–	Body fluids	≤35	[21–26]
SARS SARS-CoV-1	2002–2003	Bats, civets	≤1	[27]
Swine influenza A H1N1	2009–2010	Pigs	0.2	[28]
EBOLA Ebolavirus	2014–2016	Body fluids	≤0.1	[29]
MERS MERS-CoV	2014–2015	Bats, camelids	≤1	[30]
COVID-19 SARS-CoV-2	2019–	?	>7	[31–33]

This chapter is meant to function as a guide where the reader can find a general overview of the causative agents of key infectious diseases, what diseases they cause, the current and future treatment options, and the warning signs for future "priority pathogens."

3.2 Parasites: Protozoa and Helminths

Parasites live in or on the (human) host and survive by taking nutrients from the host, often causing diseases. There are three types of parasites: protozoa (can multiply in the human host), helminths (worm parasites that cannot multiply in the human host), and ectoparasites (which live on the human host, such as lice). In this section, we will highlight two human pathogens, the causing agent of malaria and a less-known helminth belonging to the neglected tropical diseases (NTDs) family, the causing agent of lymphatic filariasis.

3.2.1 Protozoa

This book has a chapter focusing on malaria; however, being the most important protozoa pathogen, in this chapter we will explore, albeit non-extensively, what makes the malaria-causing protozoa a high-priority pathogen. Protozoa are one-celled microscopic organisms that can be free living or parasitic. They are capable of multiplying and causing serious disease in humans. Transmission from intestine-living protozoa happens through the fecal–oral route (*e.g.*, contaminated food/water), whereas protozoa living in the human blood or tissue are usually transmitted by a vector (*e.g.*, mosquito). The latter is the case of probably the best-known human protozoan pathogen that causes malaria in humans, *Plasmodium falciparum* (*P. falciparum*) [34].

Plasmodium is the infamous causing agent of malaria, a disease that can result in a wide variety of symptoms, ranging from none/ mild symptoms to severe symptoms and even death. However, in general, if properly diagnosed and promptly treated, malaria is a curable disease. Despite this, in 2020, nearly half of the world's population was at risk from malaria. There were roughly 241 million cases and 627,000 deaths from the disease, of which 80% were children under 5 years old in sub-Saharan Africa [35].

There are several *Plasmodium* parasites (five parasite species); however, *P. falciparum* (dominant in Africa) and *Plasmodium vivax* (*P. vivax,* most common in non-African regions) present the greatest threats. These account for most malaria cases and related morbidity and mortality in humans. *P. falciparum* is responsible for most cases of severe clinical malaria, and it is virtually resistant to all drugs currently used to treat and prevent malaria. *Plasmodium* has a complex life cycle (Fig. 3.1), which makes developing anti-malaria chemotherapies complex [7, 8, 36, 37].

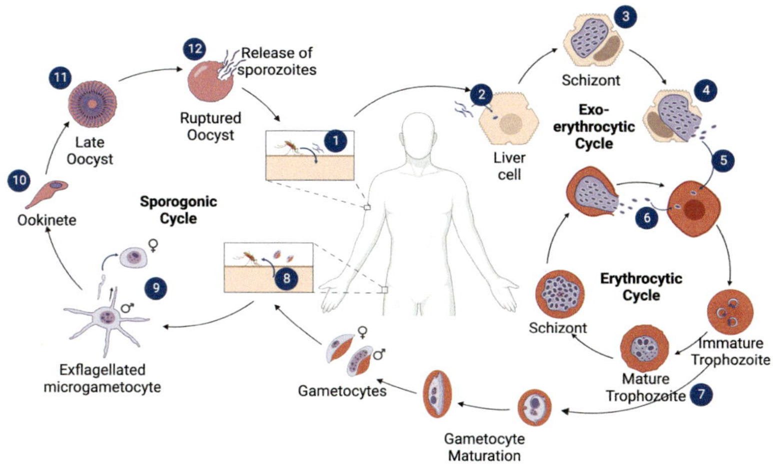

Figure 3.1 *Plasmodium falciparum* life cycle. **(1)** A malaria-infected mosquito (*Anopheles*) inoculates the human host with sporozoites. **(2)** Sporozoites infect human liver cells, **(3)** mature, and subsequently **(4)** rupture and release merozoites. **(5)** Upon replication in the liver, parasites infect and undergo asexual multiplication in erythrocytes. **(6)** These then rupture and repeat the erythrocyte infection cycle. **(7)** Some parasites differentiate and enter the blood stage, which is responsible for the clinical manifestations of the disease. **(8)** These can again be ingested by an *Anopheles* mosquito. **(9–12)** The multiplication, differentiation, and migration of the parasites into the mosquito's salivary glands allow the cycle to be perpetuated in new human hosts. Adapted from Ref. [53]. Created with BioRender.com.

Malaria prevention and treatment have involved different measures over time, and for the last two decades, WHO-recommended measurements have been effective in reducing the burden of this disease [34, 35]. Vector control, preventive chemotherapy, prompt diagnosis, and the holy grail of any infectious disease—a vaccine—have all been deployed to tackle the malaria epidemic. Vector control is an important strategy to control the perpetuation of the infection cycle by disrupting local transmission. However, emerging resistance to insecticides by *Anopheles* mosquitos has been observed in 78 countries [34, 35, 38, 39]. Another approach is to use preventive chemotherapy in infants and pregnant women, seasonal malaria chemoprevention, and mass drug administration [9, 34, 35, 40–42]. However, similar to vector control strategies, *Plasmodium* resistance to existing drugs can limit the efficacy of this approach. Early and accurate diagnosis of malaria can reduce disease and prevent death (when accompanied by treatment) and can contribute to reducing transmission. Histidine-rich protein 2 (HRP2) is the predominant target of the 419 million *P. falciparum*-detecting malaria rapid detection tests (RDTs) sold each year. Parasites that no longer express HRP2 may not be detectable by RDTs based on HRP2, and those that no longer express both HRP2 and histidine-rich protein 3 (HRP3) are completely undetectable by these RDTs [34, 35, 43–45]. These safe and cost-effective strategies should be complementary and used to impede serious disease onset and perpetuation of the infection cycle; however, all measures present serious caveats. The holy grail of infectious disease control, as the COVID-19 pandemic so clearly illustrates, is the development of an effective vaccine. However, in the case of malaria, the different stages of parasite development and corresponding different antigenic compositions and antigen variability associated with these development stages are obvious challenges for effective vaccine development. Further, although natural infection in endemic areas provides protection (hence most severe cases affecting infants and naive adults), we still do not fully understand why immunity against malaria develops slowly or how it is maintained in the absence of re-exposure [46, 47]. The immunity provided by complex antigens such as irradiated sporozoites (Fig. 3.1) and natural infection has been very difficult to replicate using purified antigens [7]. Currently, the only vaccine in use is the RTS,S vaccine, a hepatitis B virus–like particle that contains a genetically fused portion of the repeat domain and the C-terminal region of the *P. falciparum* circumsporozoite protein (CSP) [48, 49]. However, the protective effect of this vaccine is short lived, and it appears to depend on the intensity of transmission in different endemic areas; hence, it is critical to develop vaccines that induce durable sterilizing immunity [48, 49]. There are currently several vaccines in phase I–III clinical trials, so there is hope that long-lasting protection might be achieved by one of them [50, 51]. The development of new vaccines in parallel with other control measures is critical to achieving the WHO objectives for 2030 malaria control, specifically: reducing malaria case incidence by at least 90% by 2030, reducing malaria mortality rates by at least 90% by 2030, eliminating malaria in at least 35 countries by 2030, and preventing a resurgence of malaria in all countries that are malaria free [52].

3.2.2 Helminths

Malaria is not the only cause for concern in the realm of human parasites. Helminths (derived from the Greek word for worms) are large, multicellular organisms, which, like protozoa, can be free living or parasitic. Among these, there is the causing agent of lymphatic filariasis (also known as elephantiasis), which is caused by the nematodes *Wuchereria bancrofti* (~90% of the cases), *Brugia malayi* (~10% of the cases), and *Brugia timori* (can also cause the disease) [54–57]. Lymphatic filariasis is transmitted by different types of mosquitos such as *Culex*, *Anopheles*, and *Aedes* mosquitos [58].

Lymphatic filariasis is one of the several NTDs that have suffered from a lack of attention by the public health community, despite affecting more than 1 billion people worldwide. Lymphatic filariasis can be asymptomatic, acute, or chronic. While most infections are asymptomatic, chronic lymphatic filariasis can result in lymphoedema (tissue swelling) or elephantiasis (skin/tissue thickening). In its acute manifestation, lymphatic filariasis causes skin and lymph nodes/vessels inflammation and is often associated with secondary bacterial infections. Lymphatic filariasis is debilitating, may last for weeks and is a major cause of lost wages and social shame, thus often being accompanied by extreme poverty [54–58]. In 2020, ~900 million people were living in risk areas, and ~80 million people were/are affected by the disease [59]. Similar to malaria, vector control, preventive chemotherapy, and prompt diagnosis have been used as control measures and currently no vaccine is available [54, 60, 61].

3.3 Fungi

Fungi and their infections have been understudied and underrated, being surpassed by infections caused by bacteria and viruses [62, 63]. They are responsible for a wide array of diseases that range from minor allergic syndromes to superficial and life-threatening invasive fungal diseases (IFDs), affecting more than 1 billion people worldwide [64, 65]. Data on IFDs is sparse and of low quality, leading to a challenge in the evaluation of the real burden of these diseases. Nevertheless, IFDs are rising worldwide, especially among immunocompromised populations or as a co-infection. Most worryingly, there is a limitation in the diagnosis and treatment of IFDs due to the overall lack of quality diagnostics and treatment. Moreover, the emergence of antifungal resistance is rising in many settings [66].

Accordingly, in October 2022, the WHO deemed IFDs as an important health issue and released the first-ever list of fungal priority pathogens, a list of 19 fungi that represent a great threat to human health [67]. The fungal pathogens included were ranked and categorized into three priority groups: critical, high, and medium (Table 3.2). The critical group includes *Cryptococcus neoformans, Candida auris, Aspergillus fumigatus,* and *Candida albicans.* These pathogens are all associated with a serious risk of mortality and/or morbidity and are responsible for acute, subacute, and systemic fungal infections.

Table 3.2 WHO critical, high, and medium priority fungal pathogens [67]

Priority	Species	Disease/Key Facts
Critical	*Cryptococcus neoformans*	Global, opportunistic pathogenic yeast. Cryptococcosis usually affects the lungs or the CNS; cryptococcal meningitis in immunocompromised patients.
	Candida auris	Global, pathogenic yeast. Wound and ear infections, invasive candidiasis (bloodstream), usually nosocomial. Multidrug resistant.
	Aspergillus fumigatus	Global, environmental mold. Aspergillosis (pulmonary and most severely invasive) in immunocompromised patients.
	Candida albicans	Global, commensal yeast. Mouth and throat candidiasis, invasive candidiasis affecting blood (candidemia), heart, brain, eyes, bones.
High	*Nakaseomyces glabrata (Candida glabrata)*	Global, commensal yeast. Superficial mucosal infections, bloodstream infections in immunocompromised patients.
	Histoplasma spp.	Global, dimorphic fungi. Histoplasmosis with pulmonary and disseminated infection affecting lungs, mucosal surfaces, and skin in immunocompromised patients.

Priority	Species	Disease/Key Facts
	Eumycetoma causative agents	Deep tissue infection caused by various fungal pathogens, which enter the body through breaks in the skin. High amputation rates.
	Mucorales	Global, pathogenic mold. Mucormycosis affecting lungs and sinuses, can spread to the eye, GI tract and CNS in immunocompromised patients.
	Fusarium spp.	Global, but mostly tropical, pathogenic mold. Fusariosis, invasive fusariosis of the respiratory system and the eyes, but also CNS.
	Candida tropicalis	Global, commensal yeast with pathogenic potential. Invasive candidiasis of the blood, heart, CNS, eyes, bones, and internal organs.
	Candida parapsilosis	Global, commensal yeast with pathogenic potential. Invasive candidiasis of the blood, heart, CNS, eyes, bones, and internal organs; mainly immunocompromised patients.
Medium	*Scedosporium* spp.	Global, opportunistic pathogenic molds. Invasive scedosporiosis mainly of the respiratory system, also blood, CNS, systemic.
	Lomentospora prolificans	Global, opportunistic pathogenic mold. Invasive lomentosporiosis in the respiratory system, blood, CNS, others, systemic. Nosocomial, mostly immunocompromised patients.
	Coccidioides spp.	Pathogenic dimorphic fungi distributed in the Americas. Environmental mold acquired by inhalation. Coccidiomycosis affects the lungs but can expand to CNS, blood, bones, and others.
	Pichia kudriavzeveii (Candida krusei)	Global, commensal yeast with pathogenic potential. Invasive candidiasis of the blood, heart, CNS, eyes, bones, and internal organs.
	Cryptococcus gattii	Traditionally tropical and subtropical, environmental pathogenic yeast. Cryptococcosis initially affects the lungs but can spread to CNS, blood, and others.
	Talaromyces marneffei	Pathogenic dimorphic environmental fungus endemic in South-East Asia and some areas of China. Spore inhalation can infect humans, mostly immunocompromised.

(*Continued*)

Table 3.2 (*Continued*)

Priority	Species	Disease/Key Facts
	Pneumocystis jirovecii	Global, opportunistic fungal pathogen, which is acquired from person to person through the air. Mostly immunocompromised patients.
	Paracoccidioides spp.	Dimorphic environmental fungi endemic to Central and South America. Spore inhalation can infect humans, mostly immunocompromised. Paracoccidioidomycosis mainly affects the lungs, mucous membranes, and skin.

CNS: central nervous system; GI: gastrointestinal

As mentioned, these IFDs affect mostly immunocompromised patients and those with severe underlying conditions, including patients with HIV/AIDS, cancer, chronic respiratory disease, diabetes mellitus, prior tuberculosis, and patients who are critically ill in the intensive care units or take immune-suppressing medicines [64, 68]. Additionally, the COVID-19 pandemic has seen an increase in the incidence of co-infections associated with fungi, in particular aspergillosis, mucormycosis, and candidemia [69–72]. Somewhat not so surprisingly, there has also been an association between the rise and global expansion of IFDs due to climate change, as the geographical range of fungi and the adaptive landscape for resistance are changing [73, 74].

The threat to human health is heightened as these IFDs are becoming increasingly more common and resistant to treatment. In fact, there are only four classes of antifungal medicines currently available in the clinical practice (azoles, echinocandins, pyrimidines, and polyenes), and very few in the developmental pipeline [65, 75–78]. Antifungal medicines are effective, but they are also linked to a large number of unwanted side effects, particularly because the use of these medicines requires an extended knowledge on their interactions with other drugs [79]. Nevertheless, treatment failure commonly occurs due to (1) the interplay of the drug properties with the host characteristics; (2) antifungal drug properties; and (3) the inherent fungal characteristics, such as their diverse cell morphologies, and antifungal tolerance and resistance [80].

Although antifungal resistance is much less well studied than antimicrobial resistance occurring in bacteria, pathogenic fungi can acquire resistance by adaptation to drug-selective pressure, effectively changing mechanisms that directly or indirectly affect the drug-target interaction [80, 81]. Antifungal tolerance, on the other hand, is the ability of fungal cells to grow at drug concentrations above the minimum inhibitory concentration (MIC), usually associated with a general stress response [82]. Antifungal resistance and tolerance have major implications for human health, leading to prolonged treatment in longer hospital stays and therapy changes toward the more expensive and sometimes highly toxic, second-line antifungal medicines [66]. Furthermore, rapid and sensitive diagnostics for fungal diseases are lacking, and the existing ones are not globally available/affordable [65, 75].

Many fungal diseases have been shown to be acquired from our living environment, since this is where opportunistic fungi are commonly found, producing airborne spores that easily come in contact with humans. The overuse of antifungals in the environment undoubtedly affects human fungal infections, partly leading to the emergence of antifungal resistance [83]. For example, the agricultural use of azoles is responsible for the increasing numbers of azole-resistant *Aspergillus fumigatus* infections—15–20% in Europe, and over 80% in environmental samples in Asia [83–86]. Nevertheless, stopping the use of azoles in agriculture would lead to a huge effect on global food production, making this issue a true dilemma. There is an evident need for the development of new drugs, so that their uses can be clearly divided between agricultural and clinical.

3.3.1 *Candida auris*

Candida auris is part of the critical priority fungal pathogens list released by the WHO [67] and should be specifically mentioned due to its high importance as a globally emerging pathogen with multidrug resistance.

C. auris was first isolated in 2009 from the external ear canal of a patient in Japan, attributing its name, *auris* [87]. In the same year, *C. auris* was reported in South Korea in cases of chronic otitis, and in 2011 the first cases of bloodstream infection were documented in the same country [88, 89]. As of 2022, infections caused by *C. auris* have been reported in over 30 countries, in six continents [90]. Interestingly, it has been suggested that the selection of *C. auris* could be at least partially attributed to climate change [91].

C. auris infections are on the rise, essentially in nosocomial settings where it affects mostly immunocompromised and long-term hospitalized patients, with many having received previous broad-spectrum antibiotics and/or antifungals [92, 93]. *C. auris* can spread to internal organs through candidemia (systemic blood infection) and can also infect the respiratory tract, muscles, and even the CNS, with mortality rates being as high as 30–70% [94, 95].

C. auris is intrinsically resistant to fluconazole, and some isolates demonstrated elevated MICs to all available drug classes [96, 97]. This high level of multidrug resistance is what makes the emergence of *C. auris* so alarming, combined with its capacity to easily disseminate, particularly within the hospital environment. Adding to this, this yeast is commonly misidentified in the clinical laboratory, leading to the inappropriate treatment of infections [92].

Contrary to most *Candida* species, *C. auris* prefers colonizing the skin, facilitating person-to-person transmission. It can also survive on surfaces for long periods of time, increasing its risk of transmission and explaining the high number of reports of nosocomial infections [95, 98, 99]. *C. auris* has many virulence factors known to contribute to its pathogenesis, such as the production of secreted proteases and lipases, biofilm formation, and adherence capacity [100, 101].

The emergence of *C. auris* as a multidrug-resistant pathogen poses a global health threat that must be met with a global call for action, particularly in the development of novel drugs and diagnosis methods.

3.4 Viruses

There are over 250 viruses that have caused disease in humans [102]. More exist in natural reservoirs, and the current globalized, fast-paced, and fast-changing world increases the probability of "unwanted encounters" [102–104]. The severe acute respiratory syndrome coronavirus 2 (SARS-CoV-2) pandemic was a clear reminder of the challenges human viruses pose to our existence. However, this is not the first pandemic we have faced as a species, nor is this the only virus with pandemic potential, and we have not yet been able to fully address much older threats, such as the human immunodeficiency virus (HIV) or the influenza virus, for example.

In this section, we will introduce some of the viruses that have the potential to keep us, as a species, awake at night.

3.4.1 Coronaviruses

This book has a focus chapter on COVID-19, which is why here we will cover just the basic information regarding coronaviruses, and how the vaccine development against SARS-CoV-2 may help the fight against other viral pathogens. Coronaviruses are enveloped viruses, with one positive-sense, single-stranded RNA genome segment of 27–32 kb. The RNA genome is bound by the viral nucleocapsid phosphoprotein and surrounded by the viral envelope, which is composed of the viral envelope protein, the membrane protein, and the spike glycoprotein (Fig. 3.2a) [105]. Coronaviruses are

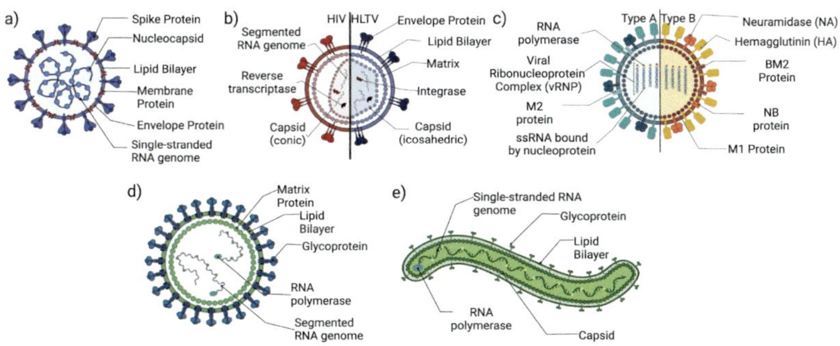

Figure 3.2 Structural organization of some high-priority viruses. **(a)** SARS-CoV-1; **(b)** HIV-1/HLTV; **(c)** Influenza A/B; **(d)** Lassa virus; and **(e)** Ebola virus. Created with BioRender.com.

highly diverse members of the family *Coronaviridae*, subfamily *Coronavirinae*, which consists of four genera—*alpha-coronavirus*, *beta-coronavirus*, *gamma-coronavirus*, and *delta-coronavirus*—based on phylogenetic relationships and genomic structures [105]. SARS-CoV-2, responsible for the pandemic that started in 2019, is not the first highly pathogenic human beta-coronavirus identified. We had previously identified SARS-CoV in 2003 and the Middle East respiratory syndrome coronavirus (MERS-CoV) in 2012. Not all coronaviruses are highly pathogenic, in fact, before SARS and MERS, coronaviruses were not considered highly pathogenic since they caused mostly mild infections in healthy people [106]. There are hundreds of coronaviruses, most of which circulate in other animals, such as pangolins, civets, camels, or bats, for example [106]. SARS-CoV and MERS-CoV likely originated in bats and were transmitted directly to humans from market civets and camels, respectively [105]. The origin of SARS-CoV-2 and the transmission link to humans remain elusive. The SARS-CoV-2 pandemic highlighted the dangers of viral infections but also the strength of human resolve and technological advances, with multiple vaccines being developed in less than 2 years [107, 108]. The vaccines against SARS-CoV-2 are highly effective but are still not able to fully impede COVID-19. Alternative/complementary strategies are required to obtain long-lasting immunity, reduce adverse disease events, and democratize access to vaccination across the globe. We have witnessed the use of several technological approaches (new and old) to develop SARS-CoV-2 vaccines from inactivated viruses to mRNA-based vaccines. The latter offers a high level of protection against COVID-19 [109]. However, as with our next high-priority viral pathogen examples, the mutation-prone nature of the virus and the coexistence of several variants affect long-lasting vaccine protection and its effectiveness.

3.4.2 Retroviruses

Retroviruses are viruses that insert their RNA into the host cell and use its own reverse transcriptase enzyme to produce double-stranded DNA (thus *retro*—backward—from the canonical DNA to RNA progression). The retrovirus family can be divided into several subfamilies, from which the best known are delta-retrovirus (*e.g.*, human T-cell leukemia virus—HTLV, Fig. 3.2b) and lentiviruses (*e.g.*, HIV, Fig. 3.2b). All encode the gag, pol, and env proteins (Fig. 3.2b), although different genomic organizations and regulatory proteins make each family unique [110]. Among these, HIV continues to be the major global health issue, unsurprisingly being one of the "Big Three." In 2020, HIV-related causes claimed between 0.5 and 1 million people, adding to the almost 50 million lives it has claimed so far [26]. There are approximately 45 million people (upper estimate) living with HIV, two-thirds of whom are living in Africa, and there are ~1.5 million new cases annually [26]. Two types of HIV have been identified: HIV-1 and HIV-2. The former, HIV-1, is more virulent and infective, thus being the cause of most HIV infections globally [26]. HIV-1, however, is characterized (as other lentiviruses) by its long incubation period and corresponding long illness duration with an average survival time (without treatment) of approximately 10 years [25, 26]. This long survival is also related to HIV-1 infection characteristics. HIV-1 does not kill the infected host; rather, it weakens the infected person's defense, making it more susceptible to several other diseases, such as infections and some types of cancer, hence the name "human immunodeficiency virus."

HIV "targets" cells expressing the CD4 receptor and the chemokine co-receptors CCR5 and CXCR4 (depending on tropism and infection stage), in T cells, monocytes, and macrophages (for example, in the spleen, lymph nodes, liver, or brain) and dendritic cells (for example, in lymphoepithelial regions) [24]. HIV destroys immune cells, and the severity and progression of the disease can be analyzed by the $CD4^+$ cell count, with infected individuals gradually becoming immunodeficient [24]. When the HIV infection is at its most advanced stage, it is referred to as the acquired immunodeficiency syndrome (AIDS), defined by the development of opportunistic diseases such as certain cancers, bacterial and fungal infections, for example [26]. HIV was discovered in 1983, and almost 30 years later, although there are several treatments that allow patients to live "normally" with the virus, there is still no vaccine. There have been several attempts to develop an HIV vaccine; however, most trials showed that antibodies induced by the vaccines did not provide protection against divergent viruses or primary viruses isolated from other patients [23, 111, 112]. This prompted many further developments regarding vaccine strategies, which, unfortunately, brought no better prospects as these failed to protect human patients against infection [113, 114]. Currently, combination antiretroviral therapy (ART) can block multiple stages of the HIV-1 life cycle to prevent progression to AIDS in people living with HIV-1. However, it cannot cure these patients as $CD4^+$ T cells are a reservoir of HIV, and to prevent viral rebound, life-long ART is needed. However, two cases hold the promise of an HIV cure: the curious cases of the "Berlin patient" and the "London patient," two patients who have been cured of HIV. In both these cases, allogeneic hematopoietic stem-cell transplantation from donors with a homozygous CCR5 delta32 mutation was performed for an associated hematological malignancy [115, 116]. This strategy, albeit promising, is not a practical tactic due to the inherent complications of such an invasive approach. CCR5- and/or CXCR4-based hematopoietic stem-cell gene therapy and/or latency revival (to "activate" HIV reservoirs) followed by drug or immune-mediated clearance of infected cells (the so-called "shock and kill" approach) may be more feasible options [22].

3.4.3 Influenza

Influenza viruses are enveloped negative single-strand RNA viruses. Influenza A and B viruses, which cause disease in humans [117, 118], contain eight genomic segments that encode for 10 essential viral proteins, and several strain-dependent accessory proteins, such as hemagglutinin (HA) and neuraminidase (NA), for example (Fig. 3.2c) [117, 118]. Influenza A viruses are divided into subtypes based on two proteins on the surface of the virus: hemagglutinin (H, H1 through H18) and neuraminidase (N, N1 through N11), the seasonal viruses being H1N1 and H3N2 [117, 118]. Influenza A viruses are the only influenza viruses known to cause flu pandemics, such as the 1918 Flu pandemic caused by an H1N1 virus of avian origin, estimated to have infected one-third of the world population (500 million) and killed at least 50 million people [31, 32]. The 1918 pandemic was, until the SARS-CoV-2 pandemic, the most famous viral pandemic. However, it is not the only one. In 1957, a new influenza A (H2N2) virus emerged in East Asia, triggering the "Asian Flu" pandemic, with an estimated 1.1 million deaths worldwide. In 1968, an influenza A virus (H3N2) killed 1 million people worldwide; this virus continues to circulate worldwide as a seasonal influenza A virus. In 2009, a new pandemic strain ("swine flu") was identified and resulted in 200,000–500,000 deaths worldwide in 1 year [119,

120]. However, influenza's socioeconomic burden is not only derived from its pandemic strains, with annual seasonal influenza epidemics causing approximately 1 billion infections, 3–5 million cases of severe illness, and 290,000–650,000 deaths worldwide [119–121]. Influenza A is under constant monitoring as its high mutation rate allows accumulating point mutations ("antigenic drift"), which enables the virus to evade immune recognition and cause repeated influenza outbreaks. Further, it can present major alterations in the HA antigen ("antigenic shift"), caused by reassortment—between animal and human subtypes—of genetic material from different A subtypes, which can result in new pandemic strains [119–121].

Influenza B viruses are classified into two lineages—B/Yamagata and B/Victoria—and subsequently into specific clades and sub-clades, for example for influenza B/Victoria, clade V1A, subclade V1A.1 [117, 118, 121–123]. Influenza B viruses generally change more slowly in terms of their genetic and antigenic properties than influenza A viruses. As such, until now, no pandemic influenza B virus has been identified albeit co-circulation of influenza B viruses from both lineages around the world [117, 118, 121–123]. The severity of infection in humans is associated with the replication of the virus in the lower respiratory tract, which usually results in severe inflammation owing to immune cell infiltration, pneumonia (viral or bacterial, so-called super-infection), and sepsis. Additional possible complications can include myocarditis, encephalitis, and multi-organ failure (*e.g.*, lung and kidney failure) [31, 32, 117–120, 124].

As mentioned earlier, influenza viruses present high mutation rates and great variability in their surface antigens, and as such immunity resulting from infection with one virus often does not protect against a different virus (regardless of whether it is from the same type/sub-type or not). This means that influenza vaccines must be designed annually to match the circulating viruses, which health authorities predict will cause future epidemics. If this "matching" is done efficiently, influenza vaccines, which usually include two influenza A subtypes (H1N1 and H3N2) and one/two influenza B lineages (Victoria and Yamagata), will provide 70–90% protection against the clinical disease in healthy adults [117–119, 124]. This decision is made twice a year (to match the flu seasons in the northern and southern hemispheres) by the WHO based on observation from HA inhibition assays [121]. However, this is not always the case and the danger of "antigenic shift" pandemic strains is always lurking, even in epidemic strains, where vaccine efficacy can vary [121–123, 125]. Thus, the search for alternative and novel anti-influenza drugs with a broad spectrum of reactivity against all viruses, better bioavailability, and easier administrative streamlined solutions is desirable. As in many priority pathogens, drug resistance is a major issue, and influenza is no different. Two classes of drugs are currently used for influenza treatment—adamantanes and NA inhibitors. The former is no longer effective, and NA-inhibitor-resistant viruses have been identified (albeit, luckily, the H1N1 resistant strain was replaced by the susceptible H1N1 2009 pandemic strain). Thus, the future demands a combination of vaccines and antiviral treatments to respond to influenza regardless of the actual virus that is dominant at any specific time [124, 126–129].

3.4.4 Viral Hemorrhagic Fever

Viral hemorrhagic fever (VHF) can lead to severe, life-threatening illnesses, which are caused by different RNA viruses (single- or double-stranded, enveloped and non-enveloped, such as Ebola or Lassa virus, for example Fig 3.2d,e and Table 3.3) [130]. These viruses cause—in different disease stages and concomitant severity degrees—fever, capillary destruction, and coagulation defects. These symptoms, associated with multi-organ system failure and often death, are the reason for the ominous disease name of VHF and great public attention when outbreaks happen [131, 132]. Transmission can occur through various routes, which are as vast as the viruses that cause VHF, from insect transmission to contact with animal excretions/fluids (Table 3.4). Person-to-person transmission is also possible for some of these viruses (Table 3.4). As mentioned before, VHF is caused by many different pathogens (Table 3.3); hence, death rates are dependent on the virus and, importantly, on the healthcare provided to patients. As such, the outbreaks of Lassa or Ebola viruses in Africa have caused death rates as high as 50%

and 90%, respectively, values, which would not be as high if access to modern healthcare was available [133–136]. This is because the most common treatment option for VHF is early and aggressive supportive care (Table 3.4), although some broad-spectrum antiviral drugs and vaccines against some viruses causing VHF are becoming available (*e.g.*, the antiviral drug remdesivir, also used for COVID-19) [137–143].

Table 3.3 Examples of VHF causative agents

Virus	Family	Disease	RNA	Ref.
Chapare	*Arenaviridae*	Chapare HF	ssbRNA	[144]
Guanarito		Venezuelan HF		[145]
Junín		Argentine HF		[146]
Lassa		Lassa fever		[147]
Lujo		Lujo HF		[144]
Machupo		Bolivian HF		[148]
Sabiá		Brazilian HF		[149]
Crimean–Congo	*Bunyaviridae*	Crimean–Congo HF	SL ssRNA	[150]
Hantavirus		HF with RS		[151]
Puumala				[152]
Rift Valley fever		Rift Valley fever		[153]
Saaremaa		HF with RS		[154]
Seoul				[155]
Sin Nombre		Hantavirus PS		[156]
SFTS Bunya		Severe fever and TS		[157]
Tula		HF with RS		[158]
Marburg	*Filoviridae*	Marburg HF	ssRNA	[159]
Ebola		Ebola disease		[160]
Dengue	*Flaviviridae*	Dengue fever		[161]
Yellow fever		Yellow fever		[162]
West Nile		West Nile disease		[163]

Single-stranded RNA (ssRNA), single-stranded bisegmented RNA (ssbRNA), hemorrhagic fever (HF), renal syndrome (RS), thrombocytopenia syndrome (TS), pulmonary syndrome (PS)

Table 3.4 Most common VHF causing agents, disease symptoms, and treatment

<table>
<tr><th>Virus Vector Transmission</th><th>Symptoms</th><th>Treatment</th><th>Ref.</th></tr>
<tr><td>• Dengue virus
• Aedes aegypti mosquito
• Mosquito bite</td><td>Nausea, vomiting, rash, aches, pain. 1 in 20 may develop severe dengue: shock and internal bleeding.</td><td rowspan="4">• Fluids and electrolytes orally or intravenously
• Medication to support blood pressure, manage fever and pain
• Avoid medications that increase the risk of bleeding (e.g., aspirin)
• Oxygenation support</td><td>[164–166]</td></tr>
<tr><td>• Ebola virus
• Animals (e.g., bat, nonhuman primate)
• Infected animal or human, body fluids, contaminated objects</td><td>Fever, aches, weakness, fatigue, sore throat, loss of appetite, vomiting, diarrhea, hemorrhaging</td><td>[167–170]</td></tr>
<tr><td>• Lassa virus
• Mastomys natalensis rodent
• Ingestion or inhalation of urine and droppings, contact with contaminated objects/food</td><td>Fever, weakness, and headache. Serious disease (<20%) hemorrhaging respiratory distress, vomiting, swelling, pain, shock, neurological problems (e.g., encephalitis)</td><td>[171–173]</td></tr>
<tr><td>• Marburg virus
• Animals (e.g., Rousettus bat)
• Contact with droppings, body fluids, contaminated objects</td><td>Fever, chills, headache, myalgia, nausea, vomiting, pain, rash</td><td>[170, 174, 175]</td></tr>
</table>

3.5 Bacteria

The concept of priority pathogen was created for bacteria. In 2017, the WHO published a list of global priority pathogens, all bacteria. Twelve species (Table 3.5), due to their antibiotic resistance profile, pose the greatest threat to human health [176, 177]. The problem is as simple as it is challenging. Quoting directly from the United Nations (UN) Ad hoc Interagency Coordination Group (IACG) on antimicrobial resistance report [178]: "Alarming levels of antimicrobial resistance have been reported in countries of all income levels. In some member countries of the Organization for Economic Cooperation and Development (OECD), about 35%

Table 3.5 WHO critical and high-priority bacterial pathogens

Priority	Name or Class (Gram +/-)	Antibiotic	Resistance Mechanism	Ref.
Critical	*Acinetobacter baumannii* (-)	Carbapenem	β-lactamases, porin, efflux pump, PBPs[3]	[195–197]
	Pseudomonas aeruginosa (-)	Carbapenem		[198–200]
	Enterobacteria[1] (-)	Carbapenem ESBL[2]		[201–203]
High	*Enterococcus faecium* (+)	Vancomycin	Target modification	[204, 205]
	Staphylococcus aureus (+)	β-lactams, Vancomycin	PBPs,[3] target modification	[206–209]
	Helicobacter pylori (-)	Clarithromycin	Target modification	[210–213]
	Campylobacter spp. (-)	Fluoroquinolone		[214, 215]
	Salmonellae (-)	Fluoroquinolone		[216, 217]
	Neisseria gonorrhoeae (-)	Cephalosporin, Fluoroquinolone	Porins, efflux pumps, PBPs,[3] target modification	[218–220]
	Shigella spp. (-)	Fluoroquinolone		[221]

[1]Family of gram-negative bacteria, *e.g., E. coli*, [2]Extended spectrum β-lactamase, [3]Penicillin-binding proteins

of common human infections are already resistant to currently available medicines, and in some low- and middle-income countries (LMICs), resistance rates are as high as 80 to 90% for some antibiotic-bacterium combinations." The consequences are severe. Again, quoting directly from the UN Ad hoc IACG on antimicrobial resistance report [178]: "Drug-resistant diseases already cause at least 700,000 deaths globally a year, including 230,000 deaths from multidrug-resistant tuberculosis, a figure that could increase to 10 million deaths globally per year by 2050 under the most alarming scenario if no action is taken. Around 2.4 million people could die in high-income countries between 2015 and 2050 without a sustained effort to contain antimicrobial resistance. The economic damage of uncontrolled antimicrobial resistance could be comparable to the shocks experienced during the 2008–2009 global financial crisis as a result of dramatically increased health care expenditures; impact on food and feed production, trade and livelihoods; and increased poverty and inequality." As mentioned before, the problem is simple; there is an urgent need for new antimicrobials. Drug resistance develops faster than new antibiotics can be developed, validated, and put into the clinic. The problem is well known: developing new antimicrobials is a complex and long process, with no guarantee of success; most molecules in development do not reach the end of the process [179, 180]. The few molecules that make it to the market are often, and understandably, used carefully to prevent the development of resistance. This increases the burden on the development pipeline, resulting in poor commercial incentive for new molecule development [179, 180].

Before antibiotics, bacterial infections were deadly. *Staphylococcus aureus,* for example, was fatal in 80% of infected wounds. Hence, it is not surprising that life expectancy was, on average, 47 years [181]. The discovery of penicillin by Sir Alexander Fleming (1928), its subsequent purification by Ernst Chain and Howard Florey (1942), and it becoming available to the public (1945) marked the beginning of the antibiotic era. This changed the world [180–184]. From then on, a wealth of new molecules were discovered until, in the 1980s, we reached what has been referred to as the "innovation gap"—a worrying concept that refers to the lack of development of new classes of antibiotics [185]. However, in 2015, some good news emerged with the discovery of the mode of action of teixobactins [186]. They belong to a new class of antibiotics that block peptidoglycan biosynthesis by binding to cell-wall precursors (*e.g.,* Lipid II) to which, so far, resistance has not been identified [186]. However, there is a dire need for more classes of antibiotics to be developed so that current classes can be "rested," and perhaps introduced again in the future, to ease the current selective pressure. The development of new chemical entities should be combined with other approaches, such as phage therapy, which has been used for decades in Eastern Europe and uses bacterial viruses (bacteriophages) to infect bacteria and, consequently, lyse them [187–189]. Although powerful, this mechanism can also give rise to bacterial resistance and comes with other pitfalls associated to, for example, the phage specificity, as these can be often strain specific, not being able to infect and kill all members of the same species [188, 190]. Another successful approach is the use of engineered endolysins, phage-encoded enzymes that efficiently and rapidly lyse and kill bacteria [191, 192]. They can be added directly to gram-positive bacteria, killing them almost instantly, and although specific enough to target a single species, they do not rely on bacterial receptors or replication to induce killing, as they bind and hydrolyze specific bonds of the bacterial cell wall. An improvement of these enzymes emerged with the development of artilysins, modified endolysins that can penetrate the gram-negative outer layer to effectively hydrolyze their cell wall and kill the bacteria [193, 194].

Although simple, the antimicrobial resistance problem is extremely complex to solve and requires a joint effort of multiple stakeholders to develop a variety of antibacterial strategies that can be used across the world regardless of healthcare conditions. This is especially true for infections that target, unsurprisingly, disproportionally developing countries, such as tuberculosis.

In this section, we focus on the bacteria that are classified as critical in the high-priority list and on *Mycobacterium tuberculosis*, the causative agent of tuberculosis. Tuberculosis, one of the "Big Three," is the second leading infectious killer worldwide, after COVID-19 [6]. However, one must not ignore the relevance of gram-positive bacteria as the infamous methicillin-resistant *Staphylococcus aureus* (MRSA). The acronym, due to the uncanny capacity of *S. aureus* to acquire antibiotic resistance, is sometimes referred to as multi-resistant *Staphylococcus aureus* and can often resist even last-resort antibiotics such as vancomycin or daptomycin [222–226].

3.5.1 *Mycobacterium tuberculosis*

M. tuberculosis belongs to the diverse family of actinobacteria, an acid-fast bacterium, undetectable by the gram stain technique [227]. It has a unique cell surface (Fig. 3.3a), which resembles both gram positive (Fig. 3.3b) and gram negative (Fig. 3.3c), with a thick peptidoglycan layer (like gram positive) and a mycolic acid/arabinogalactan outer

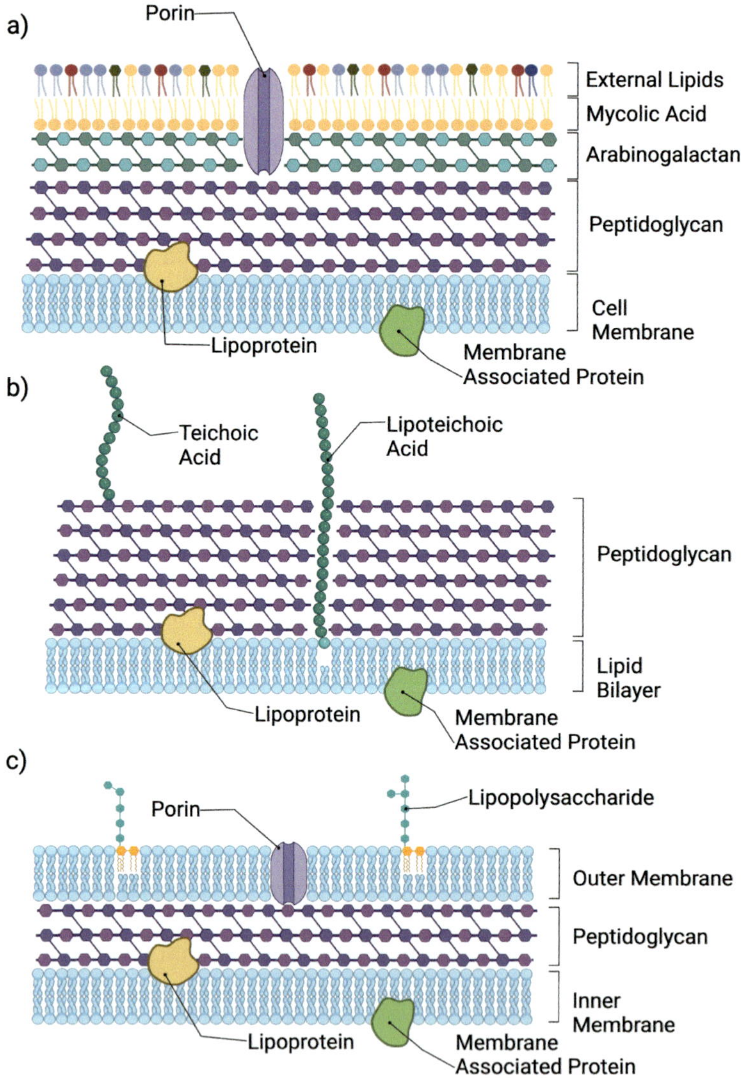

Figure 3.3 Different cell-wall organizations. **(a)** Acid-fast bacterial cell-wall organization. **(b)** Gram-positive (gram +) bacterial cell-wall organization. **(c)** Gram-negative (gram −) bacterial cell-wall organization. Created with BioRender.com

layer (like gram negative) [227]. This organization is important for *M. tuberculosis* adaptability and capacity to resist antibiotics [227]. In 2020, tuberculosis killed 1.5 million people, and an estimated 10 million people fell ill with this disease [228, 229]. However, approximately 0.5 million cases are caused by multidrug-resistant tuberculosis [228, 229]. These are striking numbers for a disease that is curable and preventable, for which there is a vaccine available (Bacillus Calmette-Guérin (BCG) vaccine). In fact, tuberculosis is a disease of poverty; the 30 countries with the highest tuberculosis burden (all low and middle income) have an approximate incidence between 183 and 400 cases per 100,000 people/year, whereas in high-income countries, this number is less than 10 cases per 100,000 people/year [230, 231].

M. tuberculosis spreads through the air from one infected person to another, via coughing, speaking, etc. However, not everyone infected with *M. tuberculosis* develops tuberculosis, giving rise to what is known as latent tuberculosis [230, 231]. Tuberculosis can infect different parts of the body; however, the classical manifestation is related to pulmonary tuberculosis and causes cough (with phlegm/blood), chest pain, fever, fatigue, wight loss, and night sweats [228–231]. The challenge of tuberculosis, especially in developing countries, is related to the treatment regime, which can take as long as 9 months of daily drug administration [228, 230]. This situation is further aggravated by the appearance of drug-resistant *M. tuberculosis*, often due to incomplete courses of treatment, poor quality of drugs, or incorrect treatment regimens [228, 230]. These can be multidrug-resistant tuberculosis (resistant to the most potent anti-tuberculosis drugs—isoniazid and rifampicin), pre-extensively drug-resistant tuberculosis (resistant to the former drugs as well as second-line injectable drugs, such as amikacin, capreomycin, and kanamycin), and extensively drug-resistant tuberculosis (resistant to isoniazid and rifampicin, a fluroquinolone, and a second-line injectable drug, or resistant to isoniazid, rifampicin, a fluroquinolone, and bedaquiline or linezolid). In 2020, approximately 50,000 cases were caused by extensively drug-resistant tuberculosis [230–232]. The treatment for these drug-resistant strains requires more aggressive drug regimens and longer treatment times, which, in developing countries, contributes to the circular nature of this problem and to the difficulty of eradicating this preventable disease [233]. Tuberculosis and HIV co-infection is relatively common (~10%), and the two pathogens accelerate the failure of the immunological capacity of an infected person. Co-infection is particularly severe in poor countries where both diseases coexist and treatment options are limited [228, 232–234]. Despite the high death rate of tuberculosis being a resultant of inequalities in access to diagnostics, treatment, and healthcare, drug-resistant tuberculosis, however, can expand the reach of this disease to underdeveloped and developed countries alike, requiring the development of new effective antimicrobials that not only tackle drug-resistant strains but also reduce the length of treatment regimens.

3.5.2 Carbapenem-Resistant Gram-Negative Bacteria

Carbapenems are viewed as the most active and potent agents against multidrug-resistant bacteria; hence, resistance in gram-negative pathogens poses a grave challenge. All WHO critical importance pathogens are carbapenem resistant (Table 3.5). Carbapenems are a class of β-lactam antibiotics, which kill bacteria via the inhibition of penicillin-binding proteins (PBPs) that catalyze the last steps of peptidoglycan biosynthesis [235]. There are several mechanisms of carbapenem resistance. One major mechanism is the hydrolysis of antibiotic molecules by carbapenemase enzymes (β-lactamases, which hydrolyze β-lactam antibiotics), often encoded on plasmids and easy to transmit across species [196, 236, 237]. β-lactamases can be of different classes: A, B, C, and D, based on the catalytic domain and substrate preference [238]. Classes A, B, and D include carbapenemase enzymes, whereas class C hydrolyzes primarily cephalosporins [196]. Bacteria can also express extended spectrum β-lactamases (ESBL), which, as the name suggests, can hydrolyze multiple β-lactam antibiotic molecules, including penicillins, cephalosporins, and the monobactam aztreonam [177, 239]. The combination of several of these enzymes can render bacteria resistant to virtually all β-lactam antibiotics [196, 236, 237]. In addition to β-lactamases, bacteria can avoid carbapenem insults by altering the permeability to these molecules. One mechanism is associated with mutations or reduced expression of genes encoding for porins (*e.g.*, OprD, OmpK26) [196, 235, 240]. Alternatively, or concomitantly, overexpression of genes encoding for efflux pumps (*e.g.*, MexAB-OprM, MexXY-OprM) has a similar outcome [196, 236, 241–243]. These mechanisms are not specific for carbapenems and thus can contribute to reduced susceptibility to other antibiotics as well [243, 244]. Lastly, albeit less frequent, mutations in PBPs can also confer some level of protection [241, 242]. To tackle this problem, nothing less than the involvement of all stakeholders in the pursuit of new molecules, new treatments, and the promotion of rational antimicrobial therapy is needed.

3.6 Future Perspectives

Tackling infectious diseases requires implementing a multi-targeted approach (Fig. 3.4) involving all stakeholders: patients and their families, health professionals, environmental and veterinary partners, scientific experts, medical and non-medical industrial actors, non-governmental organizations, and decision-makers. This approach needs to be collaborative and transdisciplinary, and it needs to recognize the interconnection between humans, nonhuman animals, plants, and their common environment, dubbed a "One Health" approach [245]. The ideal scenario in the realm of infectious diseases is, obviously, eradication. However, to date only two diseases have been considered officially eradicated by the WHO: smallpox (Table 3.1) and rinderpest (a cattle disease caused by the rinderpest virus) [246]. A fantastic tool for the reader to understand the impact of infectious diseases and the effort the scientific community has made to tackle them is the Wellcome interactive explainer, which can be explored here [247].

As it is well known, the variola virus and rinderpest virus are not extinct; samples of both viruses exist and are used in research to prevent future outbreaks. Rather, these pathogens have seen a "permanent reduction to zero of the worldwide incidence of infection (...) as a result of deliberate efforts" as defined by the WHO [248]. To achieve this, it is essential to identify the disease and its geographic/host distribution and to design strategies to contain and subsequently eliminate it. The latter is easier said than done, with the "Big Three" as striking examples. The obvious answer is a vaccine, right? It is estimated that vaccines save 2–3 million people per year [249, 250]. However, global coverage dropped from 86% in

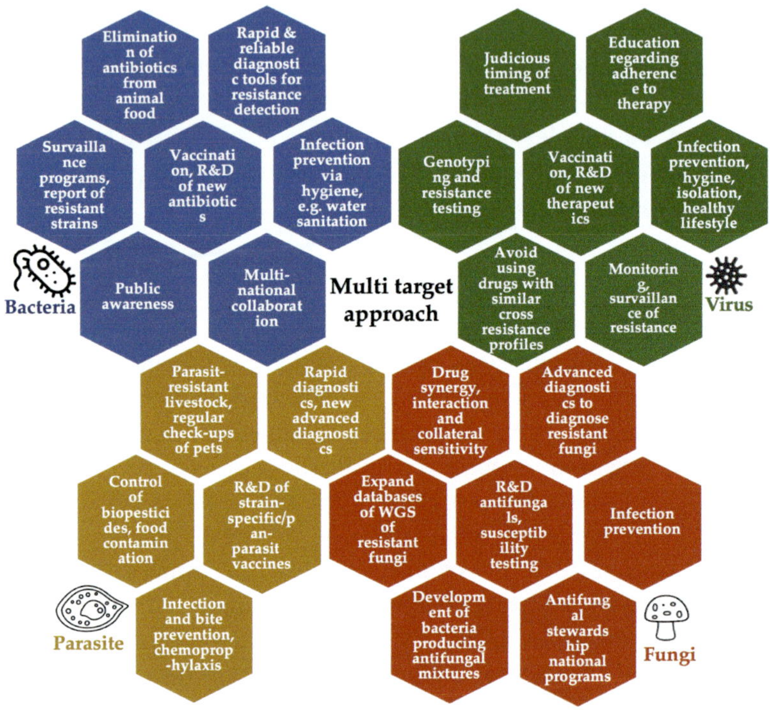

Figure 3.4 Strategies to tackle infectious diseases. A multi-target approach is needed to address the threat of infectious diseases, independent of the causative agent being bacteria (blue), virus (green), parasite (gold), or fungi (orange).

2019 to 83% in 2020 (referring to basic vaccines [250]), and there are many diseases for which no vaccine exists and in all likelihood no vaccine will exist in the near future. To this adds the vaccine confidence crisis that the COVID-19 pandemic brought to light [251, 252], and that requires a strong and pedagogic response by health professionals, scientific experts, medical and non-medical industrial actors, non-governmental organizations, and decision-makers. Regardless of the task, be it understanding, managing, fighting, or ultimately eradicating a disease, it requires global coordination, global cooperation, global health campaigns, and global economic support. Anything less results in the impossibility of mobilizing the needed resources effectively. The example of the COVID-19 pandemic is, once more, indicative of the so-needed global response. However, equally important is to monitor progress based on clear and measurable milestones, which should be framed in ambitious yet realistic objectives—a strategy that should be based on humanity's successes, such as the smallpox eradication or the near eradicated polio (as of 2020, only Pakistan and Afghanistan have active poliovirus) [253–257]. Can we implement the "one health" approach, tackle climate change, or achieve 100% vaccination rates, tackle and ultimately eradicate infectious diseases? Only time will tell; what is certain is that a global strategy is required, and this strategy involves you, dear reader, and all other stakeholders.

Funding

P.M.P. acknowledges funding by a "Maratona da Saúde" award, by a La Caixa Junior Leader Fellowship (LCF/BQ/PI20/11760012) financed by "la Caixa" Foundation (ID 100010434) and by European Union's Horizon 2020 research and innovation programme under the Marie Skłodowska-Curie grant agreement No 847648; and by the Maratona da Saúde award. P.M.P., R.P., K.B., and C.M.P.C. acknowledge funding by Science and Technology Foundation (FCT) grant (PTDC/BIA-MIC/2422/2020); Project LISBOA-01-0145-FEDER-007660 Microbiologia Molecular, Estrutural e Celular to ITQB-NOVA; MOSTMICRO-ITQB R&D Unit (UIDB/04612/2020, UIDP/04612/2020); LS4FUTURE Associated Laboratory (LA/P/0087/2020).

References

1. M. J. Selgelid, Infectious diseases, in *Encyclopedia of Applied Ethics*, edited by R. Chadwick (Elsevier, 2012), pp. 704–711.
2. From Worlds Apart to a World Prepared: Global Preparedness Monitoring Board report 2021. Geneva: World Health Organization; 2021. Licence: CC BY-NC-SA 3.0 IGO.
3. K. E. Jones, N. G. Patel, M. A. Levy, A. Storeygard, D. Balk, J. L. Gittleman, and P. Daszak, Global trends in emerging infectious diseases, *Nature* **451**, 990 (2008).

4. C. L. Ventola, The antibiotic resistance crisis: Part 1: Causes and threats, *P. T.* **40** (4), (2015).
5. R. G. Finch, Antibiotic resistance, *J. Antimicrob. Chemother.*, **42** (2), (1998).
6. World Health Organization, The top 10 causes of death, http://www.who.int/en/news-room/fact-sheets/detail/the-top-10-causes-of-death (accessed in October 2023).
7. R. Rappuoli and A. Aderem, A 2020 vision for vaccines against HIV, tuberculosis and malaria, *Nature* **473**, 463 (2011).
8. K. Bourzac, Infectious disease: Beating the Big Three, *Nature* **507**, (2014).
9. P. Makam and R. Matsa, "Big Three" infectious diseases: Tuberculosis, malaria and HIV/AIDS, *Curr. Top. Med. Chem.* **21**, 2779 (2021).
10. D. Stanisic and J. Mak, The Big 3 infectious diseases besides COVID-19 scientists are trying to find a vaccine for, https://globalbiodefense.com/2020/09/14/the-big-3-infectious-diseases-besides-covid-19-scientists-are-trying-to-find-a-vaccine-for/ (accessed in October 2023).
11. The Global Fund, Mitigating the impact of Covid-19 on countries affected by HIV, tuberculosis and malaria, 2020, https://www.theglobalfund.org/media/9819/covid19_mitigatingimpact_report_en.pdf (accessed in October 2023).
12. M. Prudêncio and J. C. Costa, Research funding after COVID-19, *Nat. Microbiol.* **5**, 986 (2020).
13. S. Sabbatani, R. Manfredi, and S. Fiorino, La Peste Di Giustiniano (Prima Parte), *Infez. Med.* **20**, 125 (2012).
14. S. N. DeWitte, Mortality risk and survival in the aftermath of the medieval black death, *PLoS One* **9**, e96513 (2014).
15. B. Bramanti, K. R. Dean, L. Walløe, and N. Chr. Stenseth, The third plague pandemic in Europe, *Proc. R. Soc. B Biol. Sci.* **286**, 20182429 (2019).
16. D. Lippi, E. Gotuzzo, and S. Caini, Cholera, *Microbiol. Spectr.* **4**, (2016).
17. J. Piret and G. Boivin, Pandemics throughout history, *Front. Microbiol.* **11**, (2021).
18. P. Berche, The enigma of the 1889 Russian flu pandemic: A coronavirus? *Presse Med.* **51**, 104111 (2022).
19. D. M. Morens and A. S. Fauci, The 1918 influenza pandemic: Insights for the 21st century, *J. Infect. Dis.* **195**, 1018 (2007).

20. J. R. Schäffr, Y. Kawaoka, W. J. Bean, J. Süss, D. Senne, and R. G. Webster, Origin of the pandemic 1957 H2 influenza A virus and the persistence of its possible progenitors in the avian reservoir, *Virology* **194**, 781 (1993).

21. C. del Rio, The global HIV epidemic: What the pathologist needs to know, *Semin. Diagn. Pathol.* **34**, 314 (2017).

22. J. Ding, Y. Liu, and Y. Lai, Knowledge from London and Berlin: Finding threads to a functional HIV cure, *Front. Immunol.* **12**, (2021).

23. M. I. Johnston and A. S. Fauci, An HIV Vaccine: Evolving concepts, *N. Engl. J. Med.* **356**, 2073 (2007).

24. S. Lucas and A. M. Nelson, HIV and the spectrum of human disease, *J. Pathol.* **235**, 229 (2015).

25. J. A. Levy, HIV pathogenesis and long-term survival, *AIDS* **7**, 1401 (1993).

26. Global progress report on HIV, viral hepatitis and sexually transmitted infections, 2021. Accountability for the global health sector strategies 2016–2021: Actions for impact. Geneva: World Health Organization; 2021. Licence: CC BY-NC-SA 3.0 IGO.

27. J. D. Cherry, The chronology of the 2002–2003 SARS mini pandemic, *Paediatr. Respir. Rev.* **5**, 262 (2004).

28. M. P. Girard, J. S. Tam, O. M. Assossou, and M. P. Kieny, The 2009 A (H1N1) influenza virus pandemic: A review, *Vaccine* **28**, 4895 (2010).

29. R. T. Kamorudeen, K. A. Adedokun, and A. O. Olarinmoye, Ebola outbreak in West Africa, 2014–2016: Epidemic timeline, differential diagnoses, determining factors, and lessons for future response, *J. Infect. Public Health* **13**, 956 (2020).

30. Y.-H. Hsieh, 2015 Middle East respiratory syndrome coronavirus (MERS-CoV) nosocomial outbreak in South Korea: Insights from modeling, *PeerJ* **3**, e1505 (2015).

31. G. E. Patterson, K. M. McIntyre, H. E. Clough, and J. Rushton, Societal impacts of pandemics: Comparing COVID-19 with history to focus our response, *Front. Public Heal.* **9**, (2021).

32. D. M. Morens, P. Daszak, H. Markel, and J. K. Taubenberger, Pandemic Covid-19 joins history's pandemic legion, *MBio* **11**, (2020).

33. World Health Organization, COVID-19 deaths, https://covid19.who.int/?mapFilter=deaths (accessed in October 2023).

34. *World Malaria Report 2021*. Geneva: World Health Organization; 2021. Licence: CC BY-NC-SA 3.0 IGO.

35. *World Malaria Report 2022*. Geneva: World Health Organization; 2022. Licence: CC BY-NC-SA 3.0 IGO.
36. W. Amelo and E. Makonnen, Efforts made to eliminate drug-resistant malaria and its challenges, *Biomed Res. Int.* **2021**, (2021).
37. S. A. Woon, L. Manning, and B. R. Moore, Antimalarials for children with plasmodium vivax infection: Current status, challenges, and research priorities, *Parasitol. Int.* **87**, (2022).
38. A. Namias, N. B. Jobe, K. P. Paaijmans, and S. Huijben, The need for practical insecticide-resistance guidelines to effectively inform mosquito-borne disease control programs, *Elife* **10**, (2021).
39. H. Vatandoost, A. A. Hanafi-Bojd, F. Nikpoor, A. Raeisi, M. R. Abai, and M. Zaim, Situation of insecticide resistance in malaria vectors in the World Health Organization of Eastern Mediterranean Region 1990–2020, *Toxicol. Res. (Camb).* **11**, 1 (2022).
40. L. M. Birkholtz, P. Alano, and D. Leroy, Transmission-blocking drugs for malaria elimination, *Trends Parasitol.* **38**, 390 (2022).
41. J. J. Moehrle, Development of new strategies for malaria chemoprophylaxis: From monoclonal antibodies to long-acting injectable drugs, *Trop. Med. Infect. Dis.* **7**, (2022).
42. C. V Plowe, Malaria chemoprevention and drug resistance: A review of the literature and policy implications, *Malar. J.* **21**, 104 (2022).
43. J. C. Mouatcho and J. P. Dean Goldring, Malaria rapid diagnostic tests: Challenges and prospects, *J. Med. Microbiol.* **62**, 1491 (2013).
44. V. N. Orish, V. F. De-Gaulle, and A. O. Sanyaolu, Interpreting rapid diagnostic test (RDT) for *Plasmodium falciparum, BMC Res. Notes* **11**, (2018).
45. K. E. Poti, D. J. Sullivan, A. M. Dondorp, and C. J. Woodrow, HRP2: Transforming malaria diagnosis, but with caveats, *Trends Parasitol.* **36**, 112 (2020).
46. J. Sang, Who cares? Psychological stress in the operating department nurse, *Br. J. Theatre Nurs.* **1**, 7 (1991).
47. T. L. Richie and A. Saul, Progress and challenges for malaria vaccines, *Nature* **415**, 694 (2002).
48. F. Zavala, RTS,S: The first malaria vaccine, *J. Clin. Invest.* **132**, (2022).
49. J. A. Stoute, M. Slaoui, D. G. Heppner, P. Momin, K. E. Kester, P. Desmons, B. T. Wellde, N. Garçon, U. Krzych, and M. Marchand, A preliminary evaluation of a recombinant circumsporozoite protein vaccine against

Plasmodium falciparum malaria. *RTS,S Malaria Vaccine Evaluation Group, N. Engl. J. Med.* **336**, 86 (1997).

50. Sanaria, Malaria eradication through vaccination, https://sanaria.com/ (accessed in October 2023).
51. L. Kurtovic, L. Reiling, D. H. Opi, and J. G. Beeson, Recent clinical trials inform the future for malaria vaccines, *Commun. Med.* **1**, 26 (2021).
52. Global technical strategy for malaria 2016–2030, 2021 update. Geneva: World Health Organization; 2021. Licence: CC BY-NC-SA 3.0 IGO.
53. T. Beck, Will malaria soon be a thing of the past? The potential of recombinant protein vaccines to control one of the world's most deadly diseases, in *Coffee Break: Tutorials for NCBI Tools*, edited by L. Dean and J. McEntyre (Bethesda (MD): National Center for Biotechnology Information (US), 1999).
54. E. L. Davis, L. J. Reimer, L. Pellis, and T. D. Hollingsworth, Evaluating the evidence for lymphatic filariasis elimination, *Trends Parasitol.* **35**, 860 (2019).
55. S. M. Hedtke, A. C. Kuesel, K. E. Crawford, P. M. Graves, M. Boussinesq, C. L. Lau, D. A. Boakye, and W. N. Grant, Genomic epidemiology in filarial nematodes: Transforming the basis for elimination program decisions, *Front. Genet.* **10**, 1282 (2019).
56. K. Deribe, D. K. Bakajika, H. M.-G. Zoure, J. O. Gyapong, D. H. Molyneux, and M. P. Rebollo, African regional progress and status of the programme to eliminate lymphatic filariasis: 2000–2020, *Int. Health* **13**, S22 (2020).
57. U. V Amazigo, S. G. A. Leak, H. G. M. Zoure, C. Okoronkwo, M. Diop Ly, S. Isiyaku, A. Crump, J. C. Okeibunor, and B. Boatin, Community-directed distributors: The "foot soldiers" in the fight to control and eliminate neglected tropical diseases, *PLoS Negl. Trop. Dis.* **15**, e0009088 (2021).
58. World Health Organization, Lymphatic filariasis, https://www.who.int/news-room/fact-sheets/detail/lymphatic-filariasis (accessed in October 2023).
59. Ending the neglect to attain the Sustainable Development Goals: A road map for neglected tropical diseases 2021–2030. Geneva: World Health Organization; 2020. Licence: CC BY-NC-SA 3.0 IGO.
60. V. P. Chavda, A. Pandya, S. Pulakkat, M. Soniwala, and V. Patravale, Lymphatic filariasis vaccine development: Neglected for how long? *Expert Rev. Vaccines* **20**, 1471 (2021).

61. R. Kalyanasundaram, V. Khatri, and N. Chauhan, Advances in vaccine development for human lymphatic filariasis, *Trends Parasitol.* **36**, 195 (2020).

62. M. C. Fisher et al., Threats posed by the fungal kingdom to humans, wildlife, and agriculture, *MBio* **11**, (2020).

63. M. L. Rodrigues and J. D. Nosanchuk, Fungal diseases as neglected pathogens: A wake-up call to public health officials, *PLoS Negl. Trop. Dis.* **14**, e0007964 (2020).

64. F. Bongomin, S. Gago, R. Oladele, and D. Denning, Global and multi-national prevalence of fungal diseases: Estimate precision, *J. Fungi* **3**, 57 (2017).

65. G. D. Brown, D. W. Denning, N. A. R. Gow, S. M. Levitz, M. G. Netea, and T. C. White, Hidden killers: Human fungal infections, *Sci. Transl. Med.* **4**, (2012).

66. M. C. Fisher et al., Tackling the emerging threat of antifungal resistance to human health, *Nat. Rev. Microbiol.* **0123456789**, (2022).

67. WHO fungal priority pathogens list to guide research, development and public health action. Geneva: World Health Organization; 2022. Licence: CC BY-NC-SA 3.0 IGO.

68. J. Wauters, I. Baar, P. Meersseman, W. Meersseman, K. Dams, R. De Paep, K. Lagrou, A. Wilmer, P. Jorens, and G. Hermans, Invasive pulmonary aspergillosis is a frequent complication of critically Ill H1N1 patients: A retrospective study, *Intensive Care Med.* **38**, 1761 (2012).

69. A. Raut and N. T. Huy, Rising incidence of mucormycosis in patients with COVID-19: Another challenge for India amidst the second wave? *Lancet Respir. Med.* **9**, e77 (2021).

70. D. Armstrong-James et al., Confronting and mitigating the risk of COVID-19 associated pulmonary aspergillosis, *Eur. Respir. J.* **56**, 2002554 (2020).

71. N. A. F. Janssen et al., Multinational observational cohort study of COVID-19–associated pulmonary aspergillosis, *Emerg. Infect. Dis.* **27**, 2892 (2021).

72. A. Arastehfar, A. Carvalho, M. H. Nguyen, M. T. Hedayati, M. G. Netea, D. S. Perlin, and M. Hoenigl, COVID-19-associated candidiasis (CAC): An underestimated complication in the absence of immunological predispositions? *J. Fungi* **6**, 211 (2020).

73. N. E. Nnadi and D. A. Carter, Climate change and the emergence of fungal pathogens, *PLOS Pathog.* **17**, e1009503 (2021).

74. A. Casadevall, D. P. Kontoyiannis, and V. Robert, On the emergence of *Candida auris*: Climate change, azoles, swamps, and birds, *MBio* **10**, (2019).

75. D. W. Denning, Antifungal drug resistance: An update, *Eur. J. Hosp. Pharm.* **29**, 109 (2022).

76. N. Osherov and D. P. Kontoyiannis, The anti-aspergillus drug pipeline: Is the glass half full or empty? *Med. Mycol.* **55**, 118 (2017).

77. J. R. Perfect, The antifungal pipeline: A reality check, *Nat. Rev. Drug Discov.* **16**, 603 (2017).

78. D. W. Denning and M. J. Bromley, How to bolster the antifungal pipeline, *Science* **347**, 1414 (2015).

79. S. Niazi-Ali, G. T. Atherton, M. Walczak, and D. W. Denning, Drug–drug interaction database for safe prescribing of systemic antifungal agents, *Ther. Adv. Infect. Dis.* **8**, 204993612110106 (2021).

80. M. C. Fisher, N. J. Hawkins, D. Sanglard, and S. J. Gurr, Worldwide emergence of resistance to antifungal drugs challenges human health and food security, *Science* **360**, 739 (2018).

81. N. Robbins, T. Caplan, and L. E. Cowen, Molecular evolution of antifungal drug resistance, *Annu. Rev. Microbiol.* **71**, 753 (2017).

82. J. Berman and D. J. Krysan, Drug resistance and tolerance in fungi, *Nat. Rev. Microbiol.* **18**, 319 (2020).

83. J. Rhodes et al., Population genomics confirms acquisition of drug-resistant *Aspergillus fumigatus* infection by humans from the environment, *Nat. Microbiol.* **7**, 663 (2022).

84. T. N. Duong et al., Azole-resistant *Aspergillus fumigatus* is highly prevalent in the environment of Vietnam, with marked variability by land use type, *Environ. Microbiol.* **23**, 7632 (2021).

85. D. Zhou et al., Extensive genetic diversity and widespread azole resistance in greenhouse populations of *Aspergillus fumigatus* in Yunnan, China, *MSphere* **6**, (2021).

86. E. Vermeulen, K. Lagrou, and P. E. Verweij, Azole resistance in *Aspergillus fumigatus*, *Curr. Opin. Infect. Dis.* **26**, 493 (2013).

87. K. Satoh, K. Makimura, Y. Hasumi, Y. Nishiyama, K. Uchida, and H. Yamaguchi, *Candida auris* Sp. Nov.: A novel ascomycetous yeast isolated from the external ear canal of an inpatient in a Japanese hospital, *Microbiol. Immunol.* **53**, 41 (2009).

88. M. Kim et al., *Candida haemulonii* and closely related species at 5 university hospitals in Korea: Identification, antifungal susceptibility, and clinical features, *Clin. Infect. Dis.* **48**, e57 (2009).
89. W. G. Lee, J. H. Shin, Y. Uh, M. G. Kang, S. H. Kim, K. H. Park, and H.-C. Jang, First three reported cases of nosocomial fungemia caused by *Candida auris*, *J. Clin. Microbiol.* **49**, 3139 (2011).
90. A. Sanyaolu, C. Okorie, A. Marinkovic, A. F. Abbasi, S. Prakash, J. Mangat, Z. Hosein, N. Haider, and J. Chan, *Candida auris*: An overview of the emerging drug-resistant fungal infection, *Infect. Chemother.* **54**, 236 (2022).
91. G. Desoubeaux, A. T. Coste, C. Imbert, and C. Hennequin, Overview about *Candida auris*: What's up 12 years after its first description? *J. Med. Mycol.* **32**, 101248 (2022).
92. E. S. Spivak and K. E. Hanson, *Candida auris*: An emerging fungal pathogen, *J. Clin. Microbiol.* **56**, (2018).
93. S. Sarma and S. Upadhyay, Current perspective on emergence, diagnosis and drug resistance in *Candida auris*, *Infect. Drug Resist.* **10**, 155 (2017).
94. P. Eggimann, J. Garbino, and D. Pittet, Epidemiology of *Candida* species infections in critically ill non-immunosuppressed patients, *Lancet Infect. Dis.* **3**, 685 (2003).
95. G. B. Ruiz and A. Lorenz, What do we know about the biology of the emerging fungal pathogen of humans *Candida auris*? *Microbiol. Res.* **242**, 126621 (2021).
96. S. R. Lockhart et al., Simultaneous emergence of multidrug-resistant *Candida auris* on 3 continents confirmed by whole-genome sequencing and epidemiological analyses, *Clin. Infect. Dis.* **64**, 134 (2017).
97. A. Chowdhary et al., A multicentre study of antifungal susceptibility patterns among 350 *Candida auris* isolates (2009–17) in India: Role of the ERG11 and FKS1 genes in azole and echinocandin resistance, *J. Antimicrob. Chemother.* **73**, 891 (2018).
98. S. Allert, D. Schulz, P. Kämmer, P. Großmann, T. Wolf, S. Schäuble, G. Panagiotou, S. Brunke, and B. Hube, From environmental adaptation to host survival: Attributes that mediate pathogenicity of *Candida auris*, *Virulence* **13**, 191 (2022).
99. R. M. Welsh, M. L. Bentz, A. Shams, H. Houston, A. Lyons, L. J. Rose, and A. P. Litvintseva, Survival, persistence, and isolation of the emerging multidrug-resistant pathogenic yeast *Candida auris* on a plastic health care surface, *J. Clin. Microbiol.* **55**, 2996 (2017).

100. E. Larkin et al., The emerging pathogen *Candida auris*: Growth phenotype, virulence factors, activity of antifungals, and effect of SCY-078, a novel glucan synthesis inhibitor, on growth morphology and biofilm formation, *Antimicrob. Agents Chemother.* **61**, (2017).

101. A. M. Borman, A. Szekely, and E. M. Johnson, Comparative pathogenicity of United Kingdom isolates of the emerging pathogen *Candida auris* and other key pathogenic *Candida* species, *MSphere* **1**, (2016).

102. N. Mollentze and D. G. Streicker, Viral zoonotic risk is homogenous among taxonomic orders of mammalian and avian reservoir hosts, *Proc. Natl. Acad. Sci. U. S. A.* **117**, 9423 (2020).

103. Z. L. Grange et al., Ranking the risk of animal-to-human spillover for newly discovered viruses, *Proc. Natl. Acad. Sci. U. S. A.* **118**, (2021).

104. R. E. Baker et al., Infectious disease in an era of global change, *Nat. Rev. Microbiol.* **20**, 193 (2022).

105. J. Cui, F. Li, and Z.-L. Shi, Origin and evolution of pathogenic coronaviruses, *Nat. Rev. Microbiol.* **17**, 181 (2019).

106. B. N. Fields, *Fields Virology*, edited by D. M. Knipe and P. M. Howley (Lippincott Williams & Wilkins, 2013).

107. L. Soraci, F. Lattanzio, G. Soraci, M. E. Gambuzza, C. Pulvirenti, A. Cozza, A. Corsonello, F. Luciani, and G. Rezza, COVID-19 vaccines: Current and future perspectives, *Vaccines* **10**, (2022).

108. P. Buchy, Y. Buisson, O. Cintra, D. E. Dwyer, M. Nissen, R. Ortiz de Lejarazu, and E. Petersen, COVID-19 pandemic: Lessons learned from more than a century of pandemics and current vaccine development for pandemic control, *Int. J. Infect. Dis.* **112**, 300 (2021).

109. S. Pecetta, D. Tortorice, F. B. Scorza, M. Pizza, G. Dougan, R. Hatchett, S. Black, D. E. Bloom, and R. Rappuoli, The trillion dollar vaccine gap, *Sci. Transl. Med.* **14**, eabn4342 (2022).

110. A.-S. Krebs, L. M. Mendonça, and P. Zhang, Structural analysis of retrovirus assembly and maturation, *Viruses* **14**, (2021).

111. M. P. Sanou, A. S. De Groot, M. Murphey-Corb, J. A. Levy, and J. K. Yamamoto, HIV-1 vaccine trials: Evolving concepts and designs, *Open AIDS J.* **6**, 274 (2012).

112. J. R. Mascola and D. C. Montefiori, The role of antibodies in HIV vaccines, *Annu. Rev. Immunol.* **28**, 413 (2010).

113. N. L. Board, M. Moskovljevic, F. Wu, R. F. Siliciano, and J. D. Siliciano, Engaging innate immunity in HIV-1 cure strategies, *Nat. Rev. Immunol.* (2021).

114. P. J. Klasse and J. P. Moore, Reappraising the value of HIV-1 vaccine correlates of protection analyses, *J. Virol.* **96**, e0003422 (2022).

115. G. Hütter et al., Long-term control of HIV by CCR5 Delta32/Delta32 stem-cell transplantation, *N. Engl. J. Med.* **360**, 692 (2009).

116. R. K. Gupta et al., HIV-1 remission following CCR5Δ32/Δ32 haematopoietic stem-cell transplantation, *Nature* **568**, 244 (2019).

117. F. Krammer et al., Influenza, *Nat. Rev. Dis. Prim.* **4**, 3 (2018).

118. J. H. Beigel, Influenza, *Crit. Care Med.* **36**, 2660 (2008).

119. World Health Organization, Pandemic Influenza Preparedness (PIP) framework, https://www.who.int/initiatives/pandemic-influenza-preparedness-framework (accessed in October 2023).

120. E. D. Kilbourne, Influenza pandemics of the 20th century, *Emerg. Infect. Dis.* **12**, 9 (2006).

121. World Health Organization, Global Influenza Programme, https://www.who.int/teams/global-influenza-programme (accessed in October 2023).

122. Centers for Disease Control and Prevention, Types of influenza viruses, https://www.cdc.gov/flu/about/viruses/types.htm (accessed in October 2023).

123. Influenza virus characterization: summary report, Europe, August 2023. Copenhagen: WHO Regional Office for Europe and Stockholm: European Centre for Disease Prevention and Control; 2023. Licence: CCBY3.0IGO

124. A. E. Fiore, A. Fry, D. Shay, L. Gubareva, J. S. Bresee, T. M. Uyeki, and Centers for Disease Control, Antiviral agents for the treatment and chemoprophylaxis of influenza: Recommendations of the Advisory Committee on Immunization Practices (ACIP), *MMWR Recomm. Rep.* **60**, 1 (2011).

125. F. Krammer et al., Influenza, *Nat. Rev. Dis. Prim.* **4**, 1 (2018).

126. C. S. Tyrrell, J. L. Y. Allen, and E. Gkrania-Klotsas, Influenza: Epidemiology and hospital management, *Medicine (Abingdon)* **49**, 797 (2021).

127. S. Rockman, B. Taylor, J. W. McCauley, I. G. Barr, R. Longstaff, and R. Bahra, Global pandemic preparedness: Optimizing our capabilities and the influenza experience, *Vaccines* **10**, (2022).

128. S. Pecetta, D. Tortorice, F. B. Scorza, M. Pizza, G. Dougan, R. Hatchett, S. Black, D. E. Bloom, and R. Rappuoli, The trillion dollar vaccine gap, *Sci. Transl. Med.* **14**, eabn4342 (2022).

129. A. Sarker, Z. Gu, L. Mao, Y. Ge, D. Hou, J. Fang, Z. Wei, and Z. Wang, Influenza-existing drugs and treatment prospects, *Eur. J. Med. Chem.* **232**, (2022).

130. J. Hidalgo, G. A. Richards, J. I. S. Jiménez, T. Baker, and P. Amin, Viral hemorrhagic fever in the tropics: Report from the Task Force on Tropical Diseases by the World Federation of Societies of Intensive and Critical Care Medicine, *J. Crit. Care* **42**, 366 (2017).

131. N. Jonkmans, V. D'Acremont, and A. Flahault, Scoping future outbreaks: A scoping review on the outbreak prediction of the WHO blueprint list of priority diseases, *BMJ Glob. Heal.* **6**, (2021).

132. Y. Botti-Lodovico et al., The origins and future of sentinel: An early-warning system for pandemic preemption and response, *Viruses* **13**, (2021).

133. S. A. M. Kularatne, Dengue fever, *BMJ* **351**, (2015).

134. M. Iannetta, A. Di Caro, E. Nicastri, F. Vairo, H. Masanja, G. Kobinger, A. Mirazimi, F. Ntoumi, A. Zumla, and G. Ippolito, Viral hemorrhagic fevers other than Ebola and Lassa, *Infect. Dis. Clin. North Am.* **33**, 977 (2019).

135. V. Rougeron, H. Feldmann, G. Grard, S. Becker, and E. M. Leroy, Ebola and Marburg haemorrhagic fever, *J. Clin. Virol.* **64**, 111 (2015).

136. D. A. Asogun, S. Günther, G. O. Akpede, C. Ihekweazu, and A. Zumla, Lassa fever: Epidemiology, clinical features, diagnosis, management and prevention, *Infect. Dis. Clin. North Am.* **33**, 933 (2019).

137. R. W. Cross et al., Quadrivalent VesiculoVax vaccine protects nonhuman primates from viral-induced hemorrhagic fever and death, *J. Clin. Invest.* **130**, 539 (2020).

138. C. D. Murin, B. M. Gunn, P. W. H. I. Parren, and G. P. Kobinger, Editorial: Antibody therapeutics for the treatment of filoviral infection, *Front. Immunol.* **13**, (2022).

139. V. Raabe, A. K. Mehta, J. D. Evans, and Science Working Group of the National Emerging Special Pathogens Training and Education Center (NETEC) Special Pathogens Research Network (SPRN), Lassa virus infection: A summary for clinicians, *Int. J. Infect. Dis.* **119**, 187 (2022).

140. H. Sudhakar, J. Bhate, and A. K. Patra, Patent landscape of novel technologies for combating category-A arenavirus infections, *Expert Opin. Ther. Pat.* **30**, 557 (2020).

141. R. D. V. Kleinert, E. Montoya-Diaz, T. Khera, K. Welsch, B. Tegtmeyer, S. Hoehl, S. Ciesek, and R. J. P. Brown, Yellow fever: Integrating current knowledge with technological innovations to identify strategies for controlling a re-emerging virus, *Viruses* **11**, (2019).

142. T. Ikegami, Candidate vaccines for human Rift Valley fever, *Expert Opin. Biol. Ther.* **19**, 1333 (2019).

143. M. R. Hickman, D. L. Saunders, C. A. Bigger, C. D. Kane, and P. L. Iversen, The development of broad-spectrum antiviral medical countermeasures to treat viral hemorrhagic fevers caused by natural or weaponized virus infections, *PLoS Negl. Trop. Dis.* **16**, (2022).

144. S. Delgado et al., Chapare virus: A newly discovered arenavirus isolated from a fatal hemorrhagic fever case in Bolivia, *PLoS Pathog.* **4**, e1000047 (2008).

145. R. B. Tesh, P. B. Jahrling, R. Salas, and R. E. Shope, Description of Guanarito virus (Arenaviridae: Arenavirus), the etiologic agent of Venezuelan hemorrhagic fever, *Am. J. Trop. Med. Hyg.* **50**, 452 (1994).

146. A. Grant, A. Seregin, C. Huang, O. Kolokoltsova, A. Brasier, C. Peters, and S. Paessler, Junín virus pathogenesis and virus replication, *Viruses* **4**, 2317 (2012).

147. V. B. Oti, A reemerging, Lassa virus: Aspects of its structure, replication, pathogenicity and diagnosis, in *Current Topics in Tropical Emerging Diseases and Travel Medicine,* edited by A. J. Rodriguez-Morales (IntechOpen, 2018).

148. J. W. Golden, B. Beitzel, J. T. Ladner, E. M. Mucker, S. A. Kwilas, G. Palacios, and J. W. Hooper, An attenuated Machupo virus with a disrupted L-segment intergenic region protects guinea pigs against lethal Guanarito virus infection, *Sci. Rep.* **7**, 4679 (2017).

149. J. P. Gonzalez, M. D. Bowen, S. T. Nichol, and R. Rico-Hesse, Genetic characterization and phylogeny of Sabiá virus, an emergent pathogen in Brazil, *Virology* **221**, 318 (1996).

150. R. Hewson et al., Crimean–Congo haemorrhagic fever virus: Sequence analysis of the small RNA segments from a collection of viruses world wide, *Virus Res.* **102**, 185 (2004).

151. M. A. Mir, Hantaviruses, *Clin. Lab. Med.* **30**, 67 (2010).

152. T. Sironen, A. Vaheri, and A. Plyusnin, Molecular evolution of Puumala Hantavirus, *J. Virol.* **75**, 11803 (2001).

153. H. Boshra, G. Lorenzo, N. Busquets, and A. Brun, Rift Valley fever: Recent insights into pathogenesis and prevention, *J. Virol.* **85**, 6098 (2011).

154. A. Plyusnin, A. Vaheri, and A. Lundkvist, Saaremaa Hantavirus should not be confused with its dangerous relative, Dobrava virus, *J. Clin. Microbiol.* **44**, 1608 (2006).

155. Q. Su, Y. Chen, M. Li, J. Ma, B. Wang, J. Luo, and H. He, Genetic characterization and molecular evolution of urban Seoul virus in Southern China, *Viruses* **11**, (2019).

156. R. Feuer, J. D. Boone, D. Netski, S. P. Morzunov, and S. C. St Jeor, Temporal and spatial analysis of sin Nombre virus quasispecies in naturally infected rodents, *J. Virol.* **73**, 9544 (1999).

157. F. Ren et al., Non-structural proteins of severe fever with thrombocytopenia syndrome virus suppress RNA synthesis in a transcriptionally active CDNA-derived viral RNA synthesis system, *Front. Microbiol.* **12**, 709517 (2021).

158. O. Vapalahti et al., Isolation and characterization of Tula virus, a distinct serotype in the genus Hantavirus, family Bunyaviridae, *J. Gen. Virol.* **77** (Pt 12), 3063 (1996).

159. W. Slenczka and H. D. Klenk, Forty years of Marburg virus, *J. Infect. Dis.* **196** S131 (2007).

160. S. T. Jacob et al., Ebola virus disease, *Nat. Rev. Dis. Prim.* **6**, 13 (2020).

161. R. J. Kuhn et al., Structure of dengue virus: Implications for flavivirus organization, maturation, and fusion, *Cell* **108**, 717 (2002).

162. G. H. Samuel, M. R. Wiley, A. Badawi, Z. N. Adelman, and K. M. Myles, Yellow fever virus capsid protein is a potent suppressor of RNA silencing that binds double-stranded RNA, *Proc. Natl. Acad. Sci. U. S. A.* **113**, 13863 (2016).

163. G. L. Campbell, A. A. Marfin, R. S. Lanciotti, and D. J. Gubler, West Nile Virus, *Lancet. Infect. Dis.* **2**, 519 (2002).

164. World Health Organization, Dengue and severe dengue, https://www.who.int/news-room/fact-sheets/detail/dengue-and-severe-dengue (accessed in October 2023).

165. European Centre for Disease Prevention and Control, Dengue, https://www.ecdc.europa.eu/en/dengue (accessed in October 2023).

166. Centers for Disease Control and Prevention, Dengue, https://www.cdc.gov/dengue/ (accessed in October 2023).

167. World Health Organization, Ebola virus disease, https://www.who.int/health-topics/ebola (accessed in October 2023).

168. European Centre for Disease Prevention and Control. Ebola and Marburg virus disease in *ECDC. Annual Epidemiological Report for 2019*. Stockholm: ECDC; 2021.

169. Centers for Disease Control and Prevention, Ebola disease, https://www.cdc.gov/vhf/ebola/index.html (accessed in October 2023).

170. Centers for Disease Control and Prevention, Marburg, https://www.cdc.gov/vhf/marburg/ (accessed in October 2023).

171. European Centre for Disease Prevention and Control, Lassa fever, https://www.ecdc.europa.eu/en/lassa-fever (accessed in October 2023).

172. World Health Organization, Lassa fever, https://www.who.int/health-topics/lassa-fever (accessed in October 2023).

173. Centers for Disease Control and Prevention, Lassa fever, https://www.cdc.gov/vhf/lassa/ (accessed in October 2023).

174. European Centre for Disease Prevention and Control. Ebola and Marburg virus disease, in *ECDC. Annual Epidemiological Report for 2019*. Stockholm: ECDC; 2021.

175. World Health Organization, Marburg virus disease, https://www.who.int/health-topics/marburg-virus-disease (accessed in October 2023).

176. E. Tacconelli et al., Discovery, research, and development of new antibiotics: The WHO priority list of antibiotic-resistant bacteria and tuberculosis, *Lancet Infect. Dis.* **18**, 318 (2018).

177. World Health Organization, WHO publishes list of bacteria for which new antibiotics are urgently needed, https://www.who.int/news/item/27-02-2017-who-publishes-list-of-bacteria-for-which-new-antibiotics-are-urgently-needed (accessed in October 2023).

178. Interagency Coordination Group in Antimicrobial Resistance, Antimicrobial Resistance Report, 2019.

179. M. Hutchings, A. Truman, and B. Wilkinson, Antibiotics: Past, present and future, *Curr. Opin. Microbiol.* **51**, 72 (2019).

180. C. Årdal, M. Balasegaram, R. Laxminarayan, D. McAdams, K. Outterson, J. H. Rex, and N. Sumpradit, Antibiotic development: Economic, regulatory and societal challenges, *Nat. Rev. Microbiol.* **18**, 267 (2020).

181. W. A. Adedeji, The treasure called antibiotics, *Ann. Ibadan Postgrad. Med.* **14**, 56 (2016).

182. R. I. Aminov, A brief history of the antibiotic era: Lessons learned and challenges for the future, *Front. Microbiol.* **1**, (2010).

183. A. Fleming, Classics in infectious diseases: On the antibacterial action of cultures of a Penicillium, with special reference to their use in the isolation of *B. influenzae* by Alexander Fleming, Reprinted from the *British Journal of Experimental Pathology* **10**:226–236, *Rev. Infect. Dis.* **2**, 129 (1980).

184. M. I. Hutchings, A. W. Truman, and B. Wilkinson, Antibiotics: Past, present and future, *Curr. Opin. Microbiol.* **51**, 72 (2019).

185. L. L. Silver, Challenges of antibacterial discovery, *Clin. Microbiol. Rev.* **24**, 71 (2011).

186. R. Shukla et al., Mode of action of teixobactins in cellular membranes, *Nat. Commun.* **11**, (2020).

187. F. L. Gordillo Altamirano and J. J. Barr, Phage therapy in the postantibiotic era, *Clin. Microbiol. Rev.* **32**, (2019).

188. K. E. Kortright, B. K. Chan, J. L. Koff, and P. E. Turner, Phage therapy: A renewed approach to combat antibiotic-resistant bacteria, *Cell Host Microbe* **25**, 219 (2019).

189. D. M. Lin, B. Koskella, and H. C. Lin, Phage therapy: An alternative to antibiotics in the age of multi-drug resistance, *World J. Gastrointest. Pharmacol. Ther.* **8**, 162 (2017).

190. G. F. Hatfull, R. M. Dedrick, and R. T. Schooley, Phage therapy for antibiotic-resistant bacterial infections, *Annu. Rev. Med.* **73**, 197 (2022).

191. M. Ur Rahman, W. Wang, Q. Sun, J. A. Shah, C. Li, Y. Sun, Y. Li, B. Zhang, W. Chen, and S. Wang, Endolysin: A promising solution against antimicrobial resistance, *Antibiotics* **10**, 1277 (2021).

192. F. Eichenseher et al., Linker-improved chimeric endolysin selectively kills *Staphylococcus aureus in vitro*, on reconstituted human epidermis, and in a murine model of skin infection, *Antimicrob. Agents Chemother.* **66**, (2022).

193. Y. Briers et al., Engineered endolysin-based "Artilysins" to combat multidrug-resistant gram-negative pathogens, *MBio* **5**, (2014).

194. H. Gerstmans, L. Rodríguez-Rubio, R. Lavigne, and Y. Briers, From endolysins to Artilysin®s: Novel enzyme-based approaches to kill drug-resistant bacteria, *Biochem. Soc. Trans.* **44**, 123 (2016).

195. M. Nguyen and S. G. Joshi, Carbapenem resistance in *Acinetobacter baumannii*, and their importance in hospital-acquired infections: A scientific review, *J. Appl. Microbiol.* **131**, 2715 (2021).

196. P. Nordmann and L. Poirel, Epidemiology and diagnostics of carbapenem resistance in gram-negative bacteria, *Clin. Infect. Dis.* **69**, S521 (2019).

197. I. Kyriakidis, E. Vasileiou, Z. D. Pana, and A. Tragiannidis, *Acinetobacter baumannii* antibiotic resistance mechanisms, *Pathogens* **10**, (2021).

198. M. Castanheira, L. M. Deshpande, A. Costello, T. A. Davies, and R. N. Jones, Epidemiology and carbapenem resistance mechanisms of carbapenem-non-susceptible *Pseudomonas aeruginosa* collected

during 2009-11 in 14 European and Mediterranean countries, *J. Antimicrob. Chemother.* **69**, 1804 (2014).

199. R. A. Bonomo and D. Szabo, Mechanisms of multidrug resistance in Acinetobacter species and *Pseudomonas aeruginosa, Clin. Infect. Dis.* **43**, (2006).

200. D. J. Buehrle, R. K. Shields, L. G. Clarke, B. A. Potoski, C. J. Clancy, and M. H. Nguyen, Carbapenem-resistant *Pseudomonas aeruginosa* bacteremia: Risk factors for mortality and microbiologic treatment failure, *Antimicrob. Agents Chemother.* **61**, (2017).

201. A. Iovleva and Y. Doi, Carbapenem-resistant Enterobacteriaceae, *Clin. Lab. Med.* **37**, 303 (2017).

202. R. F. Potter, A. W. D'Souza, and G. Dantas, The rapid spread of carbapenem-resistant Enterobacteriaceae, *Drug Resist. Updat.* **29**, 30 (2016).

203. L. K. Logan and R. A. Weinstein, The epidemiology of carbapenem-resistant Enterobacteriaceae: The impact and evolution of a global menace, *J. Infect. Dis.* **215**, S28 (2017).

204. C. A. Arias and B. E. Murray, The rise of the Enterococcus: Beyond vancomycin resistance, *Nat. Rev. Microbiol.* **10**, 266 (2012).

205. T. Raza, S. R. Ullah, K. Mehmood, and S. Andleeb, Vancomycin resistant Enterococci: A brief review, *J. Pak. Med. Assoc.* **68**, 768 (2018).

206. N. A. Turner, B. K. Sharma-Kuinkel, S. A. Maskarinec, E. M. Eichenberger, P. P. Shah, M. Carugati, T. L. Holland, and V. G. Fowler, Methicillin-resistant *Staphylococcus aureus*: An overview of basic and clinical research, *Nat. Rev. Microbiol.* **17**, 203 (2019).

207. S. Lakhundi and K. Zhang, Methicillin-resistant *Staphylococcus aureus*: Molecular characterization, evolution, and epidemiology, *Clin. Microbiol. Rev.* **31**, (2018).

208. W. A. McGuinness, N. Malachowa, and F. R. DeLeo, Vancomycin resistance in *Staphylococcus aureus, Yale J. Biol. Med.* **90**, 269 (2017).

209. A. S. Lee, H. De Lencastre, J. Garau, J. Kluytmans, S. Malhotra-Kumar, A. Peschel, and S. Harbarth, Methicillin-resistant *Staphylococcus aureus, Nat. Rev. Dis. Prim.* **4**, (2018).

210. C. Alba, A. Blanco, and T. Alarcón, Antibiotic resistance in *Helicobacter pylori, Curr. Opin. Infect. Dis.* **30**, 489 (2017).

211. A. T. Marques, J. M. B. Vítor, A. Santos, M. Oleastro, and F. F. Vale, Trends in *Helicobacter pylori* resistance to clarithromycin: From phenotypic to genomic approaches, *Microb. Genomics* **6**, (2020).

212. W. Fischbach and P. Malfertheiner, *Helicobacter-pylori*-infektion: Indikationen zu eradikation, diagnostik und therapie, *Dtsch. Arztebl. Int.* **115**, 429 (2018).

213. H.-X. Xia, X.-G. Fan, and N. J. Talley, Clarithromycin resistance in *Helicobacter pylori* and its clinical relevance, *World J. Gastroenterol.* **5**, 263 (1999).

214. S. Payot, J. M. Bolla, D. Corcoran, S. Fanning, F. Mégraud, and Q. Zhang, Mechanisms of fluoroquinolone and macrolide resistance in *Campylobacter* Spp., *Microbes Infect.* **8**, 1967 (2006).

215. J. L. Smith and P. M. Fratamico, Fluoroquinolone resistance in Campylobacter, *J. Food Prot.* **73**, 1141 (2010).

216. E. J. Threlfall, Antimicrobial drug resistance in Salmonella: Problems and perspectives in food- and water-borne infections, *FEMS Microbiol. Rev.* **26**, 141 (2002).

217. W. L. Cuypers, J. Jacobs, V. Wong, E. J. Klemm, S. Deborggraeve, and S. van Puyvelde, Fluoroquinolone resistance in Salmonella: Insights by whole-genome sequencing, *Microb. Genomics* **4**, (2018).

218. J. W. Tapsall, *Neisseria gonorrhoeae* and emerging resistance to extended spectrum cephalosporins, *Curr. Opin. Infect. Dis.* **22**, 87 (2009).

219. B. Młynarczyk-Bonikowska, A. Majewska, M. Malejczyk, G. Młynarczyk, and S. Majewski, Multiresistant *Neisseria gonorrhoeae*: A new threat in second decade of the XXI century, *Med. Microbiol. Immunol.* **209**, 95 (2020).

220. J. S. Knapp, K. K. Fox, D. L. Trees, and W. L. Whittington, Fluoroquinolone resistance in *Neisseria gonorrhoeae, Emerg. Infect. Dis.* **3**, 33 (1997).

221. M. Puzari, M. Sharma, and P. Chetia, Emergence of antibiotic resistant Shigella species: A matter of concern, *J. Infect. Public Health* **11**, 451 (n.d.).

222. M. Samura et al., Efficacy and safety of daptomycin versus vancomycin for bacteremia caused by methicillin-resistant *Staphylococcus aureus* with vancomycin minimum inhibitory concentration > 1 μg/mL: A systematic review and meta-analysis, *Pharmaceutics* **14**, 714 (2022).

223. A. D. Bai et al., *Staphylococcus aureus* bacteraemia mortality: A systematic review and meta-analysis, *Clin. Microbiol. Infect.* (2022).

224. B. L. Bilyk, V. V Panchal, M. Tinajero-Trejo, J. K. Hobbs, and S. J. Foster, An interplay of multiple positive and negative factors governs methicillin resistance in *Staphylococcus aureus, Microbiol. Mol. Biol. Rev.* e0015921 (2022).

225. Y. Zhu, W. E. Huang, and Q. Yang, Clinical perspective of antimicrobial resistance in bacteria, *Infect. Drug Resist.* **15**, 735 (2022).

226. J. V. de Oliveira Santos, S. D. da Costa Júnior, S. M. de Fátima Ramos Dos Santos Medeiros, I. D. L. Cavalcanti, J. B. de Souza, D. L. Coriolano, W. R. C. da Silva, M. H. M. E. Alves, and I. M. F. Cavalcanti, Panorama of bacterial infections caused by epidemic resistant strains, *Curr. Microbiol.* **79**, 175 (2022).

227. A. Maitra, T. Munshi, J. Healy, L. T. Martin, W. Vollmer, N. H. Keep, and S. Bhakta, Cell wall peptidoglycan in *Mycobacterium tuberculosis*: An Achilles' heel for the TB-causing pathogen, *FEMS Microbiol. Rev.* **43**, 548 (2019).

228. Global Tuberculosis Report 2021. Geneva: World Health Organization; 2021. Licence: CC BY-NC-SA 3.0 IGO.

229. World Health Organization, Tuberculosis, https://www.who.int/health-topics/tuberculosis#tab=tab_1 (accessed in October 2023).

230. G. Churchyard, P. Kim, N. S. Shah, R. Rustomjee, N. Gandhi, B. Mathema, D. Dowdy, A. Kasmar, and V. Cardenas, What we know about tuberculosis transmission: An overview, *J. Infect. Dis.* **216**, S629 (2017).

231. J. Furin, H. Cox, and M. Pai, Tuberculosis, *Lancet* **393**, 1642 (2019).

232. Centers for Disease Control and Prevention, Tuberculosis, https://www.cdc.gov/tb/ (accessed in October 2023).

233. S. R. Mase and T. Chorba, Treatment of drug-resistant tuberculosis, *Clin. Chest Med.* **40**, 775 (2019).

234. J. Bruchfeld, M. Correia-Neves, and G. Kallenius, Tuberculosis and HIV coinfection, *Cold Spring Harb. Perspect. Med.* **5**, (2015).

235. J. Birnbaum, F. M. Kahan, H. Kropp, and J. S. Macdonald, Carbapenems: A new class of beta-lactam antibiotics. Discovery and development of imipenem/cilastatin, *Am. J. Med.* **78**, 3 (1985).

236. S.-S. Jean, D. Harnod, and P.-R. Hsueh, Global threat of carbapenem-resistant gram-negative bacteria, *Front. Cell. Infect. Microbiol.* **12**, (2022).

237. Y. Doi, Treatment options for carbapenem-resistant gram-negative bacterial infections, *Clin. Infect. Dis.* **69**, S565 (2019).

238. R. P. Ambler, The structure of beta-lactamases, *Philos. Trans. R. Soc. Lond. B. Biol. Sci.* **289**, 321 (1980).

239. D. L. Paterson and R. A. Bonomo, Extended-spectrum β-lactamases: A clinical update, *Clin. Microbiol. Rev.* **18**, 657 (2005).

240. L. Martínez-Martínez, Extended-spectrum β-lactamases and the permeability barrier, *Clin. Microbiol. Infect.* **14**, 82 (2008).

241. A. Vahhabi, A. Hasani, M. A. Rezaee, B. Baradaran, A. Hasani, H. Samadi Kafil, F. Abbaszadeh, and L. Dehghani, A plethora of carbapenem resistance in *Acinetobacter baumannii*: No end to a long insidious genetic journey, *J. Chemother.* **33**, 137 (2021).

242. M. Tilahun, Y. Kassa, A. Gedefie, and M. Ashagire, Emerging carbapenem-resistant *Enterobacteriaceae* infection, its epidemiology and novel treatment options: A review, *Infect. Drug Resist.* **14**, 4363 (2021).

243. A. R. Losito, F. Raffaelli, P. Del Giacomo, and M. Tumbarello, New drugs for the treatment of *Pseudomonas aeruginosa* infections with limited treatment options: A narrative review, *Antibiot.* **11**, (2022).

244. B. Moyá, A. Beceiro, G. Cabot, C. Juan, L. Zamorano, S. Alberti, and A. Oliver, Pan-β-lactam resistance development in *Pseudomonas aeruginosa* clinical strains: Molecular Mechanisms, penicillin-binding protein profiles, and binding affinities, *Antimicrob. Agents Chemother.* **56**, 4771 (2012).

245. J. R. Sinclair, Importance of a one health approach in advancing global health security and the Sustainable Development Goals, *Rev. Sci. Tech.* **38**, 145 (2019).

246. B. Greenwood, The contribution of vaccination to global health: Past, present and future, *Philos. Trans. R. Soc. B Biol. Sci.* **369**, (2014).

247. Wellcome Trust, Wellcome Explainer, https://wellcome.org/news/how-science-combatting-infectious-disease (accessed in October 2023).

248. W. R. Dowdle, The principles of disease elimination and eradication, *Bull. World Health Organ.* **76**, 22 (1998).

249. S. Vanderslott, B. Dadonaite, and M. Roser, Vaccination, https://ourworldindata.org/vaccination (accessed in October 2023).

250. World Health Organization, Immunization coverage, https://www.who.int/news-room/fact-sheets/detail/immunization-coverage (accessed in October 2023).

251. J. S. Barrett, S. Y. Yang, K. Muralidharan, V. Javes, K. Oladuja, M. S. Castelli, N. Clayton, J. Liu, and A. Ramos, Considerations for addressing anti-vaccination campaigns: How did we get here and what can we do about it? *Clin. Transl. Sci.* (2022).

252. È. Dubé, J. K. Ward, P. Verger, and N. E. Macdonald, Vaccine hesitancy, acceptance, and anti-vaccination: Trends and future prospects for public health, *Annu. Rev. Public Health* **42**, 175 (2020).

253. M. Schleiff, A. Olateju, E. Decker, A. H. Neel, R. Oke, M. A. Peters, A. Rao, and O. Alonge, A multi-pronged scoping review approach to understanding the evolving implementation of the smallpox and polio eradication programs: What can other global health initiatives learn? *BMC Public Health* **20**, 1698 (2020).

254. R. Davis and W. B. Mbabazi, Challenges to global measles eradication: Is it all in the timing? *Pan Afr. Med. J.* **27**, (2017).

255. A. Mohammed, O. Tomori, and J. N. Nkengasong, Lessons from the elimination of poliomyelitis in Africa, *Nat. Rev. Immunol.* **21**, 823 (2021).

256. V. Kayser and I. Ramzan, Vaccines and vaccination: History and emerging issues, *Hum. Vaccines Immunother.* **17**, 5255 (2021).

257. World Health Organization, Poliomyelitis, https://www.who.int/health-topics/poliomyelitis (accessed in October 2023).

Chapter 4

Overview of Nanotherapeutics for Viral Infections

Valeria Nele
Department of Pharmacy, University of Naples Federico II, Via D. Montesano, 49, 80131 Naples, Italy
valeria.nele@unina.it

4.1 Introduction

Infectious diseases represent a serious burden on the global healthcare system and are characterized by high mortality rates, especially in developing countries, which often lack infrastructure and have poor access to drugs and treatments [1]. Preventive treatment (*i.e.*, vaccination) has proven effective against some viral diseases such as hepatitis and measles [2]. However, this approach has been unsuccessful with viruses characterized by a high mutation rate such as influenza, which requires the development of new vaccines every year [3]. In the last three decades, the viral outbreaks due to the human immunodeficiency virus (HIV), Ebola virus, Zika

Nanotherapeutics for Infectious Diseases
Edited by Íris Luz Batalha

ISBN 978-981-5129-36-6 (Hardcover), 978-1-003-56678-6 (eBook)
www.jennystanford.com

virus, and coronaviruses have severely impacted human economy and well-being and have emphasized the lack of effective measures to fight these diseases [4].

Viruses are extremely effective at infecting potential hosts to proliferate and survive; typically, the virus structure comprises a protein coating (capsid) that protects the viral genetic material and facilitates viral entry in the host cell [5]. Viral capsids can exhibit various morphologies (*i.e.*, helical, icosahedral, and complex shapes, which are often a combination of helical and icosahedral structures) [6]. Furthermore, viral capsids are often coated with a lipidic bilayer that contains fusogenic proteins able to promote viral fusion with the host cell membrane [7]; in this case, viruses are referred to as "enveloped viruses" (Fig. 4.1).

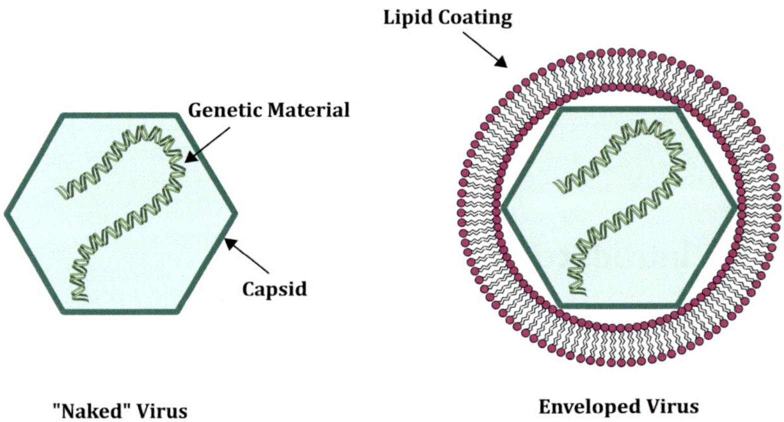

Figure 4.1 Schematic of the structures of a "naked" and an enveloped virus. Partially created using Servier Medical Art licensed under a Creative Commons Attribution 3.0 Unported License (https://creativecommons.org/licenses/by/3.0/).

Enveloped viruses can enter the host cells either via direct fusion with the host cell membrane or via endocytosis and subsequent endosomal escape [8]. Viruses typically exploit the host cell machinery to replicate. For example, viruses whose genome is encoded by deoxyribonucleic acid (DNA) contain double-stranded DNA, which is transcribed into mRNA by the host cell machinery, leading to the production of viral proteins. Newly synthesized viral proteins are then packed into new viral particles and are released from the host cell via an outward budding mechanism [9].

Common human viruses include the HIV, which induces acquired immunodeficiency syndrome (AIDS) and depletes $CD4^+$ T cells, making the host organism vulnerable to any infection [10]. Other common human viruses are the human influenza A virus, which causes approximately 250,000–500,000 deaths per year; the hepatitis viruses, responsible for liver damage-induced deaths; the Zika virus; and the Ebola virus, which causes hemorrhagic fever and has a death rate of 50%.

4.2 Antiviral Therapeutic Strategies

Numerous antiviral therapeutic strategies have emerged in the last few decades; these therapies can target either specific viral components or a phase of the viral life cycle. An example is the development of small molecules such as peptides, which can disrupt viral envelopes; this approach has been proposed for the treatment of hepatitis C [11], HIV, and measles [12]. Alternatively, it is possible to block viral entry in the host cell by directly inhibiting viral binding to the host cell membrane receptors via antibodies. Monoclonal antibodies have been proven effective against the influenza virus H7N9 [13] and the Zika virus [14]. More recently, multivalent antibodies that can target multiple epitopes on the viral surface have been introduced [15, 16].

Therapeutic approaches targeting viral replication have also been developed; in this case, small interfering RNA (siRNA) or micro RNA (miRNA) able to interfere with the synthesis of viral proteins can be used [17]. Other small molecules that target viral replication comprise nucleoside, pyrophosphate, acyclic guanosine, and acyclic nucleoside phosphonate analogues [18]. For example, remdesivir, a nucleoside analogue, can be converted into an active triphosphate metabolite following cellular uptake; this metabolite inhibits RNA polymerase and has been shown to be effective in the treatment of the Ebola virus [19]. Further therapeutic strategies include the inhibition of virion assembly [18], which exploits protease inhibitors able to halt the synthesis of new viral components and can be used in combination with drugs targeting viral replication [20], and the reduction of inflammation. In this case, drugs capable of exerting an immunomodulatory activity by targeting cytokines such as tumor necrosis factor (TNF) can be used [21].

4.3 Nanotherapeutics to Tackle Viral Infections

In the last few decades, several therapeutic approaches that leverage nanotechnology to treat viral infections have emerged. Nanoparticles can effectively encapsulate antiviral drugs and improve their pharmacokinetic profile; furthermore, they can be designed to improve organ specificity, prevent systemic toxicity, and exhibit intrinsic antiviral activity.

4.3.1 Lipid Nanoparticles

As mentioned above, RNA interference is a potential antiviral approach. However, RNA suffers from poor stability in serum and is prone to rapid degradation; as such, it requires the use of nanocarriers for optimal performance. Lipid-based nanoparticles are one of the most widely investigated systems for RNA delivery due to their ability to encapsulate and deliver RNA payloads with high efficiencies.

Lipid-based nanoparticles comprise solid lipid nanoparticles (SLN), nanostructured lipid nanoparticles (NLC), liposomes, and lipid nanoparticles (LNP). SLNs comprise a solid lipid core protected by a surfactant layer, exhibit sizes between 50 and 1000 nm, and can be prepared with cationic lipids able to complex negatively charged RNA [22]. NLCs are second-generation SLNs prepared from a mixture of solid and liquid lipids and surfactants and are particularly suited for dermal drug delivery applications [23]. Liposomes comprise an aqueous core enclosed by a lipid shell and can be prepared with mixtures of cationic and zwitterionic lipids for RNA complexation to form lipoplexes [24].

LNPs have received enormous attention in the last few years given their use as vaccines against Covid-19 infections; these nanoparticles contain an ionizable lipid that becomes positively charged at acidic pH, structural lipids such as cholesterol and phosphatidylcholines, and PEGylated lipids [24].

Given their versatility, lipid-based nanoparticles encapsulating RNA have been widely investigated to treat viral infections. For example, LNPs loaded with anti-hepatitis B (HBV) siRNA have been shown to reduce HBV titters up to 7 days post intravenous administration in mice [25]. Similarly, lipoplexes comprising a cationic lipid, cholesterol, and protamine were loaded with siRNA targeted to the 5'UTR of the herpes complex virus (HCV) genome; a significant inhibition of viral replication was observed upon systemic injection in a liver tumor-xenotransplant mouse model of HCV [26]. Lipid-based nanoparticles can also be functionalized with specific moieties to achieve targeted drug delivery. Kim et al. were able to functionalize the surface of cationic liposomes with a monoclonal antibody against lymphocyte function-associated antigen-1 (LFA-1) integrin for the targeted delivery of liposomes to human lymphocytes and monocytes. Liposomes were loaded with an siRNA targeting the chemokine receptor CCR5 and injected intravenously in mice; a reduction in viral load and in the loss of $CD4^+$ T cells was observed when challenged with HIV plasma [27]. The chemokine receptor CCR5 that functions as a co-receptor for macrophage-tropic strains of HIV is an attractive target for therapeutic ablation as its natural mutation is well tolerated and provides protection from HIV.

More recently, lipid-based nanoparticles have also been investigated for antiviral vaccine applications. In this case, nanoparticles are loaded with messenger RNA (mRNA) encoding viral antigens. LNPs loaded with a nucleoside-modified mRNA elicited protective immunity against the Zika virus after a single intramuscular injection in rhesus macaques [28] and were proven successful also against other viruses such as Ebola, influenza, and hepatitis [29]. Clinical studies are ongoing to assess the efficacy and the safety profile of mRNA-LNP vaccines; promising results have been obtained with mRNA-LNP vaccines against the influenza virus H10N8, which induced good immune response and mild to moderate side effects in most patients after intramuscular injection. However, increased reactogenicity was observed when a second vaccine dose was administered, highlighting the need to further optimize the LNP composition in order to minimize side effects [29].

Lipid-based nanoparticles can also be used to deliver small-molecule antiviral drugs such as antiretrovirals to prolong their circulation time and enhance cellular uptake. Antiretroviral therapy is the current gold standard for HIV treatment; however, this therapy poses some challenges associated with adverse effects, the development of viral resistance, poor stability and reduced shelf-life, and low patient compliance [30]. SLNs were able to ensure the sustained release of encapsulated antiretrovirals such as lopinavir [31] and darunavir [32] and to enhance their bioavailability. In other examples, multiple antiretrovirals could be encapsulated within a single lipid nanoparticle, enabling a multi-drug therapeutic approach [33, 34].

In these examples, the nanoparticles were administered via intravenous injection, which may not be ideal in terms of patient compliance. As such, there has been an increasing interest in nanocarriers suitable for oral administration for antiretroviral delivery. The main challenges associated with oral administration are the first-pass metabolism effect and potential drug degradation by enzymes and acidic pH [35]; these require careful design of nanocarriers to ensure successful drug delivery. Coating of the liposome surface with biotin moieties has been shown to improve the uptake of antiretrovirals in the intestine; alternatively, permeation enhancers such as Transcutol® and biliary salts can improve the oral delivery of antiretrovirals [35]. Intranasal administration has also been investigated for antiretroviral therapy; in this case, the nanoparticle surface can be functionalized with muco-penetrating moieties such as polyethylene glycol (PEG) and fatty acids [35].

4.3.2 Polymeric Nanoparticles

Polymeric nanoparticles can be obtained by using natural polymers such as chitosan, dextran, and alginate, or by using synthetic polymers such as polylactic acid (PLA), polylactide-co-glycolic acid (PLGA), and poly(Ɛ-caprolactone) (PCL). Chitosan is a cationic polysaccharide derived from chitin and has been extensively investigated for drug delivery applications due to its biodegradability, hydrophilicity, and biocompatibility [36]. Chitosan nanoparticles loaded with a potent antiretroviral drug (*i.e.*, saquinavir) for the treatment of HIV ensured greater control of viral proliferation compared to the free drug with two different HIV strains [37]. Chitosan nanoparticles have also been proposed for acyclovir delivery against HSV-1 and showed a potent antiviral activity with minimal cytotoxicity toward host cells [38].

Polymeric nanoparticles based on polyglycerols have shown intrinsic virucidal activity against HSV-2 and respiratory syncytial virus (RSV) without significant cell toxicity. These nanoparticles can irreversibly bind viral surface receptors, making viral DNA accessible to enzymatic degradation and permanently damaging the viral envelope [39]. Eshaghi et al. have recently shown the potential of PLA-PLGA virus-mimicking nanoparticles loaded with two antiretrovirals to inhibit the HIV-1 infection of primary human macrophages over 35 days [40]. A similar approach was demonstrated by Shibata et al., who developed PLGA nanoparticles loaded with a combination of efavirenz and lopinavir, which could efficiently inhibit HIV-1 infection [41]. PLGA nanoparticles have also been coated with the plasma membrane of natural $CD4^+$ T lymphocyte cells to promote binding to the HIV envelope glycoprotein gp120 and to inhibit the HIV-induced apoptosis of native $CD4^+$ T lymphocyte cells [42].

4.3.3 Inorganic Nanoparticles

Metal-based nanoparticles have been extensively investigated for their intrinsic antiviral properties. Among these, silver nanoparticles (AgNPs) have been shown to block viral entry in the host cell and intracellular viral replication. AgNPs with sizes ranging from 30 to 50 nm were able to prevent the interaction between the HIV-1 virus and the CD4 receptors of HeLa-CD4-LTR-β-gal cells *in vitro* [43], while 10 nm AgNPs could effectively inhibit hemagglutination due to the H1N1 influenza virus [44]. The ability of AgNPs to block intracellular replication has been demonstrated for both the HBV and the HIV-1 viruses [45, 46]. It has been hypothesized that AgNPs exert antiviral activity by binding to viral coat proteins and inhibiting the interaction of the virus with target cells or by interacting with the genomic material of the virus, which leads to the inhibition of genome replication inside the host cell [47].

Zinc-based nanomaterials can also be exploited for antiviral applications. Zirconia nanoparticles had a protective effect on mice treated with a lethal dose of the H5N1 influenza virus [48], while in other examples, PEGylated zinc oxide nanoparticles achieved inhibition rates higher than 90% for the H1N1 and the HSV-1 viruses, respectively [49, 50]. Silica nanoparticles (SiNPs) can also have neutralizing activity against HIV and RSV infections; coating the surface of SiNPs with (3-aminopropyl)triethoxysilane (APTES) enhanced the antiviral activity of these nanoparticles [51, 52].

Gold nanoparticles (AuNPs) are another class of metal nanoparticles with antiviral activity. AuNPs can block the interaction between viral glycoproteins, such as gp120, and CD4 receptors, preventing viral entry in the host cell. AuNPs coated with a silicon dioxide shell have been shown to inhibit adenovirus activity with minimal cellular toxicity [53]. In another example, the surface of AuNPs was functionalized with siRNA against the dengue virus; the nanoparticles could reduce the replication of the dengue virus serotype 2 (DENV-2) and the release of infectious virions [54]. Recently, porous AuNPs were shown to exhibit enhanced antiviral activity against the influenza A virus compared to nonporous AuNPs, potentially due to an increased surface area [55].

4.3.3.1 Magnetic nanoparticles

Magnetic nanoparticles (MNPs) have emerged as a novel strategy in the treatment of viral infections due to their versatility in terms of the variety of mechanisms of antiviral action. MNPs can be made of different elements such as cobalt, iron, platinum, nickel, or metal alloys. Among these, iron-based MNPs comprising maghemite (Fe_2O_3, γ-Fe_2O_3) or magnetite (Fe_3O_4) have been largely explored for nanomedicine applications [56]. Their magnetic properties can be leveraged for targeted drug delivery, imaging applications, and hyperthermia-induced drug delivery. In the context of viral infections, MNPs can be engineered to block viral entry into host cells, interfere with viral replication, or deliver drugs that can, in turn, suppress viral replication, translation, and transcription.

MNPs coated with aziridines, poly-(hexamethylene biguanide) (PHMBG), and low molecular weight ethyleneimines were able to interact with viral glycoproteins and exerted antiviral activity by preventing viral entry into host cells [57]. In another example, the electrostatic interactions between negatively charged virions and poly-cationic MNPs were leveraged to exert antiviral activity against HSV-1, bacteriophage MS2, enveloped viral hemorrhagic septicemia virus (VHSV), and non-enveloped infectious pancreatic necrosis virus (IPNV) [57]. Iron oxide–based MNPs coated with polymers such as PEG and polyvinylpyrrolidine (PVP) exhibited antiviral activity against the H1N1 virus as well as improved biocompatibility compared to bare nanoparticles [58]. In another example, iron oxide–based MNPs were coated with glycine and tested against the H1N1 virus, showing a size-dependent antiviral activity [59].

The surface of MNPs can also be engineered with antiviral drugs for drug delivery applications. For example, therapeutic antisense nucleotides ("DNAzymes") able to inhibit the expression of viral genes were covalently attached to the surface of iron oxide nanoparticles for the treatment of HCV. The nanoparticle fate could be monitored by magnetic resonance imaging due to the presence of the magnetic core [60]. In another example, Javant et al. used a layer-by-layer assembly approach to coat iron oxide nanoparticles with a latency-breaking agent (vorinostat) and tenofovir for the treatment of HIV. Drug delivery across the blood–brain barrier was achieved by the application of a magnetic field [61].

The application of external alternating magnetic fields on MNPs causes the transfer of magnetic energy as heat, a phenomenon known as magnetic hyperthermia. This phenomenon can be leveraged to treat viral infections, as proposed by Williams et al., who used superparamagnetic MNPs to kill HIV-infected T cells upon exposure to an alternating magnetic field [62].

4.3.4 Hybrid Organic–Inorganic Nanoparticles

Hybrid organic–inorganic nanoparticles have received increasing attention as antiviral therapeutics due to the possibility of combining various properties within a single nanoparticle. It has been shown that AgNPs coated with mercaptoethane sulfonate can block HSV-1 entry due to competitive binding with cellular heparan sulfate [63]. HSV-1 entry could also be inhibited by mercaptoethane sulfonate-coated AuNPs, which offered superior performance to the free mercaptoethane sulfonate [64]. A similar concept has been exploited to design tannic acid–coated AgNPs, which could neutralize HSV-2 in a size-dependent manner [65]. By tuning the length of the conjugated sulfonate linkers, Cagno et al. were able to further improve the virucidal activity of AuNPs; the longer linkers could destroy the viral capsids and induce irreversible viral damage. The nanoparticles exhibited broad-spectrum antiviral activity against HSV, human papillomavirus, RSV, dengue virus, and lentivirus [66].

4.4 Conclusion and Future Perspective

The recent viral outbreaks have highlighted the need for novel therapeutic strategies to treat viral infections. In this context, nanoparticles can offer high versatility and flexibility since their physicochemical properties can be tuned to achieve site-specific and effective antiviral activity. Furthermore, nanoparticles can ameliorate the pharmacokinetic profile of several antiviral drugs when their circulation time and bioavailability are increased. Metal-based nanoparticles are typically characterized by excellent virucidal properties; however, further studies are needed to better understand their safety profile and *in vivo* biodistribution. Recent advances in the development of lipid-based nanoparticles encapsulating nucleic acids have enabled the success of lipid nanoparticle mRNA vaccines and will pave the way for the clinical approval of other lipid nanoparticle-based antiviral therapeutics.

References

1. G. A. Roth et al., Global, regional, and national age-sex-specific mortality for 282 causes of death in 195 countries and territories, 1980–2017: A systematic analysis for the global burden of disease study 2017, *Lancet* **392**, 1736 (2018).
2. J. B. Ulmer, U. Valley, and R. Rappuoli, Vaccine manufacturing: Challenges and solutions, *Nat. Biotechnol.* **24**, 1377 (2006).
3. R. G. Pebody, M. Zambon, and M. Ramsay, Flu vaccines: An annual challenge, *Br. Med. J.* **361**, k2705 (2018).
4. B. Allegranzi, C. Kilpatrick, J. Storr, E. Kelley, B. J. Park, and L. Donaldson, Global infection prevention and control priorities 2018–22: A call for action, *Lancet Glob. Health* **5**, e1178 (2017).
5. J. Mercer, M. Schelhaas, and A. Helenius, Virus entry by endocytosis, *Annu. Rev. Biochem.* **79**, 803 (2010).
6. J. Johnson, Structures of virus and virus-like particles, *Curr. Opin. Struct. Biol.* **10**, 229 (2000).
7. F. A. Rey and S.-M. Lok, Common features of enveloped viruses and implications for immunogen design for next-generation vaccines, *Cell* **172**, 1319 (2018).
8. V. Más and J. A. Melero, Entry of enveloped viruses into host cells: Membrane fusion, *Subcell Biochem.* **68**, 467 (2013).
9. J. A. den Boon, A. Diaz, and P. Ahlquist, Cytoplasmic viral replication complexes, *Cell Host Microbe* **8**, 77 (2010).
10. S. G. Deeks, J. Overbaugh, A. Phillips, and S. Buchbinder, HIV infection, *Nat. Rev. Dis. Primers* **1**, 15035 (2015).

11. N.-J. Cho, H. Dvory-Sobol, A. Xiong, S.-J. Cho, C. W. Frank, and J. S. Glenn, Mechanism of an amphipathic α-helical peptide's antiviral activity involves size-dependent virus particle lysis, *ACS Chem. Biol.* **4**, 1061 (2009).
12. M. D. Bobardt, G. Cheng, L. de Witte, S. Selvarajah, U. Chatterji, B. E. Sanders-Beer, T. B. H. Geijtenbeek, F. V. Chisari, and P. A. Gallay, Hepatitis C virus NS5A anchor peptide disrupts human immunodeficiency virus, *Proc. Natl. Acad. Sci. U.S.A.* **105**, 5525 (2008).
13. K. Tharakaraman et al., A Broadly neutralizing human monoclonal antibody is effective against H7N9, *Proc. Natl. Acad. Sci. U.S.A.* **112**, 10890 (2015).
14. D. M. Magnani et al., Neutralizing human monoclonal antibodies prevent zika virus infection in macaques, *Sci. Transl. Med.* **9** (2017).
15. L. M. Walker and D. R. Burton, Passive immunotherapy of viral infections: "Super-antibodies" enter the fray, *Nat. Rev. Immunol.* **18**, 297 (2018).
16. A. F. Labrijn, M. L. Janmaat, J. M. Reichert, and P. W. H. I. Parren, Bispecific antibodies: A mechanistic review of the pipeline, *Nat. Rev. Drug Discov.* **18**, 585 (2019).
17. F. L. Tan and J. Q. Yin, RNAi, a new therapeutic strategy against viral infection, *Cell Res.* **14**, 460 (2004).
18. E. de Clercq and G. Li, Approved antiviral drugs over the past 50 years, *Clin. Microbiol. Rev.* **29**, 695 (2016).
19. T. K. Warren et al., Therapeutic efficacy of the small molecule GS-5734 against Ebola virus in rhesus monkeys, *Nature* **531**, 381 (2016).
20. R. M. Gulick et al., Treatment with indinavir, zidovudine, and lamivudine in adults with human immunodeficiency virus infection and prior antiretroviral therapy, *N. Engl. J. Med.* **337**, 734 (1997).
21. K. J. Szretter, S. Gangappa, X. Lu, C. Smith, W.-J. Shieh, S. R. Zaki, S. Sambhara, T. M. Tumpey, and J. M. Katz, Role of host cytokine responses in the pathogenesis of avian H5N1 influenza viruses in mice, *J. Virol.* **81**, 2736 (2007).
22. Y. Duan, A. Dhar, C. Patel, M. Khimani, S. Neogi, P. Sharma, N. Siva Kumar, and R. L. Vekariya, A brief review on solid lipid nanoparticles: Part and parcel of contemporary drug delivery systems, *RSC Adv.* **10**, 26777 (2020).
23. I. Chauhan, M. Yasir, M. Verma, and A. P. Singh, Nanostructured lipid carriers: A groundbreaking approach for transdermal drug delivery, *Adv. Pharm. Bull.* **10**, 150 (2020).

24. R. Tenchov, R. Bird, A. E. Curtze, and Q. Zhou, Lipid nanoparticles—From liposomes to MRNA vaccine delivery, a landscape of research diversity and advancement, *ACS Nano* **15**, 16982 (2021).

25. D. V. Morrissey et al., Potent and persistent *in vivo* anti-HBV activity of chemically modified SiRNAs, *Nat. Biotechnol.* **23**, 1002 (2005).

26. P. K. Chandra, A. K. Kundu, S. Hazari, S. Chandra, L. Bao, T. Ooms, G. F. Morris, T. Wu, T. K. Mandal, and S. Dash, Inhibition of hepatitis C virus replication by intracellular delivery of multiple SiRNAs by nanosomes, *Mol. Ther.* **20**, 1724 (2012).

27. S.-S. Kim et al., RNAi-mediated CCR5 silencing by LFA-1-targeted nanoparticles prevents HIV infection in BLT mice, *Mol. Ther.* **18**, 370 (2010).

28. N. Pardi et al., Zika virus protection by a single low-dose nucleoside-modified MRNA vaccination, *Nature* **543**, 248 (2017).

29. M.-G. Alameh, D. Weissman, and N. Pardi, Messenger RNA-based vaccines against infectious diseases, *Curr. Top. Microbiol. Immunol.* **440**, 111 (2022).

30. M. J. Faria, C. M. Lopes, J. das Neves, and M. Lúcio, Lipid nanocarriers for anti-HIV therapeutics: A focus on physicochemical properties and biotechnological advances, *Pharmaceutics* **13**, 1294 (2021).

31. M. R. Aji Alex, A. J. Chacko, S. Jose, and E. B. Souto, Lopinavir loaded solid lipid nanoparticles (SLN) for intestinal lymphatic targeting, *Eur. J. Pharm. Sci.* **42**, 11 (2011).

32. M. Bhalekar, P. Upadhaya, and A. Madgulkar, Formulation and characterization of solid lipid nanoparticles for an anti-retroviral drug darunavir, *Appl. Nanosci.* **7**, 47 (2017).

33. J. P. Freeling, J. Koehn, C. Shu, J. Sun, and R. J. Y. Ho, Long-acting three-drug combination anti-HIV nanoparticles enhance drug exposure in primate plasma and cells within lymph nodes and blood, *AIDS* **28**, 2625 (2014).

34. J. P. Freeling, J. Koehn, C. Shu, J. Sun, and R. J. Y. Ho, Anti-HIV drug-combination nanoparticles enhance plasma drug exposure duration as well as triple-drug combination levels in cells within lymph nodes and blood in primates, *AIDS Res. Hum. Retroviruses* **31**, 107 (2015).

35. M. J. Faria, C. M. Lopes, J. das Neves, and M. Lúcio, Lipid nanocarriers for anti-HIV therapeutics: A focus on physicochemical properties and biotechnological advances, *Pharmaceutics* **13**, 1294 (2021).

36. H. Boroumand et al., Chitosan-based nanoparticles against viral infections, *Front. Cell Infect. Microbiol.* **11** (2021).

37. L. N. Ramana, S. Sharma, S. Sethuraman, U. Ranga, and U. M. Krishnan, Evaluation of chitosan nanoformulations as potent anti-HIV therapeutic systems, *Biochim. Biophys. Acta, Gen. Subj.* **1840**, 476 (2014).

38. M. Donalisio, F. Leone, A. Civra, R. Spagnolo, O. Ozer, D. Lembo, and R. Cavalli, Acyclovir-loaded chitosan nanospheres from nano-emulsion templating for the topical treatment of herpesviruses infections, *Pharmaceutics* **10**, 46 (2018).

39. E. Mohammadifar, M. Gasbarri, V. Cagno, K. Achazi, C. Tapparel, R. Haag, and F. Stellacci, Polyanionic amphiphilic dendritic polyglycerols as broad-spectrum viral inhibitors with a virucidal mechanism, *Biomacromolecules* **23**, 983 (2022).

40. B. Eshaghi, J. Fofana, S. B. Nodder, S. Gummuluru, and B. M. Reinhard, Virus-mimicking polymer nanoparticles targeting CD169 + macrophages as long-acting nanocarriers for combination antiretrovirals, *ACS Appl. Mater. Interfaces* **14**, 2488 (2022).

41. A. Shibata, E. McMullen, A. Pham, M. Belshan, B. Sanford, Y. Zhou, M. Goede, A. A. Date, and C. J. Destache, Polymeric nanoparticles containing combination antiretroviral drugs for HIV type 1 treatment, *AIDS Res. Hum. Retroviruses* **29**, 746 (2013).

42. X. Wei, G. Zhang, D. Ran, N. Krishnan, R. H. Fang, W. Gao, S. A. Spector, and L. Zhang, T-cell-mimicking nanoparticles can neutralize HIV infectivity, *Adv. Mat.* **30**, 1802233 (2018).

43. H. H. Lara, N. V Ayala-Nuñez, L. Ixtepan-Turrent, and C. Rodriguez-Padilla, Mode of antiviral action of silver nanoparticles against HIV-1, *J. Nanobiotechnol.* **8**, 1 (2010).

44. D. Xiang, Q. Chen, L. Pang, and C. Zheng, Inhibitory effects of silver nanoparticles on H1N1 influenza A virus *in vitro*, *J. Virol. Methods* **178**, 137 (2011).

45. L. Lu, R. W.-Y. Sun, R. Chen, C.-K. Hui, C.-M. Ho, J. M. Luk, G. K. Lau, and C.-M. Che, Silver nanoparticles inhibit hepatitis B virus replication, *Antivir. Ther.* **13**, 253 (2008).

46. R. W.-Y. Sun, R. Chen, N. P.-Y. Chung, C.-M. Ho, C.-L. S. Lin, and C.-M. Che, Silver nanoparticles fabricated in Hepes buffer exhibit cytoprotective activities toward HIV-1 infected cells, *Chem. Commun. (Camb.)* 5059 (2005).

47. Z. A. Ratan, F. R. Mashrur, A. P. Chhoan, S. Md. Shahriar, M. F. Haidere, N. J. Runa, S. Kim, D.-H. Kweon, H. Hosseinzadeh, and J. Y. Cho, Silver nanoparticles as potential antiviral agents, *Pharmaceutics* **13**, 2034 (2021).

48. C. Huo, J. Xiao, K. Xiao, S. Zou, M. Wang, P. Qi, T. Liu, and Y. Hu, Pretreatment with zirconia nanoparticles reduces inflammation induced by the pathogenic H5N1 influenza virus, *Int. J. Nanomed.* **15**, 661 (2020).

49. A. Tavakoli, A. Ataei-Pirkooh, G. M. Sadeghi, F. Bokharaei-Salim, P. Sahrapour, S. J. Kiani, M. Moghoofei, M. Farahmand, D. Javanmard, and S. H. Monavari, Polyethylene glycol-coated zinc oxide nanoparticle: An efficient nanoweapon to fight against herpes simplex virus type 1, *Nanomedicine* **13**, 2675 (2018).

50. H. Ghaffari et al., Inhibition of H1N1 influenza virus infection by zinc oxide nanoparticles: Another emerging application of nanomedicine, *J. Biomed. Sci.* **26**, 70 (2019).

51. J. M. de Souza e Silva, T. D. M. Hanchuk, M. I. Santos, J. Kobarg, M. C. Bajgelman, and M. B. Cardoso, Viral inhibition mechanism mediated by surface-modified silica nanoparticles, *ACS Appl. Mater. Interfaces* **8**, 16564 (2016).

52. L. A. Osminkina et al., Porous silicon nanoparticles as scavengers of hazardous viruses, *J. Nanopart. Res.* **16**, 2430 (2014).

53. V. Lysenko et al., Nanoparticles as antiviral agents against adenoviruses, *Adv. Nat. Sci. Nanosci. Nanotechnol.* **9**, 025021 (2018).

54. A. M. Paul, Y. Shi, D. Acharya, J. R. Douglas, A. Cooley, J. F. Anderson, F. Huang, and F. Bai, Delivery of antiviral small interfering RNA with gold nanoparticles inhibits dengue virus infection *in vitro*, *J. Gen. Virol.* **95**, 1712 (2014).

55. J. Kim et al., Porous gold nanoparticles for attenuating infectivity of influenza A virus, *J. Nanobiotechnol.* **18**, 54 (2020).

56. A. Akbarzadeh, M. Samiei, and S. Davaran, Magnetic nanoparticles: Preparation, physical properties, and applications in biomedicine, *Nanoscale Res. Lett.* **7**, 144 (2012).

57. L. Bromberg, D. J. Bromberg, T. A. Hatton, I. Bandín, A. Concheiro, and C. Alvarez-Lorenzo, Antiviral properties of polymeric aziridine- and biguanide-modified core–shell magnetic nanoparticles, *Langmuir* **28**, 4548 (2012).

58. S. R. Kumar, M. Paulpandi, M. Manivel Raja, D. Mangalaraj, C. Viswanathan, S. Kannan, and N. Ponpandian, An *in vitro* analysis of H1N1 viral inhibition using polymer coated superparamagnetic Fe_3O_4 nanoparticles, *RSC Adv.* **4**, 13409 (2014).

59. R. Kumar et al., Iron oxide nanoparticles based antiviral activity of H1N1 influenza A virus, *J. Infect. Chemother.* **25**, 325 (2019).

60. S.-R. Ryoo, H. Jang, K.-S. Kim, B. Lee, K. B. Kim, Y.-K. Kim, W.-S. Yeo, Y. Lee, D.-E. Kim, and D.-H. Min, Functional delivery of DNAzyme with iron oxide nanoparticles for hepatitis C virus gene knockdown, *Biomaterials* **33**, 2754 (2012).

61. R. Jayant, V. Atluri, M. Agudelo, V. Sagar, A. Kaushik, and M. Nair, Sustained-release nanoART formulation for the treatment of neuroAIDS, *Int. J. Nanomed.* 1077 (2015).

62. J. Williams, P. Southern, A. Lissina, H. Christian, A. Sewell, R. Phillips, Q. Pankhurst, and J. Frater, Application of magnetic field hyperthermia and superparamagnetic iron oxide nanoparticles to HIV-1-specific T-cell cytotoxicity, *Int. J. Nanomed.* 2543 (2013).

63. D. Baram-Pinto, S. Shukla, N. Perkas, A. Gedanken, and R. Sarid, Inhibition of herpes simplex virus type 1 infection by silver nanoparticles capped with mercaptoethane sulfonate, *Bioconjug. Chem.* **20**, 1497 (2009).

64. D. Baram-Pinto, S. Shukla, A. Gedanken, and R. Sarid, Inhibition of HSV-1 attachment, entry, and cell-to-cell spread by functionalized multivalent gold nanoparticles, *Small* **6**, 1044 (2010).

65. P. Orlowski et al., Tannic acid modified silver nanoparticles show antiviral activity in herpes simplex virus type 2 infection, *PLoS One* **9**, e104113 (2014).

66. V. Cagno et al., Broad-spectrum non-toxic antiviral nanoparticles with a virucidal inhibition mechanism, *Nat. Mater.* **17**, 195 (2018).

Chapter 5

Formulation Approaches for COVID-19 Vaccines

Mohammed M. Al Qaraghuli,[a,b] Mohammad A. Obeid,[c] Ahmed A. Abdulhussein Al-Ali,[d,e] Judith Kuntsche,[d] Carsten Uhd Nielsen,[d] and Paul A. Mulheran[f]

[a]*Strathclyde Institute of Pharmacy and Biomedical Sciences, University of Strathclyde, 161 Cathedral Street, Glasgow, G4 0RE, UK*
[b]*EPSRC Future Manufacturing Research Hub for Continuous Manufacturing and Advanced Crystallisation (CMAC), University of Strathclyde, 99 George Street, Glasgow, G1 1RD, UK*
[c]*Department of Pharmaceutics and Pharmaceutical Technology, Faculty of Pharmacy, Yarmouk University, Irbid, Jordan*
[d]*Department of Physics, Chemistry and Pharmacy, University of Southern Denmark, Campusvej 55, 5230 Odense M, Denmark*
[e]*Department of Pharmacy, Al-Kunooze University College, Al-Zaytoon District, Basra, Iraq*
[f]*Department of Chemical and Process Engineering, University of Strathclyde, 75 Montrose Street, Glasgow, G1 1XJ, UK*
mohammed.al-qaraghuli@strath.ac.uk

COVID-19 is a respiratory disease caused by a new type of coronavirus known as severe acute respiratory syndrome coronavirus 2 (SARS-CoV-2). In 2019, this disease was discovered in Wuhan, China,

Nanotherapeutics for Infectious Diseases
Edited by Íris Luz Batalha

ISBN 978-981-5129-36-6 (Hardcover), 978-1-003-56678-6 (eBook)
www.jennystanford.com

and since then it has pandemically infected and, unfortunately, in many cases taken the lives of millions of people globally. Since the emergence of COVID-19, the entire biopharmaceutical research community has focused on the swift development of vaccines that can hinder the spread of the virus and reduce the shockingly high number of deaths. These efforts have led to the approval of a new class of vaccines known as messenger RNA (mRNA) vaccines, which were independently developed by Pfizer-BioNTech and Moderna. The mRNA vaccines are formulated as nanoparticles (NPs) where the fragile mRNA is encapsulated to stabilize the mRNA and enhance efficacy. Through accelerated approval of regulatory authorities, there are currently 10 approved vaccines, 153 candidate vaccines in clinical development, and 196 vaccines in preclinical development. The clinical candidates include seven NP-based vaccines, which are mainly based on protein subunits or RNA-based platforms. This chapter will provide insights into SARS-CoV-2 structure, potential vaccine approaches, the currently approved COVID-19 vaccines, and the candidates in preclinical and clinical development. The advancement in nanotechnology has enabled the utilization of various forms of NPs in vaccine development, such as liposomes, lipid NPs, protein NPs, and nanobodies. Therefore, this chapter will also detail the different types of potential NPs available for vaccine development and the formulation of these vaccines in terms of active ingredients, excipients, and storage.

5.1 Overview of COVID-19 and SARS-CoV-2 Virus

The World Health Organization (WHO) was alerted in December 2019 regarding an outbreak of pneumonia in Wuhan, China, and it formally named the disease, on February 11, 2020, as Coronavirus Disease-2019 (COVID-19) [1]. COVID-19 is caused by SARS-CoV-2, which belongs to the *Coronaviridae* family.

This disease can mainly spread from person to person by droplets generated by an infected person [2]. COVID-19 is generally associated with symptoms such as dry cough, fever, loss of smell and taste, fatigue, and dyspnea in severe cases [3]. Many children and young adults are often asymptomatic, while elderly people or patients with comorbidities are at a higher risk of severe disease and death. These underlying medical conditions include diabetes; hypertension; lung, liver, and kidney disease; cancer patients on chemotherapy; transplant recipients; and patients receiving steroids [4]. The incubation period of COVID-19 is ~5 days. Severe disease usually develops ~8 days after symptom onset, and life-threatening disease and death occur after ~16 days [3]. As of April 29, 2022, there have been 509,531,232 cumulative cases and 6,230,357 cumulative deaths globally [5].

SARS-CoV-2 is a single-stranded RNA-enveloped virus [6], which encodes 9860 amino acids and is 29,903 base pairs in length [7]. SARS-CoV-2 shares 79% genome sequence identity with the previous versions of SARS-CoV that caused severe acute respiratory syndrome (SARS) and 50% with MERS-CoV that produced the Middle East respiratory syndrome (MERS) [6].

SARS-CoV-2 is composed of four structural proteins (Fig. 5.1): spike protein (S), envelope protein (E), nucleocapsid protein (N), and membrane protein (M), and 16 non-structural proteins (nsp1–16) [8]. The S protein extends from the viral membrane and is arranged as trimers on the virion surface to reveal the appearance of a crown, or corona in Latin [9].

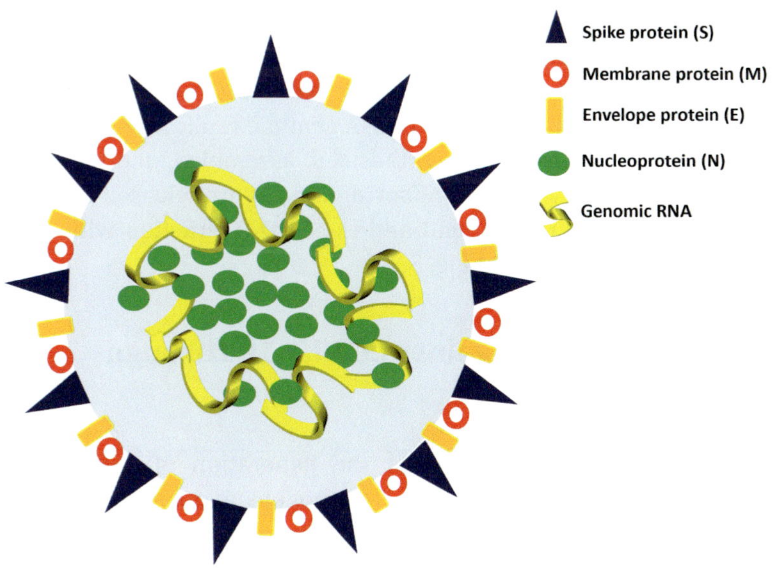

Figure 5.1 Schematic view of the SARS-CoV-2 virus. This schematic figure is modified from Farouq et al. 2021 [10].

Structural studies of coronaviruses have shown that a large number of glycosylated S proteins cover the surface of SARS-CoV-2 and bind to the host cell receptor angiotensin-converting enzyme 2 (ACE2), mediating viral cell entry [11]. The spike receptor-binding domain (RBD) of coronaviruses can fold independently of the rest of the S protein and comprises all of the structural information for host receptor binding [12]. The S protein shape was determined by cryo-electron microscopy, revealing different conformations of the S RBD domain in opened and closed states and its corresponding functions [13, 14].

The size of S protein is 180–200 kDa, and it consists of an extracellular N-terminus, a transmembrane (TM) domain attached to the viral membrane, and a short intracellular C-terminal segment [15]. The SARS-CoV-2 S gene encodes 22 N-linked glycans, which likely play a role in protein folding and immune evasion [16]. The S protein is composed of 1273 amino acids and consists of a signal peptide comprising the N-terminus (residues 1–13), the S1 subunit that includes 672 amino acids (residues 14–685), and the S2 subunit composed of 588 amino acids (residues 686–1273) [13]. The S1 RBD is represented by residues 319–541.

Once the S protein binds to the receptor, the transmembrane serine protease 2 (TMPRSS2) enables virus entry into the cell by activating the S protein [15]. This enables the virus to enter the cell, releasing the viral RNA and enabling polyprotein translation from the RNA genome, as well as viral RNA genome replication [17]. Then the structural proteins are synthesized, assembled, and packed in the host cell, after which the offspring viruses are released [18]. The S protein trimer arrangement and the RBD interaction with ACE2 receptor are illustrated in Fig. 5.2.

5.2 Protection against COVID-19 through Vaccination

Vaccines are used to stimulate the generation of a protective immune response against various pathogens and can be generally prepared using either the causative agent of a disease (attenuated or inactivated) or a synthetic substitute. The main aim of many researchers and biopharmaceutical companies since the start of the COVID-19 pandemic has been to design a vaccine that can generate antibodies against the S protein or the RBD [19–21]. These exceptional efforts have produced 153 candidate vaccines in clinical development, distributed across 11 platforms, and 196 vaccines in preclinical development [22]. The developed vaccines utilize either traditional approaches such as inactivated viruses, recombinant proteins, vectored vaccines, or new approaches such as RNA and DNA vaccines (Table 5.1).

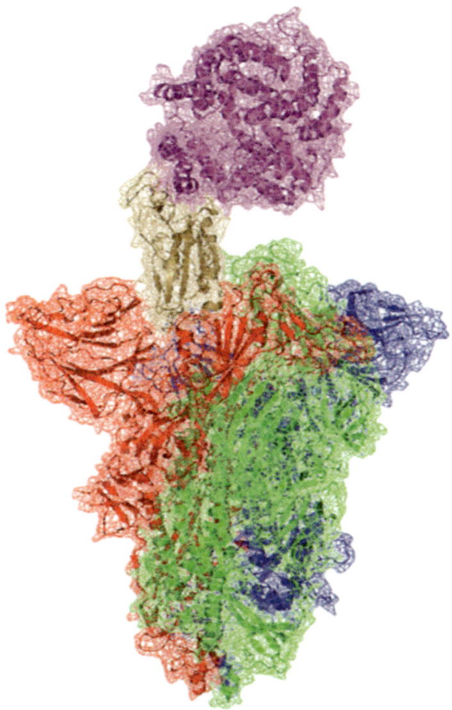

Figure 5.2 SARS-CoV-2 S protein complexed with ACE2 receptor. The 7DF4 crystal structure was retrieved from the Protein Data Bank (PDB). The complex is shown as loops, and the surface is displayed as a mesh. The S protein monomers are colored red, blue, and green. The RBD (brown) of the S protein (blue monomer) binds to the human ACE2 receptor (magenta) to promote attachment and fusion. The structure was visualized by PyMOL (PyMOL Molecular Graphics System, Version 1.7.4 Schrödinger, LLC.).

Table 5.1 COVID-19 candidate vaccines in clinical development

Platform	**Candidate Vaccines (Number and Percentage)**	
Protein subunit	51	33.3%
Viral vector (nonreplicating)	21	13.7%
DNA	16	10.5%
Inactivated virus	21	13.7%
RNA	28	18.3%
Viral vector (replicating)	4	2.6%
Virus-like particle	6	3.9%
Viral vector (replicating) + antigen-presenting cell	2	1.3%
Live-attenuated virus	2	1.3%
Viral vector (nonreplicating) + antigen-presenting cell	1	0.7%
Bacterial antigen-spore expression vector	1	0.7%
Total	**153**	**100%**

Note: Data were retrieved from the WHO COVID-19 vaccine tracker and landscape, April 2022 [22].

With exceptional development and approval processes, 10 vaccines have already been licensed or approved for emergency use, as summarized in Table 5.2. The approved vaccines have revealed the predominance of three vaccine classes: inactivated virus, viral vector, and RNA-based vaccine. In addition, the wide implementation of the vaccination process has demonstrated both safety and efficacy of these vaccines.

Virus-based vaccines include either weakened or inactivated pathogens [23]. Live-attenuated virus vaccines essentially contain several antigenic components derived from the attenuated live viruses, which are immunogenic but not infectious [24]. The inactivated virus vaccines contain inactivated SARS-CoV-2, usually by chemicals, so that the antigenicity of the viral surface components are preserved [25]. Virus-based vaccines are generally well-established due to previously approved methodologies, regulation policies, and established manufacturing facilities [26]. CoronaVac, WIBP-CorV, and BBIBP-CorV vaccines are examples in this class.

Table 5.2 List of approved COVID-19 vaccines

Vaccine Name (Tradename)	Platform	Number of Doses	Route of Administration	Developer
CoronaVac (Sinovac)	Inactivated virus	2	Intramuscular	Sinovac Research and Development Co., Ltd.
WIBP-CorV (Sinopharm WIBP COVID-19 vaccine)	Inactivated virus	2	Intramuscular	Sinopharm, CNBG, and WIPB
BBIBP-CorV Vaccine (Sinopharm BIBP COVID-19)	Inactivated virus	2	Intramuscular	Sinopharm, CNBG, WIPB, and BIBP
ChAdOx1-S-(AZD1222) (Vaxzevria/ Covishield)	Viral vector (nonreplicating)	2	Intramuscular	AstraZeneca and University of Oxford
Ad5-nCoV (Convidecia)	Viral vector (nonreplicating)	1	Intramuscular	CanSino Biological Inc. and BIB
Ad26.CoV2.S (Janssen COVID-19 vaccine)	Viral vector (nonreplicating)	1–2	Intramuscular	Janssen Pharm (Johnson & Johnson)
mRNA-1273 (Spikevax)	RNA-based vaccine	2	Intramuscular	Moderna and NIAID
BNT162B2 (Comirnaty)	RNA-based vaccine	2	Intramuscular	Pfizer, BioNTech, and Fosun Pharma
MVC-COV1901 (MVC COVID-19 vaccine)	Protein subunit	2	Intramuscular	Medigen Vaccine Biologics, Dynavax, and NIAID
mRNA-1273.351. Prefusion stabilized S protein of the SARS-COV-2 B.1.351 variant	RNA-based vaccine	3	Intramuscular	Moderna and NIAID

Note: Data were retrieved from the WHO COVID-19 vaccine tracker and landscape, April 2022 [22].

Viral vector–based vaccines depend on the ability to insert viral antigen-encoding DNA sequences into host cells, which generate an expression of viral antigens by the host cells [27]. The vector is an innocuous, viable, and nonreplicating carrier virus that emulates infection caused by SARS-CoV-2 [28]. Retroviruses, adenovirus, and vaccinia viruses are conventionally used as carrier vehicles in this type of vaccine [29]. Adenoviruses are widely implemented because of their ability to infect different cell types, the absence of genome integration, effective transgene expression, and high *in vitro* growth [30]. This type of vaccine requires additional harmonization of efficacy and safety [31]. Three approved vaccines were developed within this class: ChAdOx1-S-(AZD1222), Ad5-nCoV, and Ad26. CoV2.S.

Protein subunit vaccines rely on the utilization of viral protein antigens to trigger immunity [32]. Both full-length proteins and protein domains can be used; however, the former can induce the development of antibodies against various epitopes [33]. In addition, protein subunit vaccines require effective adjuvants to generate a robust immune response [24]. This class of vaccine contains one approved vaccine (MVC-COV1901) and 51 COVID-19 vaccine candidates that are currently under clinical investigation (Table 5.1).

Nucleic acid (DNA and RNA) vaccines are based on inserting specific viral genes into human cells and incorporating these within NPs to protect the delicate genetic materials and to enhance the transfection efficacy [34]. Nucleic acid vaccines based on mRNA have shown huge potential during the COVID-19 pandemic, and they have been developed promptly, due to rapid and low-cost manufacturing routes and high safety profile [35]. In order to ensure the stability and efficacy of this type of vaccines, special formulation strategies, such as the use of NPs and storage at a very low temperature, are generally required [36], as will be discussed in the following sections. The three approved vaccines from this class are mRNA-1273, BNT162b2, and mRNA-1273.351.

5.3 Types of NPs Used in COVID-19 Vaccine Development

NP vaccines offer an exceptional prospect to advance vaccine development and offer solutions to the current COVID-19 pandemic. NPs are materials that range in size from 1 to 100 nm, which can imitate the structural features of viruses. NP vaccines can be generally classified into two groups [37]: (i) those where NPs encapsulate vaccine antigens or nucleic acid cargos and (ii) those with NPs that display vaccine antigens on their surface.

In response to the rapid spread of the COVID-19 pandemic, several types of NPs have been investigated for the development of COVID-19 vaccines. These include lipid nanoparticles (LNPs), liposomes, protein nanoparticles (PNPs), nanobodies (Nbs), and many others. These NPs were used for the delivery of protein subunits, nucleic acids, or other vaccine components. There are currently seven NP-based vaccine candidates in clinical trials, and they are mainly based on protein subunits or RNA-based vaccines (Table 5.3).

Table 5.3 List of nanoparticle-based vaccines against COVID-19 in preclinical and clinical development

Candidate Vaccine	Description	Platform	Developer	Phase	Clinical Trial ID
NVX-COV2373 vaccine	SARS-CoV-2 rS/Matrix M1-Adjuvant (full-length recombinant SARS-CoV-2 glycoprotein NPs vaccine adjuvanted with Matrix M)	Protein subunit	Novavax	Phase 3	NCT04611802 EUCTR2020-004123-16-GB NCT04583995 NCT05249816
CoV-2 SAM (LNP) vaccine	A self-amplifying mRNA (SAM) LNP platform + spike antigen	RNA-based vaccine	GSK	Phase 1	NCT04758962
SARS-CoV-2-Spike-Ferritin-NPs (SpFN) vaccine	SpFN uses spike proteins with a liposomal formulation QS21 (ALFQ) adjuvant	Protein subunit	WRAIR	Phase 1	NCT04784767

(*Continued*)

Table 5.3 (*Continued*)

Candidate Vaccine	Description	Platform	Developer	Phase	Clinical Trial ID
HDT-301 vaccine	Self-replicating mRNA vaccine formulated as an LNP	RNA-based vaccine	SENAI CIMATEC	Phase 1	NCT04844268
LNP-nCOV saRNA-02 vaccine	Self-amplifying RNA (saRNA) encapsulated in LNP	RNA-based vaccine	MRC/UVRI and LSHTM	Phase 1	NCT04934111
mRNA-1273.211 vaccine	A multivalent booster candidate combining mRNA-1273 plus mRNA-1273.351	RNA-based vaccine	ModernaTX, Inc.	Phase 2/3	NCT04927065
PepGNP-SARS-CoV-2 vaccine	A CD8 T-cell priming adaptive vaccine composed of a coronavirus-specific peptide mounted on a gold NP	Protein subunit	Emergex Vaccines Holding Limited	Phase 1	NCT05113862
RBD protein vaccine	Delivered in mannose-conjugated chitosan NPs	Protein subunit	Ohio State University/ KazNAU	Preclinical	Preclinical
Recombinant protein NPs	Recombinant protein subunit based on S protein and other epitopes	Protein subunit	SPbSRIVS	Preclinical	Preclinical
NPS vaccine (AKS-452)	Based on Akston's proprietary Fc fusion protein platform	Protein subunit	LakePhar/ Akston Biosciences	Preclinical	Preclinical

Note: Data were retrieved from the WHO COVID-19 vaccine tracker and landscape [22].

In the development of NP-based vaccines for COVID-19, NPs can improve the outcomes of these vaccines based on their ability to enhance the uptake by the target cells, binding specificity to the viral genome, activation of signaling pathways at the mitochondrial level, along with their ability to modulate the viral transcription [38]. In this section, the characteristics of the most investigated NPs for COVID-19 vaccine development will be discussed with a focus on the vaccines that have been approved by the regulatory authorities.

5.3.1 Protein Nanoparticles

PNPs can be used for the development of COVID-19 vaccines. They are a virus-like particle-delivery system that forms through spontaneous self-assembly of proteins into nanometer-sized particles. These types of NPs are considered safe and effective in the development of vaccines [39].

In the context of COVID-19 vaccine development, PNPs can display multiple copies of COVID-19 antigens on their surface, such as the RBD or the S protein where these antigens can be covalently or noncovalently attached to the PNP surface. In another type of PNP, the NPs along with the target antigens can be encoded in a single peptide chain, which then self-assembles to form antigen-presenting NPs [40]. With this unique structure, the PNPs resemble the structure of COVID-19 viruses, and this is what the mammalian immune systems recognize to start producing the required antibodies and other immune responses related to the antigen presented on the surface of the PNPs. This results in the development of proper protection against the COVID-19 infection [41]. In contrast to the live-attenuated virus vaccines, which may recover virulence or drive infection, PNP-based vaccines do not have any viral genetic materials, which enables them to induce an immune response while simultaneously being safe at the same time. Moreover, the antigens on the live-attenuated virus vaccines might be destroyed during the course of viral inactivation in contrast to the PNPs that carry the antigens in their original form [39].

Several researchers have attempted this approach in the development of COVID-19 vaccines. In the work of He et al., self-assembling PNP formulations were prepared using the SpyTag/SpyCatcher system carrying either RBD or a stabilized spike protein on their surfaces. Following administration into BALB/c mice models, the RBD-PNPs showed a more potent neutralizing antibody response compared to the administration of the RBD alone and the spike protein alone. Moreover, one of the PNP formulations bearing spike protein resulted in up to tenfold higher neutralizing antibody response compared to the free spike protein administration. This was accompanied by the proper induction of the T cells, demonstrating a promising vaccine candidate [42].

Powell et al. developed PNP vaccine models using ferritin NPs decorated with one of the two SARS-CoV-2 spikes on their surfaces, either full-length ectodomain (S-Fer) or a C-terminal 70 amino-acid deletion (SΔC-Fer). Single immunization of mice with either of the two PNP formulations resulted in the production of neutralizing antibodies, which was at least twofold higher than the antibodies presented in the plasma of patients recovered from COVID-19 infection [43]. Moreover, in the work of Geng et al., PNPs bearing 120 copies of SARS-CoV-2 RBD on their surfaces were prepared and resulted in the production of neutralizing antibodies five times more than the naked RBD vaccines following immunization of mice models. These neutralizing antibodies induced by the PNP-RBD vaccine lasted for two months after immunization [41].

5.3.2 Inorganic Nanoparticles

Several inorganic NPs have been applied recently for various biological and medical applications. These include silver, iron oxide, zinc oxide, nonmetallic oxide, and gold NPs, and many others as drug delivery systems. While several molecules can be attached to these types of NPs for various therapeutic effects such as vaccines, some of these NPs can themselves act as adjuvants by boosting the immune response in immunization [10]. Gold NPs, for example, have been studied extensively as an adjuvant in various viral infections such as influenza A [44]. Iron oxide NPs are also promising NPs that have been approved by the US Food and Drug Administration (FDA) for the treatment of anemia. These NPs also possess an antiviral activity *in vitro*, and many researchers employed them in the development of COVID-19 therapeutics. Iron oxide NPs are believed to interact with the spike protein of the RDB of SARS-CoV-2, and this interaction results in viral inactivation through the induction of various conformational changes in the viral proteins [45].

Thalhauser et al. employed silica NPs for the development of an effective vaccine against the human immunodeficiency virus (HIV) by attaching the viral trimeric envelop to silica NPs. This complex increased the uptake by the target antigen-presenting cells by four- to fivefold [46].

The use of silver NPs has also been extensively studied as adjuvants and therapeutic agents in viral infections. These NPs are believed to have antiviral activity by preventing the viral entry into the target cells by specific interaction with the virus surface [47]. This could be a promising effect in the development of COVID-19 vaccines either when administered alone or in combination with other therapeutic agents such as bronchodilators [48]. Multivalent NPs have been investigated for their use against COVID-19. These types of NPs have a high density of binding sites on their surfaces and can form multivalent pairs between the ligand and its receptor. This, in theory, could be used for COVID-19 treatment based on the similarities between these NPs and the virus, which could be useful for improving the therapeutic effect for the currently existing antiviral agents against COVID-19 infections [49].

5.3.3 Nanobodies

Nanobodies are single-binding domain antibodies that can be obtained from, for example, Camelidae [50]. Although nanobodies are not regarded as NPs, they are related to nanotechnology based on their nanometer size, which is around 4 × 2.5 nm. Their small size allows these molecules to bind with high affinity to hidden areas of antigens (cryptic epitopes) that conventional antibodies cannot reach. Moreover, nanobodies can be generated easily at a large scale using various expression systems such as bacteria, phage, or yeast, and these generated nanobodies are stable at various harsh conditions such as extreme temperature and pH. These characteristics make nanobodies a worthy candidate in the development of vaccines against various viral infections, including COVID-19 [51]. In this, nanobodies are believed to bind to specific antigens such as the RBD or S1 or S2 of the SARS-CoV-2, and this binding will block the interaction between the RBD and the ACE2 receptors to prevent the viral entry into the target cells. This would result in an effective way of preventing the spread of SARS-CoV-2 infections [39]. A comparison between the nanobody structure and the conventional monoclonal antibody (mAb) structure is illustrated in Fig. 5.3.

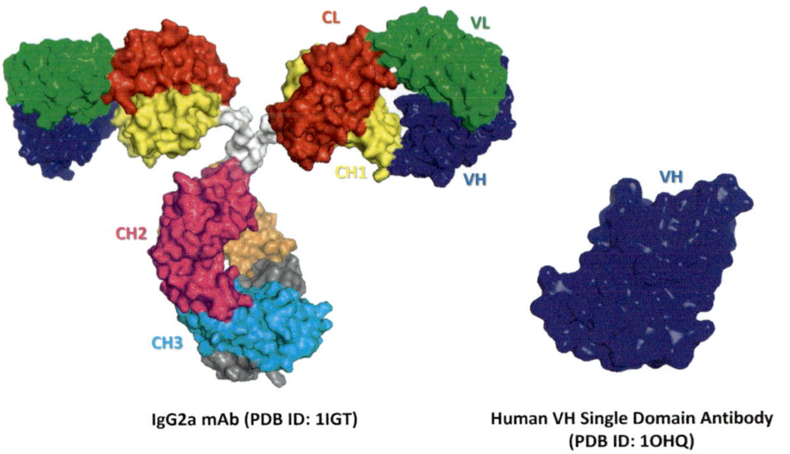

Figure 5.3 Structure of an mAb and nanobody. Illustrative view of an IgG2a mAb (PDB ID: 1IGT) and a nanobody (PDB ID: 1OHQ). The antibody domains were color coded as follow—variable light domain (VL): green, variable heavy domain (VH): blue, constant light domain (CL): red, constant heavy domain 1 (CH1): yellow, CH2: magentas/orange, CH3: cyan/gray, and Linker: light gray. The nanobody is a human VH single-domain antibody colored as blue. The structures were visualized by PyMOL (PyMOL Molecular Graphics System, Version 1.7.4, Schrödinger, LLC.).

5.3.4 Polymeric Nanoparticles

Polymeric NPs are widely used in the field of drug delivery, and as the name implies, these NPs are composed of a wide range of synthetic or natural polymers with single or branched chains [52]. Poly(lactic-co-glycolic acid) (PLGA) and polylactic acid (PLA) are among the most commonly used polymers in the construction of polymeric NPs. These NPs are versatile in terms of their conjugations, which means various compounds can be attached to them for therapeutic or diagnostic purposes. This conjugation will result in a higher drug stability and bioavailability with improved therapeutic outcomes

[53].

PLGA NPs have been extensively investigated for the development of vaccines against hepatitis B virus, tetanus toxoid, and *Mycobacterium tuberculosis* in which they carry and deliver specific antigens related to these infections [54]. Other types of polymeric NPs, such as chitosan NPs and poly-γ-glutamic acid NPs, have also been applied in the development of vaccines against the influenza virus (H1N1), bovine parainfluenza 3 virus (BPI3V), swine influenza virus (H1N2), and respiratory syncytial virus (RSV) [55]. Moreover, polymeric NPs can also be employed for the delivery of nucleic acid–based vaccines, where the loading of the negatively charged nucleic acids can be improved by the electrostatic interactions with the cationic groups of the polymers [56]. In the context of COVID-19, Li et al. developed a PLGA-based multifunctional alveolar macrophage (AM)-like NP that can bind to the SARS-CoV-2 virus at a faster rate than the binding of the virus to the cell membrane. This binding of the virus with the NPs is to be followed by the photothermal deactivation of the virus, which can protect from the COVID-19 infections [57]. Moreover, PLGA-based polymeric NPs have been extensively investigated in the development of diagnostic tools for the determination of COVID-19 infection [56].

5.3.5 Liposomes

Liposomes are lipid-based NPs that are composed of a membrane bilayer structure encapsulating an aqueous moiety. This vascular structure was first reported by Bangham et al. and employed widely for the delivery of various hydrophilic or hydrophobic molecules [58]. Within the liposome structure, hydrophilic therapeutics are encapsulated in the aqueous part, while the hydrophobic agents are typically embedded in the membrane bilayer structure [59]. The major constituents of liposomes are phospholipids, which are amphiphilic molecules with hydrophilic heads and hydrophobic tails. Currently, there are different types of natural or synthetic phospholipids such as phosphatidylcholine, phosphatidylethanolamine, phosphatidylglycerol, and phosphatidylserine, which can be used in the preparation of liposomes. By changing the type of phospholipid used, the overall characteristics of the prepared liposomes, such as particle size and surface charge, can be tailored.

Cholesterol is another major component in liposome formulations, positioned within the membrane bilayer, and is used to render the membrane more rigid and less permeable. This helps increase the stability of the prepared particles and minimizing the drug leakage upon storage. Other components can also be used in the preparation of liposomes, such as charge-inducing lipids [60]. This unique structure of liposomes and the different types of lipids available make them the most studied types of NPs, as they can be engineered in different compositions and different ratios to optimize the characteristics of the prepared particles [61, 62].

5.3.6 Lipid Nanoparticles

LNPs are novel drug delivery systems used widely in pharmaceutical applications due to their high safety and biocompatibility along with their ability to penetrate cell membranes. LNPs are being employed to encapsulate mRNA vaccines and protect them from degradation as naked mRNA is highly unstable and rapidly degraded by nucleases (Fig. 5.4). Therefore, the use of LNPs increases the stability of these nucleic acid vaccines and ensures effective delivery to the target cells [63]. In various *in vivo* models such as mice and hamsters, many of these LNP-based mRNA vaccines were able to induce sufficient immunogenicity against SARS-CoV-2 infection [64, 65].

The currently approved mRNA-based COVID-19 vaccines all utilize LNP delivery systems. The development of these successful vaccines was believed to be based on the experience behind the successful development of Onpattro™ by Alnylam Pharmaceuticals, which is an LNP-based short-interfering RNA (siRNA) drug approved by the FDA for the treatment of polyneuropathies induced by hereditary transthyretin amyloidosis [66].

Upon reaching the target cells, these LNP-based mRNA vaccines are internalized by endocytosis followed by the release of the mRNA molecules into the cytoplasm and then translated to the target protein encoded by the mRNA sequences. This translated target protein is recognized by the immune system, resulting in the activation of T and B cells. This activation results in the production of virus-neutralizing antibodies and the killing of virus-infected cells in addition to the formation of immunity by memory B cells [67].

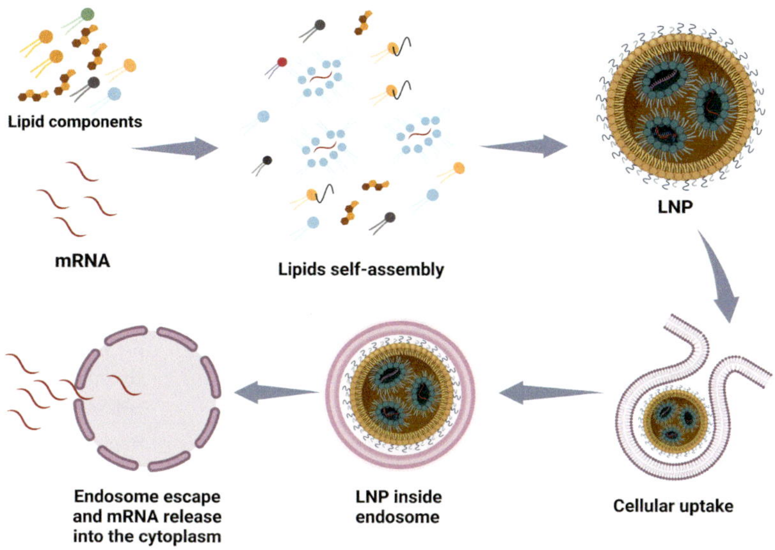

Figure 5.4 Schematic diagram of mRNA encapsulation in an LNP, leading to cellular uptake and cargo release.

LNPs are composed of different components that need to be optimized to be suitable for mRNA delivery. These components involve the use of an ionizable cationic lipid and neutral phospholipids, polyethylene glycol (PEG) with a suitable chain length, and cholesterol (for example, see Table 5.4). Although all these components are important in the formation of stable LNPs, the ionizable and PEG lipids need to be carefully chosen and optimized in terms of their lipid ratios. A key difference between LNPs and liposomes is the presence of lipids in the core, though water can also be present to some extent [68].

The role of the ionizable cationic lipids is to form lipoplexes with the negatively charged nucleic acids and to facilitate endosome escape after the LNP uptake by the endo-lysosomal pathway. Inside the endosome compartments, these pH-sensitive lipids become protonated and positively charged, interacting with the negatively charged endosome membranes and enhancing the endosome membrane rupture with the subsequent release of the LNP's load into the cytoplasm [69].

Table 5.4 Specifications and compositions of marketed mRNA LNP vaccines against COVID-19

	Pfizer Vaccine	**Moderna Vaccine**
Product name	Comirnaty	Spikevax
Generic name (INN)	Tozinameran	Elasomeran
Other names	BNT162b2	mRNA-1273
Active ingredient	Nucleoside-modified mRNA encoding the viral spike glycoprotein (S) of SARS-CoV-2	Nucleoside-modified mRNA encoding the viral spike glycoprotein (S) of SARS-CoV-2
Ionizable lipid	((4-Hydroxybutyl)azanediyl)bis(hexane-6,1-diyl)bis(2-hexyldecanoate) (ALC-0315) 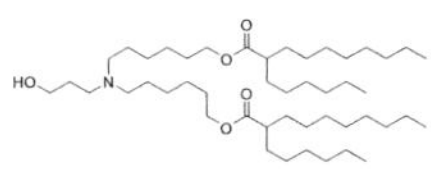	Heptadecan-9-yl 8-{(2-hydroxyethyl) [6-oxo-6-(undecyloxy)hexyl] amino} octanoate (SM-102) 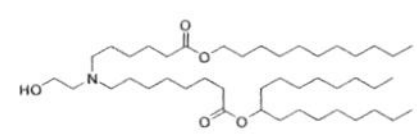
PEGylated lipid	2-[(Polyethylene glycol)-2000]-N,N-ditetradecylacetamide (ALC-0159) 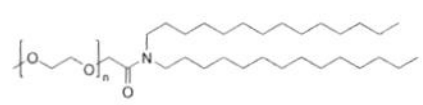 n=45	1,2-Dimyristoyl-rac-glycero-3-methoxypolyethylene glycol-2000 (PEG2000-DMG) 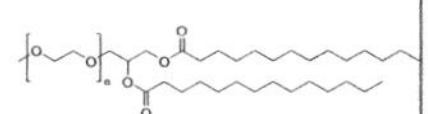 n=45
Sterol	Cholesterol 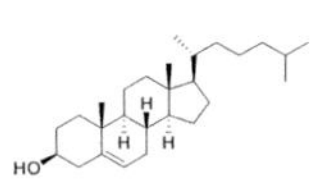	
Phospholipid	1,2-Distearoyl-sn-glycero-3-phosphocholine (DSPC) 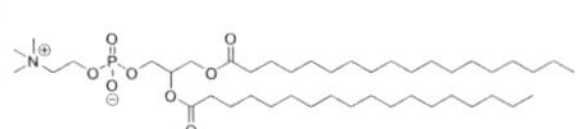	

	Pfizer Vaccine			Moderna Vaccine
Ratio of ionizable lipid: PEGylated lipid: cholesterol: phospholipid (mol%)	46.3:1.6:42.7:9.4			50:1.5:38.5:10
Vial cap color	Purple cap	Grey cap	Orange cap	Red cap
mRNA per dose (age group)	30 μg (12 years and older)	30 μg (12 years and older)	10 μg (5 to 11 years)	50 μg (6–11 years) 100 μg (12 years and older)
Dosage volume	0.3 mL	0.3 mL	0.2 mL	0.25 or 0.5 mL
Buffer type	PBS	Tris	Tris	Tris
pH	6.9–7.9			7.0–8.0
Non-lipid excipients	- Potassium chloride - Potassium dihydrogen phosphate - Sodium chloride - Disodium phosphate dihydrate - Sucrose - Water for injections - Sodium hydroxide - Hydrochloric acid	- Trometamol - Trometamol hydrochloride - Sucrose - Water for injections	- Trometamol - Trometamol hydrochloride - Sucrose - Water for injections	- Trometamol - Trometamol hydrochloride - Acetic acid - Sodium acetate trihydrate - Sucrose - Water for injections
Pharmaceutical form	Concentrate for injection, dispersion	Dispersion for injection	Concentrate for injection, dispersion	Dispersion for injection
Dilute before use (mL)	Yes with 1.8 mL NaCl 0.9% w/v	No	Yes with 1.3 mL NaCl 0.9% w/v	No
Shelf life				
Original vial −90°C to −60°C −25°C to −15°C 2°C to 8°C	 9 months 1 month	 9 months 10 weeks	 9 months 10 weeks	 9 months 30 days
Opened vial 2°C to 30°C	6 hours	12 hours	12 hours	19 hours

The development of effective LNPs for mRNA delivery requires the screening of different types of lipids and evaluating their efficacy in the formation of NPs that can carry, protect, and deliver the payload. This means that it might be necessary to test thousands of preliminary LNP formulations before determining the optimal one. The LNPs are composed of a surface lipid bilayer formed by the self-assembly of phospholipids, which surrounds an amorphous, dense core. This core mainly consists of water pores encapsulated by organized cationic lipids or lipid mixtures [70].

Among the currently approved COVID-19 vaccines, both vaccines developed by BioNTech/ Pfizer and Moderna are based on LNPs and will be discussed in the following section (see Table 5.4). Several other types of NPs have been employed for mRNA delivery for COVID-19 vaccines, such as dendrimers, cationic nanoemulsions, liposomes, or polysaccharide particles [71]. If these systems were successful in delivering their mRNA load to the target cells, then the outcomes from these vaccines are thought to be comparable to the currently approved LNP-based vaccines.

5.4 Formulation Aspects of the Approved COVID-19 Vaccines

5.4.1 mRNA LNP Vaccines

mRNA vaccines encapsulated in LNP carriers have been the most administered vaccines to the population in response to the coronavirus pandemic in Europe and the USA. As of April 13, 2022, approximately 1 billion doses of the mRNA vaccines Comirnaty and Spikevax have been administered to the population in Europe [72]. Comirnaty was developed and produced in a collaboration between BioNTech in Germany and Pfizer in the USA and is referred to as the "Pfizer vaccine" in the following text. Spikevax was developed and produced in a collaboration between Moderna, the National Institute of Allergy and Infectious Diseases (NIAID), and the Biomedical Advanced Research and Development Authority (BARDA), and this vaccine is referred to as the "Moderna vaccine" below.

Several advantages of using mRNA vaccines have been noted, including a high efficacy of approximately 95% in clinical trials to prevent SARS-CoV-2 infection, as well as to reduce the severity of infection and mortality rate [73]. mRNA vaccines do not have concerns regarding endotoxins and infections in comparison to live-attenuated and inactivated vaccines. Importantly, there is no risk of genomic integration of the mRNA vaccines as they do not require nuclear entry to exert their activities [74]. Furthermore, mRNA vaccines have high specificity as the mRNA encodes essential antigens that are needed to develop adaptive immunity. Thus, mRNA vaccines seem to have a good safety profile [74]. Moreover, the mRNA vaccines can be synthesized in a short period of time at large scales and are, therefore, an efficient option in response to the pandemic [74]. Overall, mRNA vaccines consist of an active pharmaceutical ingredient (API), *i.e.*, the mRNA sequence, and a number of pharmaceutical excipients, *e.g.*, lipids, tonicity and pH-adjusting agents, and a vehicle (Table 5.4).

5.4.2 Active Ingredients of Pfizer and Moderna Vaccines

Pfizer and Moderna vaccines contain nucleoside-modified mRNA that encodes the viral spike glycoprotein (S) of SARS-CoV-2 referred to as Tozinameran and Elasomeran, respectively. Tozinameran and Elasomeran are single-strand RNA consisting of 4284 and 4101 nucleotide residues, respectively [75, 76]. The mRNA of the vaccines is produced using a cell-free *in vitro* transcription procedure from the corresponding DNA templates, encoding the viral spike glycoprotein (S) of SARS-CoV-2 [77, 78]. The mRNA has a 5′-cap structure and a poly-A tail in which all uridine units are replaced by N1-methylpseudouridine (m1Ψ). These modifications in the mRNA protect the RNA from degradation, drive high level of translation and protein expression, and reduce unwanted recombination during plasmid production [79]. Single doses for adults of 0.3 mL Pfizer vaccine contain 30 μg mRNA, while 0.5 mL of Moderna vaccine contains 100 μg mRNA [77, 78].

5.4.3 LNPs of Pfizer and Moderna Vaccines

The *in vivo* administration of unencapsulated mRNA, *i.e.*, without a carrier, is a challenge in practice, especially when vaccinating a large population, as the naked mRNA requires special storage and handling conditions, and well-trained staff to administer the vaccine to the target site in the body. Moreover, if a dose of naked mRNA is administered *in vivo*, the mRNA is rapidly degraded by extracellular ribonucleases (RNase). Additionally, due to the negative charge of mRNA, it might not be possible for the mRNA to permeate the negatively charged cell membrane to reach the cytoplasm [80, 81]. Therefore, the encapsulation of mRNA in a carrier such as LNPs can protect the mRNA from degradation and enable cellular entry of the API.

LNPs have shown promising results in the delivery of siRNA (Onpattro™) to patients. Onpattro™ was approved in 2018 for the treatment of hereditary transthyretin amyloidosis at a dose of 0.3 mg per kg every three weeks [82]. Onpattro™, which is an LNP formulation with entrapped siRNA, is designed to target a sequence within the untranslated region (30UTR) of transthyretin mRNA [82]. Therefore, the same strategy using LNPs as carriers to deliver mRNA vaccines was investigated *in vitro* and *in vivo* by the pioneers of mRNA vaccines, BioNTech and Moderna. The LNP suspensions used for Pfizer and Moderna vaccines contain two main groups of excipients: lipids forming the NPs and non-lipid excipients that ensure dispersion stability and pharmaceutical quality.

5.4.4 Lipid Excipients

The lipids for NPs used for mRNA delivery often fall within four types: (i) Ionizable or cationic lipids, which are also known as pH-sensitive lipids because they have negligible charge at a physiological pH (7.4) but become positively charged in an acidic environment such as in the endosome where the pH is approximately 6.5. These positively charged lipids may form a membrane-destabilizing ion pair within the endosomal membrane, and this may assist in mRNA escape from the endosome. Moreover, due to the inverted hexagonal structure of the charged lipids, the cell membranes can be easily disrupted to facilitate endocytosis [70]; (ii) PEGylated lipids where polyethylene glycol (PEG) is covalently bound to a lipid anchor. PEGylated lipids are used to decrease nonspecific binding to proteins, enhance the circulation time by promoting escape from first-pass metabolism or the reticuloendothelial system, increase half-life, reduce particle size, and prevent particle aggregation [83]; (iii) sterols (cholesterol) to stabilize the LNPs and to enhance cellular membrane fusion [81]; and (iv) phospholipids, which are also referred to as helper lipids and are used to stabilize the LNP structure and promote the formation of the lipid bilayer structure of NPs [81].

The LNPs of Pfizer and Moderna vaccines consist of mainly these four types of lipids (Table 5.4), however with a few differences in the selected lipids and/or the molar ratio if the same lipid is used in the vaccines. In both vaccines, the ionizable lipids and cholesterol are the main components of LNPs with approximately 90% [70], although the two vaccines use different ionizable lipids (Table 5.4). The phospholipid is the same in Pfizer and Moderna vaccines, although slightly different in quantity (9.4% and 10%, respectively) [70]. The PEGylated lipid is also different in the two vaccines.

5.4.5 Non-lipid Excipients

Non-lipid excipients are essential to prepare the LNP carriers of the Pfizer and Moderna vaccines. These excipients have buffering, osmotic, and cryoprotective functions in the mRNA vaccines (Table 5.4). Despite the difference in the buffer systems that are used by Pfizer (original retired formulation, purple cap vial) and Moderna, using phosphate buffered saline (PBS) and Tris-acetate buffer, respectively, both vaccines are able to maintain the pH in the formulation at approximately 7–8 [77, 78]. In the recent optimized formulation of Pfizer vaccines (orange and grey cap vials), the PBS buffer in the original formulation (purple cap vials) was replaced by a Tris buffer [84]. This change in the buffering systems has improved the storage properties of the Pfizer vaccines without affecting their safety and effectiveness, when the doses of the vaccines from orange and grey cap vials administered to the age groups of 5–11 years and 12 years and above, respectively [78, 84]. In both vaccines, sucrose is used as a cryoprotectant to maintain the physical properties of LNPs during freezing and thawing conditions [82].

5.4.6 Storage of Pfizer and Moderna Vaccines

One of the obstacles in the storage and distribution of Pfizer and Moderna vaccines is the ultralow storage temperature, which is needed to preserve the activity of the mRNA. In addition, the shelf life of the vaccines changes according to the temperature used during transport and storage. For example, the shelf life of Pfizer vaccine is 9 months when stored at −90°C to −60°C; however, the shelf life is limited to only 10 weeks when stored at 2°C to 8°C [78]. Another issue is that the vaccine cannot be re-frozen once it has been thawed, thus the shelf life needs to be changed on the vial container to 10 weeks instead of 9 months [78].

Pfizer has introduced three types of vaccine formulations provided in vials with different cap colors (Table 5.4). The original formulation of Pfizer vaccine is provided in purple cap vials, which was administered to the age group of 12 years and above [78]. Pfizer has created a transitional plan to replace the purple cap vials with grey cap vials, since the latter formulation does not need to be diluted before use and can be stored for 10 weeks in a refrigerator instead of only one month [84]. Moreover, after puncturing a grey cap vial, the content must be used within 12 h, while a shorter period (6 h only) is offered for the content to be used in the case of purple cap vials [85]. The orange cap vials are used to vaccinate the age group of 5 to 11 years [85]. From Table 5.4, it can be noted that similar non-lipid excipients are used for orange and grey caps.

A Moderna vaccine vial can be stored for 9 months at −25°C to −15°C and for 30 days at 2°C to 8°C. After the vial is punctured, the doses of Moderna vaccine should be administered within 19 h. Even though the transport and storage of the Moderna vaccine is more convenient than the Pfizer vaccine, the most used vaccine in Europe is still the Pfizer vaccine, as of April 13, 2022 [72].

5.5 Future Perspectives

The stability of the mRNA strand in LNP vaccines and the storage conditions required for transport, handling, and administration are the most important issues that need to be resolved. Here, nanoparticle technology is seen to play a critical role. Researchers have reported that the mRNA stability in aqueous solution is less than that in LNP formulations [68]. However, Pfizer and Moderna formulations also contain water, necessitating the storage conditions mentioned above. So, while the mRNA in these vaccines is partially protected against hydrolysis by incorporation into the LNPs, hydrolysis cannot be completely circumvented. Lyophilization (freeze drying) has, therefore, been suggested as a solution to preserve the stability of mRNA vaccine formulations. Lyophilization is possible for both LNP formulations and naked mRNA solutions. Recently, it has been suggested that stability could be improved if the vaccine formulation is prepared by mixing reconstituted naked mRNA with the LNPs just prior to use [68]. This area of research is still ongoing and requires more attention of the scientific field.

LNP formulations for RNA delivery require different types of lipids and other excipients. This is a rich topic requiring much more investigation. It is worth noting that changing the buffer system from PBS to Tris in the Pfizer vaccine resulted in an extended shelf life of the vaccine from 1 month to 10 weeks at 2°C to 8°C, as well as prolongation of the period the vaccine can be stored after opening from 6 h to 12 h. This indicates that there is much more to be optimized and understood to advance and improve NP delivery systems.

References

1. P. Sun, X. Lu, C. Xu, W. Sun, and B. Pan, Understanding of COVID-19 based on current evidence, *J. Med. Virol.* **92**, 548 (2020).
2. J. Wang and G. Du, COVID-19 may transmit through aerosol, *Ir. J. Med. Sci.* **189**, 1143 (2020).
3. B. Hu, H. Guo, P. Zhou, and Z.-L. Shi, Characteristics of SARS-CoV-2 and COVID-19, *Nat. Rev. Microbiol.* **19**, 3 (2021).
4. A. Sanyaolu, C. Okorie, A. Marinkovic, R. Patidar, K. Younis, P. Desai, Z. Hosein, I. Padda, J. Mangat, and M. Altaf, Comorbidity and its impact on patients with COVID-19, *Sn Compr. Clin. Med.* **2**, 1069 (2020).
5. WHO Coronavirus (COVID-19) Dashboard, https://covid19.who.int.
6. R. Lu et al., Genomic characterisation and epidemiology of 2019 novel coronavirus: Implications for virus origins and receptor binding, *Lancet Lond. Engl.* **395**, 565 (2020).
7. NIH, Severe acute respiratory syndrome coronavirus 2 isolate Wuhan-Hu-1, complete genome, http://www.ncbi.nlm.nih.gov/nuccore/MN908947.3 (accessed in September 2022).

8. M.-Y. Wang, R. Zhao, L.-J. Gao, X.-F. Gao, D.-P. Wang, and J.-M. Cao, SARS-CoV-2: Structure, biology, and structure-based therapeutics development, *Front. Cell. Infect. Microbiol.* **10**, (2020).
9. L. Zhang, C. B. Jackson, H. Mou, A. Ojha, E. S. Rangarajan, T. Izard, M. Farzan, and H. Choe, The D614G mutation in the SARS-CoV-2 spike protein reduces S1 shedding and increases infectivity, BioRxiv 2020.06.12.148726 (2020).
10. M. A. H. Farouq, M. M. Al Qaraghuli, K. Kubiak-Ossowska, V. A. Ferro, and P. A. Mulheran, Biomolecular interactions with nanoparticles: Applications for coronavirus disease 2019, *Curr. Opin. Colloid Interface Sci.* **54**, 101461 (2021).
11. M. Letko, A. Marzi, and V. Munster, Functional assessment of cell entry and receptor usage for SARS-CoV-2 and other lineage B betacoronaviruses, *Nat. Microbiol.* **5**, 562 (2020).
12. F. Li, W. Li, M. Farzan, and S. C. Harrison, Structure of SARS coronavirus spike receptor-binding domain complexed with receptor, *Science* **309**, 1864 (2005).
13. L. Duan, Q. Zheng, H. Zhang, Y. Niu, Y. Lou, and H. Wang, The SARS-CoV-2 spike glycoprotein biosynthesis, structure, function, and antigenicity: Implications for the design of spike-based vaccine immunogens, *Front. Immunol.* **11**, (2020).
14. D. Wrapp, N. Wang, K. S. Corbett, J. A. Goldsmith, C.-L. Hsieh, O. Abiona, B. S. Graham, and J. S. McLellan, Cryo-EM structure of the 2019-NCoV spike in the prefusion conformation, *Science* **367**, 1260 (2020).
15. Y. Huang, C. Yang, X. Xu, W. Xu, and S. Liu, Structural and functional properties of SARS-CoV-2 spike protein: Potential antivirus drug development for COVID-19, *Acta Pharmacol. Sin.* **41**, 1141 (2020).
16. Y. Watanabe, J. D. Allen, D. Wrapp, J. S. McLellan, and M. Crispin, Site-specific glycan analysis of the SARS-CoV-2 spike, *Science* **369**, 330 (2020).
17. P. V'kovski, A. Kratzel, S. Steiner, H. Stalder, and V. Thiel, Coronavirus biology and replication: Implications for SARS-CoV-2, *Nat. Rev. Microbiol.* **19**, 3 (2021).
18. A. R. Fehr and S. Perlman, Coronaviruses: An overview of their replication and pathogenesis, *Methods Mol. Biol. Clifton NJ* **1282**, 1 (2015).
19. D. E. Speiser and M. F. Bachmann, COVID-19: Mechanisms of vaccination and immunity, *Vaccines* **8**, 3 (2020).

20. J. R. Teijaro and D. L. Farber, COVID-19 vaccines: Modes of immune activation and future challenges, *Nat. Rev. Immunol.* **21**, 4 (2021).
21. N. C. Kyriakidis, A. López-Cortés, E. V. González, A. B. Grimaldos, and E. O. Prado, SARS-CoV-2 vaccines strategies: A comprehensive review of phase 3 candidates, *Npj Vaccines* **6**, 1 (2021).
22. World Health Organization, COVID-19 vaccine tracker and landscape, https://www.who.int/publications/m/item/draft-landscape-of-covid-19-candidate-vaccines (accessed in September 2022).
23. M. Lu et al., A safe and highly efficacious measles virus-based vaccine expressing SARS-CoV-2 stabilized prefusion spike, *Proc. Natl. Acad. Sci.* **118**, e2026153118 (2021).
24. J. Zhang, H. Zeng, J. Gu, H. Li, L. Zheng, and Q. Zou, Progress and prospects on vaccine development against SARS-CoV-2, *Vaccines* **8**(2), 153 (2020).
25. L. P. Samaranayake, C. J. Seneviratne, and K. S. Fakhruddin, Coronavirus disease 2019 (COVID-19) vaccines: A concise review, *Oral Dis.* (2021).
26. X. Zhou, X. Jiang, M. Qu, G. E. Aninwene, V. Jucaud, J. J. Moon, Z. Gu, W. Sun, and A. Khademhosseini, Engineering antiviral vaccines, *ACS Nano* **14**(10), 12370–12389 (2020).
27. R. Vanaparthy, G. Mohan, D. Vasireddy, and P. Atluri, Review of COVID-19 viral vector-based vaccines and COVID-19 variants, *Infez. Med.* **29**, 328 (2021).
28. J. Wang, Y. Peng, H. Xu, Z. Cui, and R. O. Williams, The COVID-19 vaccine race: Challenges and opportunities in vaccine formulation, *AAPS PharmSciTech* **21**, 225 (2020).
29. S. Rauch, E. Jasny, K. E. Schmidt, and B. Petsch, New vaccine technologies to combat outbreak situations, *Front. Immunol.* **9**, 1963 (2018). doi: 10.3389/fimmu.2018.01963.
30. M. Robert-Guroff, Replicating and non-replicating viral vectors for vaccine development, *Curr. Opin. Biotechnol.* **18**, 546 (2007).
31. T. Ura, K. Okuda, and M. Shimada, Developments in viral vector-based vaccines, *Vaccines* **2**, 3 (2014).
32. N. Lurie, M. Saville, R. Hatchett, and J. Halton, Developing Covid-19 vaccines at pandemic speed, *N. Engl. J. Med.* **382**, 1969 (2020).
33. S. Xiaojie, L. Yu, Y. lei, Y. Guang, and Q. Min, Neutralizing antibodies targeting SARS-CoV-2 spike protein, *Stem Cell Res.* **50**, 102125 (2021).

34. E. H. Pilkington et al., From influenza to COVID-19: Lipid nanoparticle MRNA vaccines at the frontiers of infectious diseases, *Acta Biomater.* **131**, 16 (2021).

35. E. N. Gary and D. B. Weiner, DNA vaccines: Prime time is now, *Curr. Opin. Immunol.* **65**, 21 (2020).

36. W. Ho, M. Gao, F. Li, Z. Li, X.-Q. Zhang, and X. Xu, Next-generation vaccines: Nanoparticle-mediated DNA and MRNA delivery, *Adv. Healthc. Mater.* **10**, 2001812 (2021).

37. M. N. Vu, H. G. Kelly, S. J. Kent, and A. K. Wheatley, Current and future nanoparticle vaccines for COVID-19, *EBioMedicine* **74**, (2021).

38. T. A. Tabish and M. R. Hamblin, Multivalent nanomedicines to treat COVID-19: A slow train coming, *Nano Today* **35**, 100962 (2020).

39. L. Du, Y. Yang, X. Zhang, and F. Li, Recent advances in nanotechnology-based COVID-19 vaccines and therapeutic antibodies, *Nanoscale* **14**, 1054 (2022).

40. C. Keech et al., Phase 1–2 trial of a SARS-CoV-2 recombinant spike protein nanoparticle vaccine, *N. Engl. J. Med.* **383**, 2320 (2020).

41. Q. Geng et al., Novel virus-like nanoparticle vaccine effectively protects animal model from SARS-CoV-2 infection, *PLOS Pathog.* **17**, e1009897 (2021).

42. L. He, X. Lin, Y. Wang, C. Abraham, C. Sou, T. Ngo, Y. Zhang, I. A. Wilson, and J. Zhu, Single-component, self-assembling, protein nanoparticles presenting the receptor binding domain and stabilized spike as SARS-CoV-2 vaccine candidates, *Sci. Adv.* **7**, eabf1591 (2021).

43. A. E. Powell, K. Zhang, M. Sanyal, S. Tang, P. A. Weidenbacher, S. Li, T. D. Pham, J. E. Pak, W. Chiu, and P. S. Kim, A single immunization with spike-functionalized ferritin vaccines elicits neutralizing antibody responses against SARS-CoV-2 in mice, *ACS Cent. Sci.* **7**, 183 (2021).

44. J. Kim et al., Porous gold nanoparticles for attenuating infectivity of influenza A virus, *J. Nanobiotechnology* **18**, 54 (2020).

45. Y. Abo-zeid, N. S. M. Ismail, G. R. McLean, and N. M. Hamdy, A molecular docking study repurposes FDA approved iron oxide nanoparticles to treat and control COVID-19 infection, *Eur. J. Pharm. Sci.* **153**, 105465 (2020).

46. S. Thalhauser, D. Peterhoff, R. Wagner, and M. Breunig, Presentation of HIV-1 envelope trimers on the surface of silica nanoparticles, *J. Pharm. Sci.* **109**, 911 (2020).

47. R. S. El-Mohamady, T. A. Ghattas, M. F. Zawrah, and Y. G. M. Abd El-Hafeiz, Inhibitory effect of silver nanoparticles on bovine herpesvirus-1, *Int. J. Vet. Sci. Med.* **6**, 296 (2018).

48. D. S. Sarkar, Silver nanoparticles with bronchodilators through nebulisation to treat COVID 19 patients, *J. Curr. Med. Res. Opin.* **3**, 04 (2020).

49. P. Hassanzadeh, Nanotheranostics against COVID-19: From multivalent to immune-targeted materials, *J. Controlled Release* **328**, 112 (2020).

50. S. Muyldermans, A guide to: Generation and design of nanobodies, *FEBS J.* **288**, 2084 (2021).

51. G. Zhao et al., A novel nanobody targeting Middle East respiratory syndrome coronavirus (MERS-CoV) receptor-binding domain has potent cross-neutralizing activity and protective efficacy against MERS-CoV, *J. Virol.* **92**, e00837 (2018).

52. A. A. Yetisgin, S. Cetinel, M. Zuvin, A. Kosar, and O. Kutlu, Therapeutic nanoparticles and their targeted delivery applications, *Mol. Basel Switz.* **25**, E2193 (2020).

53. D. Yang, Application of nanotechnology in the COVID-19 pandemic, *Int. J. Nanomedicine* **16**, 623 (2021).

54. P. Bhardwaj, E. Bhatia, S. Sharma, N. Ahamad, and R. Banerjee, Advancements in prophylactic and therapeutic nanovaccines, *Acta Biomater.* **108**, 1 (2020).

55. S. Al-Halifa, L. Gauthier, D. Arpin, S. Bourgault, and D. Archambault, Nanoparticle-based vaccines against respiratory viruses, *Front. Immunol.* **10**, 22 (2019).

56. L. E. Charelli, G. C. de Mattos, A. de Jesus Sousa-Batista, J. C. Pinto, and T. A. Balbino, Polymeric nanoparticles as therapeutic agents against coronavirus disease, *J. Nanoparticle Res.* **24**, 12 (2022).

57. B. Li et al., Antiviral and anti-inflammatory treatment with multifunctional alveolar macrophage-like nanoparticles in a surrogate mouse model of COVID-19, *Adv. Sci.* (2021).

58. A. D. Bangham, M. W. Hill, and N. G. A. Miller, Preparation and use of liposomes as models of biological membranes, in *Methods in Membrane Biology: Volume 1*, edited by E. D. Korn (Springer US, Boston, MA, 1974), pp. 1–68.

59. M. A. Obeid, T. Teeravatcharoenchai, D. Connell, K. Niwasabutra, M. Hussain, K. Carter, and V. A. Ferro, Examination of the effect of niosome preparation methods in encapsulating model antigens on the vesicle

characteristics and their ability to induce immune responses, *J. Liposome Res.* **31**, 195 (2021).

60. M. A. Obeid, R. J. Tate, A. B. Mullen, and V. A. Ferro, Lipid-based nanoparticles for cancer treatment, in *Lipid Nanocarriers for Drug Targeting*, edited by A. M. Grumezescu (William Andrew Publishing, 2018), pp. 313–359.

61. G. Gregoriadis, Liposomes and MRNA: Two technologies together create a COVID-19 vaccine, *Med. Drug Discov.* **12**, 100104 (2021).

62. A. Gebril, M. A. Obeid, E. M. Bennett, A. Pujol, M. L. Chovel, T. Mahy, R. Acevedo, and V. A. Ferro, Mucosal and systemic immune responses following mucosal immunisation of tetanus toxoid entrapped in lipid nanoparticles prepared by microwave reactor, *Eur. J. Pharm. Biopharm.* **171**, 11 (2022).

63. M. A. Obeid, C. Dufès, S. Somani, A. B. Mullen, R. J. Tate, and V. A. Ferro, Proof of concept studies for SiRNA delivery by nonionic surfactant vesicles: *In vitro* and *in vivo* evaluation of protein knockdown, *J. Liposome Res.* **29**, 229 (2019).

64. K. S. Park et al., Lipid-based vaccine nanoparticles for induction of humoral immune responses against HIV-1 and SARS-CoV-2, *J. Control. Release* **330**, 529 (2021).

65. D. Laczkó et al., A single immunization with nucleoside-modified MRNA vaccines elicits strong cellular and humoral immune responses against SARS-CoV-2 in mice, *Immunity* **53**, 724 (2020).

66. A. Akinc et al., The onpattro story and the clinical translation of nanomedicines containing nucleic acid-based drugs, *Nat. Nanotechnol.* **14**, 12 (2019).

67. A. K. Blakney, S. Ip, and A. J. Geall, An update on self-amplifying MRNA vaccine development, *Vaccines* **9**, 2 (2021).

68. L. Schoenmaker, D. Witzigmann, J. A. Kulkarni, R. Verbeke, G. Kersten, W. Jiskoot, and D. J. A. Crommelin, MRNA-lipid nanoparticle COVID-19 vaccines: Structure and stability, *Int. J. Pharm.* **601**, 120586 (2021).

69. M. A. Obeid, H. Alyamani, H. Amawi, A. A. A. Aljabali, M. Rezigue, S. N. Abdeljaber, and V. A. Ferro, SiRNA delivery to melanoma cells with cationic niosomes, *Methods Mol. Biol. Clifton NJ* **2265**, 621 (2021).

70. I. Pantelić, T. Ilić, I. Nikolić, and S. Savić, Lipid nanoparticles employed in MRNA-based COVID-19 vaccines: An overview of materials and processes used for development and production, *Arch. Pharm.* **72**, 1 (2022).

71. S. Ranjbar, Y. Fatahi, and F. Atyabi, The quest for a better fight: How can nanomaterials address the current therapeutic and diagnostic obstacles in the fight against COVID-19? *J. Drug Deliv. Sci. Technol.* **67**, 102899 (2022).

72. COVID-19 Vaccine Tracker | European Centre for Disease Prevention and Control, https://vaccinetracker.ecdc.europa.eu/public/extensions/covid-19/vaccine-tracker.html#distribution-tab (accessed in April 2022).

73. A. I. Francis, S. Ghany, T. Gilkes, and S. Umakanthan, Review of COVID-19 vaccine subtypes, efficacy and geographical distributions, *Postgrad. Med. J.* **98**(1159), 389–394 (2021). doi: 10.1136/postgradmedj-2021-140654.

74. J. Kim, Y. Eygeris, M. Gupta, and G. Sahay, Self-assembled MRNA vaccines, *Adv. Drug Deliv. Rev.* **170**, 83 (2021).

75. JAN Elasomeran: Japanese Accepted Names for Pharmaceuticals, https://jpdb.nihs.go.jp/jan/DetailList_en.aspx?submit=all_alp+Search&keyword=Elasomeran (accessed in September 2022).

76. JAN Tozinameran: Japanese Accepted Names for Pharmaceuticals, https://jpdb.nihs.go.jp/jan/DetailList_en.aspx?submit=all_alp+Search&keyword=Tozinameran (accessed in September 2022).

77. European Commission, *Moderna, Annex I Summary of Product Characteristics*, 2022. Access date: 4/10/2022.

78. European Commission, *Pfizer, Annex I Summary of Product Characteristics*, 2022. Access date: 4/10/2022.

79. K. D. Nance and J. L. Meier, Modifications in an emergency: The role of N1-methylpseudouridine in COVID-19 vaccines, *ACS Cent. Sci.* **7**, 748 (2021).

80. N. Pardi, M. J. Hogan, F. W. Porter, and D. Weissman, MRNA vaccines: A new era in vaccinology, *Nat. Rev. Drug Discov.* **17**, 261 (2018).

81. Y. Cao and G. F. Gao, mRNA vaccines: A matter of delivery, *EClinicalMedicine* **32**, 100746 (2021).

82. Y. Suzuki and H. Ishihara, Difference in the lipid nanoparticle technology employed in three approved SiRNA (Patisiran) and MRNA (COVID-19 vaccine) drugs, *Drug Metab. Pharmacokinet.* **41**, 100424 (2021).

83. A. Khurana, P. Allawadhi, I. Khurana, S. Allwadhi, R. Weiskirchen, A. K. Banothu, D. Chhabra, K. Joshi, and K. K. Bharani, Role of nanotechnology behind the success of MRNA vaccines for COVID-19, *Nano Today* **38**, 101142 (2021).

84. FDA, Emergency Use Authorization (EUA) for an Unapproved Product Review Memorandum, 2022. Access date: 4/10/2022.

85. CVDvaccine, Vaccine Formulation/Presentation Guide, 2022. Access date: 4/10/2022.

Chapter 6

Overview of Nanotherapeutics for Bacterial Infections

José Muñoz-López, Cátia D.F. Lopes, and Íris Luz Batalha
Institute for Bioengineering of Catalonia (IBEC), Barcelona Institute of Science and Technology (BIST), Carrer de Baldiri Reixac 10-12, 08028 Barcelona, Spain
jmunoz@ibecbarcelona.eu, ibatalha@ibecbarcelona.eu

Antibiotic resistance is one of the biggest threats to global health, food security, and development, which affects people regardless of age, gender, and origin. Despite being a natural process, the misuse of antibiotics has speeded up the development of the phenotypic and genotypic drug-resistance mechanisms through which bacteria can guarantee their survival. These mechanisms include limitation of drug uptake, drug target modification, drug inactivation, and active drug efflux. As a consequence, conventional antibiotics are becoming less effective and a growing number of bacterial infections, including the ones causing pneumonia, tuberculosis, gonorrhea, and salmonellosis, are becoming harder to treat [1, 2].

On the lookout for non-conventional approaches to surpass and mitigate the effects of antibiotic resistance, nanomaterial-based

Nanotherapeutics for Infectious Diseases
Edited by Íris Luz Batalha

ISBN 978-981-5129-36-6 (Hardcover), 978-1-003-56678-6 (eBook)
www.jennystanford.com

systems render new opportunities in the therapeutics field as drug carriers or, even, as inherent antimicrobial agents. Nanoparticles (NPs) have tunable size, shape, and surface chemistry, and a high surface-to-volume ratio that confers them unique physicochemical properties (*e.g.*, electrical, magnetic, optical, etc.). Hence, nanomaterials represent an opportunity to create novel and cutting-edge therapeutic strategies to combat both planktonic bacteria and biofilms as they can circumvent existing resistance mechanisms by employing an array of bactericidal pathways. For example, positively charged NPs can engage in electrostatic interactions with the negatively charged groups on bacterial surfaces, resulting in damage to the membrane and leakage of cytoplasmic contents. Nanomaterials can also bind to various intracellular components such as ribosomes, proteins, and DNA, disrupting their normal functions. Additionally, nanomaterials possessing catalytic properties can boost the production of reactive oxygen species (ROS) like hydroxyl radicals and superoxides, leading to oxidative stress within the bacterial cells. Furthermore, nanomaterials can serve as carriers for therapeutic agents, some of which readily penetrate bacterial cells through membrane fusion, thus facilitating the delivery of their cargo (Fig. 6.1) [3–5].

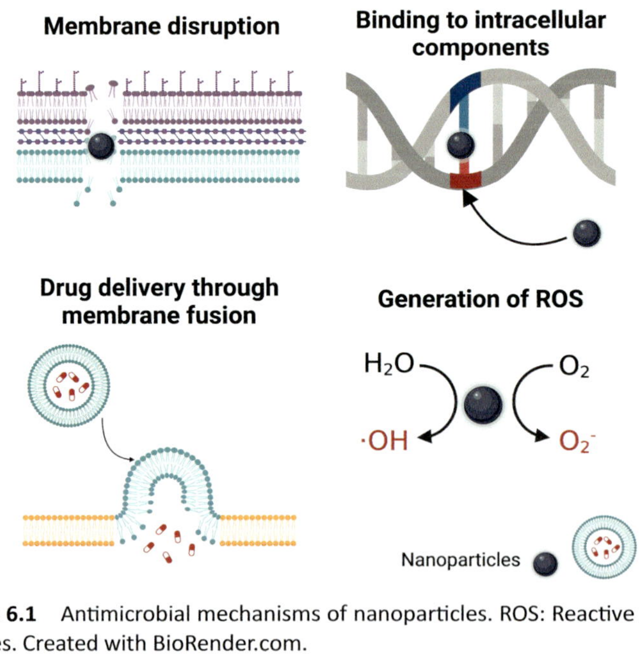

Figure 6.1 Antimicrobial mechanisms of nanoparticles. ROS: Reactive Oxygen Species. Created with BioRender.com.

6.1 Inorganic Nanomaterials

Inorganic NPs encompass a wide range of materials, including transition metals (*e.g.*, gold (Au), silver (Ag), copper (Cu)), metal oxides (*e.g.*, iron oxides (FeO, Fe_2O_3, Fe_3O_4), magnesium oxide (MgO), titanium dioxide (TiO_2), aluminum oxide (Al_2O_3), zinc oxide (ZnO)), metalloids (*e.g.*, silica (SiO_2)), carbon (*e.g.*, carbon nanotubes), and more. Their antimicrobial effect relies on certain characteristics such as size, shape, ζ-potential, presence of ligands, pH, roughness, stability, crystal structure, and chemical composition [6].

Inorganic NPs composed of transition metals can form strong coordination bonds with non-bonding orbitals characteristic of heteroatoms (*i.e.*, N, O, and S) abundant in biomolecules (*e.g.*, proteins and nucleic acids), effectively interfering with their biological function.

Positively charged metal-based NPs can also establish strong electrostatic interactions with the negatively charged bacterial cell wall, regardless of its composition (*i.e.*, gram-positive or gram-negative bacteria), resulting in its disruption and increased permeability [7]. In addition, NPs can also release free metal ions into the extracellular space, which can enter the cell and interfere with metabolic pathways [3, 7].

Another antimicrobial mechanism utilized by this class of NPs is the production of ROS, such as free radicals and peroxides, leading to the oxidative stress of bacterial cells. Indeed, ROS production is a natural product of cellular respiration, and low levels of ROS (which act as redox-signaling messengers) are needed for the normal physiological functioning of cells [8]. In addition, bacteria have a plethora of antioxidant enzymes (*e.g.*, glutathione (GSH), superoxide dismutase, and catalase) which protect biological macromolecules from oxidative stress [9]. Nevertheless, excessive ROS production caused by metal NPs can break this balance and eventually kill the bacteria by damaging the cellular membrane, DNA, and mitochondria. In fact, this is the main antimicrobial mechanism triggered by various metal-based NPs, such as Au, Cu, ZnO, and TiO_2. Therefore, this class of NPs can be adopted for antimicrobial purposes with broad-spectrum activity. One major shortcoming is that this lack of specificity in their mechanisms of action also means that they can damage pathogenic bacteria and mammalian cells alike [10].

In particular, silver NPs (AgNPs) have a strong antimicrobial activity due to the high affinity of Ag^+ ions to amines, phosphates, and thiols, functional groups widely found in proteins and nucleic acids [10]. In fact, AgNPs are the most commercialized engineered nanomaterials, accounting for over 50% of global nanomaterial consumer products, being extensively used in personal care products, textiles, food packaging, and healthcare products, due to their recognized antibacterial, antifungal, and antiviral properties [11, 12]. As such, AgNPs have been widely studied and used alone or in combination with other nanomaterials, in the form of nanocomposites or hybrid materials [13–20].

Metallic copper NPs offer an affordable alternative to silver-based nanomaterials; however, they have major limitations such as the requirement for higher dosage concentrations and instability due to the rapid oxidation of metallic copper to cupric oxide (CuO) and cuprous oxide (Cu_2O) upon exposure to air. In fact, the formation of an oxide layer at the surface of copper NPs is inevitable due to the greater thermodynamical stability of the oxide phases [21]. In addition, the release of high concentrations of Cu^{2+} (and its reduced form Cu^+) is toxic to both bacterial and mammalian cells, due to the generation of excessive ROS [10, 22, 23]. Despite these limitations, the antimicrobial properties of copper have been recognized for centuries, with various forms of copper compounds employed by ancient civilizations for a range of purposes. For instance, the ancient Egyptians used metals such as silver and copper to sterilize drinking water and treat wounds. Romans documented numerous medicinal applications of copper for different ailments. The Aztecs utilized copper to alleviate sore throats, while in Persia and India, copper was employed in the treatment of eye infections and venereal ulcers. In the past decades, Cu-based nanomaterials with low or non-cytotoxic profiles have been successfully developed against both gram-positive and gram-negative bacteria, such as *Escherichia coli, Staphylococcus aureus, Pseudomonas aeruginosa, Salmonella enterica,* and *Staphylococcus epidermis,* among many other pathogenic microorganisms [24].

The greater stability of copper oxide phases has driven research toward the development of copper oxide nanomaterials, as an alternative to metallic copper. Several studies reported the antibacterial activity of copper oxides against gram-positive and gram-negative bacteria, such as *Lactobacillus brevis, P. aeruginosa, E. coli*, and especially against methicillin-resistant *S. aureus* (*MRSA*) [25–29]. In addition, copper oxide NPs, especially Cu_2O, have been reported to show lower host cell cytotoxicity, good environmental acceptability, and remarkable broad-spectrum antibacterial activity. However, the major limitation of copper oxide NPs when compared to metallic copper is the lack of sufficient stability due to their strong tendency to aggregate [30].

Conversely to AgNPs and CuNPs, AuNPs are much more inert and biocompatible, yet they have significantly lower antibacterial activity. Despite this drawback, AuNPs can be easily functionalized with antibiotics, cationic moieties, or even enzymes, to increase their antibacterial effect [31]. For example, a hybrid system composed of ampicillin conjugated to lysozyme-capped gold nanoclusters was developed to treat *MRSA* infections, increasing ampicillin antibacterial effect against this strain and other non-resistant bacterial strains through different mechanisms, which include: (i) increasing ampicillin concentration at the target site; (ii) allowing for multivalent presentation and increased permeation of ampicillin; (iii) lysozyme-mediated cell wall lysis; (iv) bacterial efflux pump dysfunction; and (v) Au ion–mediated destabilization [32].

In a different example, amoxicillin and ofloxacin antibiotics were individually loaded in Au-silica core-shell mesoporous NPs and tested against *P. aeruginosa*. The NPs allowed a 2-fold (for ofloxacin) and 20-fold (for amoxicillin) decrease in the quantity of antibiotics needed to inhibit bacterial growth [33].

Metal oxide NPs, such as ZnO, TiO_2, and Al_2O_3 NPs, are another group of nanomaterials with potent antibacterial activity. Among these, ZnO NPs have entered the spotlight due to their low toxicity and highest biocompatibility [12, 27–29]. In addition, ZnO NPs also render high antibacterial effectiveness at low concentrations (0.16–5.00 mmol/L) against a wide range of strains and at a relatively low cost [34]. For example, gelatin-coated ZnO NPs showed not only antibacterial activity but also inhibition of biofilm formation against *P. aeruginosa* and *Enterococcus faecalis* at a concentration of 50 μg/mL, though higher activity was observed for the gram-negative bacteria over the gram-positive [35]. The green synthesis of ZnO NPs yields coated NPs with enhanced antimicrobial activities. For example, ZnO NPs prepared in *Azadirachta indica* aqueous leave extract have a surface layer composed of terpenoids, flavonoids, phenolic acid, and proteins, which increases the concentration of H_2O_2 at the surface. Therefore, they present increased ROS activity and consequently better antimicrobial activity than bare ZnO produced using synthetic methods. Indeed, these NPs presented antimicrobial activity against *S. aureus, P. aeruginosa, E. coli, Proteus mirabillis,* and *Bacillus subtilis* at micromolar concentrations [36].

6.2 Polymeric Nanomaterials

Polymeric nanomaterials can be composed of natural or synthetic polymers and generally present good biocompatibility, stability, and the ability to encapsulate both hydrophilic and hydrophobic drugs. They can be assembled into a varied array of three-dimensional nanostructures, such as micelles, vesicles, dendrimers, nanofibers, and nanogels.

6.2.1 Natural Polymers

Natural polymers are materials that occur in nature and can be extracted from a variety of plant, animal, and microbial sources. They are biocompatible, biodegradable, and eco-friendly [37, 38]. Indeed, most of these polymers are characterized by having a heteroatom backbone (*e.g.,* –C–O–, –C–N–, C–S–, among many others), which facilitates hydrolysis and bond cleavage. This kind of chemistry not only endows natural polymeric nanomaterials with control over drug release, but also facilitates their excretion from the body [39]. Natural polymers fall into three major classes: polysaccharides (*e.g.,* chitosan, cellulose, starch, hyaluronan, dextran), proteins (*e.g.,* collagen, fibrin, elastin, soy protein), and bacterial polyesters (*e.g.,* poly-3-hydroxybutarate (P3HB), poly-4-hydroxybutarate (P4HB), poly-3-hydrovalerate (PHV), and copolymer poly-3-hydroxybutarate-*co*-poly3-hydrovalerate (PHBV)) [38]. Due to their physicochemical properties, such as solubility, hygroscopicity, viscosity, and easy derivatization, polysaccharides are widely used in the production of nanomaterials for therapeutics and drug delivery purposes [40].

Polysaccharides are widely found across different biological organisms, spanning from the simplest prokaryotic cell to complex organisms such as plants and animals. These natural polymers consist of monosaccharide units linked by glycosidic bonds. The most typical examples in biomedicine are chitosan, cellulose, hyaluronic acid, and dextran. However, the diversity of chemical structures that these polymers provide (different stereoisomerism, branching degree, degree of N-acetylation, degree of polymerization, O-glycosidic bond type, etc.) and their ionic nature (cellulose and dextran are an exception for the latter feature) confer a broad spectrum of biological properties and activities (Fig. 6.2) [41].

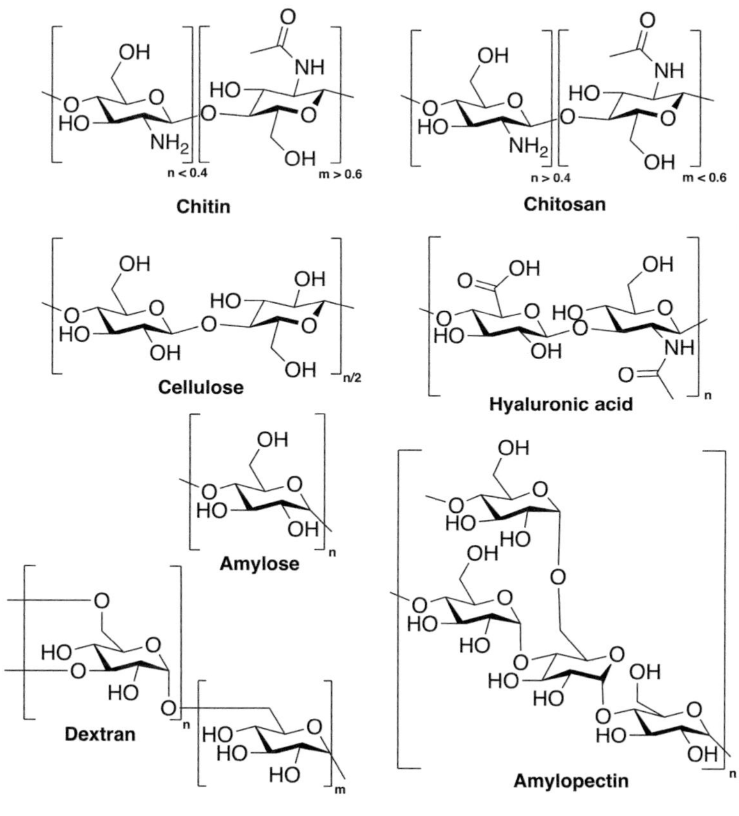

Figure 6.2 Chemical structures of some polysaccharides used in the synthesis of nanomaterials. *n* and *m* indicate the degree of polymerization. Created with ChemDraw 21.0.0.28.

Chitin is the primary component of the exoskeleton of crustaceans, and it is also the second most abundant polysaccharide across living organisms. However, its insolubility in water and other common solvents limits its use in the field of nanomedicine. Chitosan, a chitin derivative obtained from partial *N*-deacetylation, is more commonly used since its solubility in acidic conditions is increased not only due to the protonation of the basic amino groups but also due to the decrease in the polymer's molecular weight (from 1000–2500 kDa in chitin to 100–500 kDa in chitosan) [41]. In line with this, chitosan has good biocompatibility and biodegradability, and due to its greater positive net charge in mildly acidic conditions, it provides a greater antibacterial effect than chitin. The positively charged amines can electrostatically interact with negatively charged phospholipids, proteins, and carbohydrates present on the surface of the bacterial cell and inhibit bacterial growth [42].

Chitosan can be used as an adjuvant or delivery carrier, protecting the carried molecules from enzymatic degradation [37]. For example, poly(lactic-*co*-glycolic acid) (PLGA)-chitosan composite NPs enhanced the antibacterial and antibiofilm performance of ciprofloxacin against *E. faecalis* [43]. On the other hand, composites of chitin/chitosan with metallic and metal oxide NPs (*e.g.*, chitin/Ni, chitin/Ag, chitosan/Ag, chitin/Cu, chitosan/ZnO, and chitosan/TiO_2) showed antimicrobial activity against both gram-positive and gram-negative bacteria [44–51].

Hyaluronic acid, naturally found in the extracellular matrix, is a hydrophilic linear polysaccharide composed of alternately linked D-glucuronic and *N*-acetyl-D-glucosamine units via β-1,3 and β-1,4 glycosidic bonds. Opposite to chitosan, hyaluronic acid is a negatively charged polysaccharide that undergoes rapid enzymatic degradation [52]. Bacteria often have the ability to adhere to wounds and plastic materials, but for further spreading, which enables penetration into deeper tissues, the use of enzymes as invasion factors is required. Many pathogenic gram-positive bacteria (*e.g.*, *S. aureus*) produce hyaluronidase, an enzyme capable of degrading the hyaluronic acid present in the extracellular matrix and greatly enhancing bacterial infiltration. Baier et al. showed that hyaluronic acid–based nanocapsules containing the antimicrobial agent polyhexanide could be efficiently cleaved in the presence of bacterial hyaluronidase and efficiently killed the pathogenic bacteria [53]. When a sustained cargo release is desired, the conjugation of synthetic polymers, such as polyethylene glycol (PEG), to hyaluronic acid slows down and controls its degradation [54, 55]. This approach, however, has been more extensively studied in the delivery of anticancer drugs since hyaluronic acid targets CD44 receptors, which are overexpressed in cancer cells [56–60].

Starch is a nontoxic complex polysaccharide formed by two different sugars: amylose and amylopectin. The former is a linear polymer of α-(1,4)-linked D-glucose units, while the latter is a highly branched polymer composed of short chains of α-(1,4)-linked D-glucose with branches formed by α-(1,6) linkages at the branch positions [41]. Due to its insolubility, starch is generally used in combination with other polysaccharides, synthetic polymers, or metal ions, and the use of such composites shows great promise for biocide purposes in the food packaging industry due to abundant sourcing and low price [61–64]. In the biomedical field, the starch biopolymer is often used in the form of hydrogels decorated with inorganic NPs [57–61]. For instance, the incorporation of CuO NPs in carboxymethylated starch hydrogels yielded antibacterial activity against both gram-positive and gram-negative bacteria. In particular, hydrogels containing 2 wt.% CuO NPs also displayed low cytotoxicity and good biocompatibility [65]. Similarly, starch-based hydrogels reinforced with silica-coated copper NPs presented antimicrobial activity against *E. coli* and *S. aureus* for at least four cycles of use [66]. The possibility of using these hydrogel composites dermally paves the way for the development of wound dressing applications.

Cellulose is the most abundant polysaccharide and the primary constituent of plant cell walls, where it is mainly extracted from. Its linear polysaccharide structure is formed by β-(1,4)-linked D-glucose units, giving it a flat ribbon-like conformation. This β-glycosidic linkage enables the formation of inter- and intra-chain hydrogen bonds leading to a network that provides extra stability. At the nanoscale, cellulose finds numerous applications in drug delivery due to its unique physical properties (*e.g.*, good mechanical strength and stiffness), large specific surface area, low cytotoxicity, and renewability [63–65]. This configuration of cellulose coined the term nanocellulose, and it can be found across the four main natural sources (*i.e.*, bacteria, plants, algae, and animals) in the form of fibrils or crystals [67–69].

Dextran is mostly obtained from *Leuconostoc mesenteroides*, and its basic structure consists of main chains formed by (1,6)-α-D-glucose with various ratios of linkages and branches. Contrary to cellulose and starch, dextran is highly water soluble, which facilitates its use in the development of novel nanotherapeutics that can be taken orally or even by injection. Indeed, dextran sulfate sodium NPs have shown inherent antibacterial effects at minimum inhibitory concentrations (MICs) below 250 μg/mL against both gram-positive and gram-negative bacteria, including *E. coli*, *Streptococcus pyogenes*, *B. subtilis*, *Klebsiella pneumoniae*, *P. aeruginosa*, *S. aureus*, and *Proteus vulgaris*, due to their ability to passively diffuse through the bacterial cell pores. The antimicrobial effect of the particles was almost equivalent to the standard antibiotic ciprofloxacin [70]. However, the main drawback of dextran is that larger molecular weights of 60 kDa are poorly excreted from the kidney and remain in the blood before being completely metabolized, so its clinical use should be restricted under this cut-off value [41]. In another example, amphiphilic dextran–decanoate NPs loaded with ciprofloxacin presented antibacterial activity against the oral bacteria *E. faecalis* at an MIC of 0.7 μg/mL, showing potential for the treatment and prevention of root canal infections [71].

Polyhydroxyalkanoates (PHA) are natural polyesters produced through enzyme hydrolysis by certain types of bacteria and other organisms as means of energy and carbon source during periods of detrimental conditions, such as lack of macroelements, trace elements, or oxygen, for their thriving [72, 73]. From a chemical point of view, PHAs are constituted of *R*-3-hydroxy and *R*-4-hydroxy fatty acids with different chain lengths, spanning from 3 to 14 carbon atoms [13, 74] (Fig. 6.3). The degradation of these compounds produces the corresponding *R*-hydroxyalkanoic acids (such as 3-hydroxybutyric, 3-hydroxyvaleric, and 4-hydroxybutyric acids), which are recognized as degradation products and are conveniently removed from the body. In fact, 3-hydroxybutyric acid is naturally found in the human bloodstream [75]. It is noteworthy to mention that some PHAs show antimicrobial activity and their potency seems to be correlated with the chain length of the monomeric fatty acid. Although the mechanism still remains unclear, it has been attributed to their detergent-like properties, which alter the permeability of the bacterial cell membrane [76–78]. These biogenic polymers outperform many others in the field of nanomedicine due to the lack of toxicity, biocompatibility, biodegradability, as well as high drug-loading capacity [72].

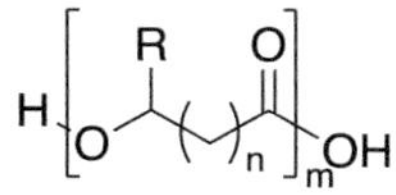

Figure 6.3 Schematic representation of polyhydroxyalkanoates (PHAs), where *R* represents an alkyl group, *n* the chain length of the repeating unit (up to *n* = 8), and *m* the degree of polymerization. Created with ChemDraw 21.0.0.28.

Among PHAs, poly-3-hydroxybutyrate (PHB) has several advantages, including biocompatibility and the ability to degrade *in vivo* and *in vitro* without the generation of harmful by-products, as it happens with the acidic degradation products of synthetic polymers, such as PLGA, which could lead to an inflammatory response [72]. Using electrospinning techniques, PHB can be tailored to form nanofibrous scaffolds as carriers for antibiotic drugs. For example, PHB nanofiber mats surface-loaded with kanamycin sulfate showed antimicrobial activity against *S. aureus* by a sustained release of over 95% of the antibiotic drug within 8 h [79]. Similarly, different porous morphologies of PHB electrospun meshes have been assessed for the release of levofloxacin against *Micrococcus luteus, Serratia marcescens*, and *E. coli.* In nanofiber mats loaded with levofloxacin, a slower release was observed with 30.4% to 32.5% of the incorporated antibiotic being released during 24 h and displaying bacteria-killing efficacy against gram-negative and gram-positive bacteria [15]. Hence, sustained drug release with enhanced antimicrobial activity can be achieved for different fibrous nanostructures of the PHB polymer.

Poly(3-hydroxybutyrate-*co*-3-hydroxyvalerate) (PHBV) polymer has been proven as an effective material for bone tissue engineering due to its biocompatibility and slow degradation rate. Nanocomposites formulated with PHBV, nanodiamond, and nanohydroxyapatite loaded with vancomycin have been developed for bone fixation devices with antibacterial activity. When tested for the treatment and prophylaxis of bone infection, these nanocomposites guaranteed a sustained *in vitro* release of the antibiotic over 22 days [80].

The combination of PHA nanomaterials with inorganic NPs not only brought about an alternative to antibiotics for the treatment of infections but has also shown enhanced antimicrobial activity in some cases. For example, PHBV nanofibers had negligible activity against *S. aureus* and *K. pneumoniae* bacteria, but showed antibacterial activity against both strains when loaded with 5–13 nm AgNPs [16]. In addition, PHBV polymer extracted from a mixed bacterial culture has been demonstrated to be a good stabilizing agent for the *in situ* preparation of AgNPs, since the antimicrobial activity of the NPs is dependent on their size, size distribution, and aggregation propensity. The AgNP–PHBV materials presented strong antibacterial activity against *S. enterica* at low concentration (0.1–1 ppm) [80].

6.2.2 Synthetic Polymers

As opposed to natural polymers, which are found in plant, animal, and microbial sources, chemically synthesized polymers are non-naturally occurring polymers obtained by different synthetic methods. In the field of medical materials, polylactic acid (PLA), PLGA, polyurethane (PU), poly(methyl methacrylate) (PMMA), polyvinylpyrrolidone (PVP), silicone rubber, and polyvinyl alcohol (PVA) polymers are the most used due to their biological compatibility (Fig. 6.4) [37].

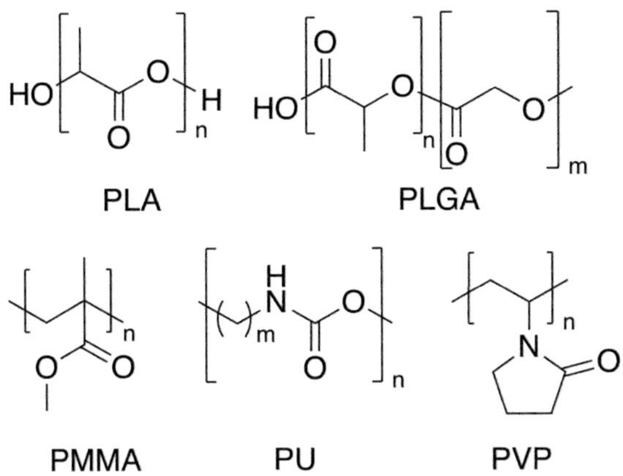

Figure 6.4 Chemical structures of the synthetic polymers most used in nanomedicine. PLA: polylactic acid; PLGA: poly(lactic-co-glycolic acid); PMMA: poly(methyl methacrylate); PU: polyurethane; PVP: polyvinylpyrrolidone. *n* and *m* indicate the degree of polymerization. Created with ChemDraw 21.0.0.28.

PLA NPs can be synthesized by a wide range of processes that produce solid colloidal particles ranging from nanometer to micrometer in size. For example, double emulsion and salting-out methods are suitable to produce micro-sized PLA particles, whereas nanoprecipitation and dialysis enable the formation of PLA NPs. Such methods, among many others, allow the encapsulation of bioactive molecules, such as genes, proteins, vaccines, and small molecule drugs [81]. PLA, PGLA, and their copolymers are recognized by the U.S. Food and Drug Administration (FDA) as nontoxic, biocompatible, and completely biodegradable. When used as carriers, due to their high structural integrity, the slow degradation of PLA and PLGA NPs enables the controlled release of drugs. For example, PLA NPs loaded with rifampicin and functionalized with a cationic peptide (poly-L-lysine) have shown to interact strongly with *S. aureus*, under planktonic and biofilm modes of growth [18]. On the other hand, PLGA NPs loaded with 1,3-bis[3,5-bis(trifluoromethyl)phenyl] urea have recently been demonstrated to be a potent antimicrobial system against the cariogenic bacterial strain *Streptococcus mutans*, responsible for dental caries. The drug-loaded PLGA NPs not only significantly inhibited the growth and lactic acid production of planktonic *S. mutans*, but also *S. mutans* biofilms, while presenting low cytotoxicity [82].

The exceptional mechanical properties of PU materials (*e.g.*, elasticity, strength, and resilience) enable their use in the biomedical field, such as in the production of artificial organs, catheter interventions, and polymeric drug capsules. PU nanomaterials do not present inherent antimicrobial activity, yet such properties can be rendered by their functionalization with quaternary ammonium salts or with metal-based NPs [83–85]. For example, antimicrobial PU catheters doped with AgNPs have shown good antimicrobial activity against *E. coli* and *S. aureus*, eradicating up to 80% of the bacterial population when challenged in a killing test [86]. Notwithstanding, although PU materials are non-teratogenic and non-allergenic and present low cytotoxicity, they cannot be naturally degraded, which implies environmental pollution [30]. To overcome this drawback, biodegradable PU materials, such as starch and cellulose derivatives of PU, have been developed. In fact, urethane-functionalized starch is stated to improve the properties of native starch such as pore size, volume, and binding sites [37, 87]. Desai et al. developed cross-linked starch polyurethanes as nanocarriers for anti-tuberculosis drugs and showed that streptomycin-loaded and isoniazid-loaded nanocrystals were 42 and 7 times more active against the *Mycobacterium tuberculosis* H37Rv strain than the respective free drugs. A pH-dependent drug-release study showed that the antibiotic-loaded starch-derived nanopolyurethanes have a sustained release profile for isoniazid, rifampicin, and pyrazinamide, whereas burst release was observed for streptomycin, suggesting that the drug is present in the nanosphere's outer layer [87]. Several other hybrid starch-based PU nanocomposites in combination with inorganic NPs have been designed and tested satisfactorily against gram-positive and gram-negative bacteria [88, 89].

PVP is a water-soluble polymer made from the *N*-vinylpyrrolidone monomer. Approved by the FDA, PVP attracts considerable attention because of its excellent chemical and physical properties. In this sense, PVP has been widely studied as a stabilizer of AgNPs since it has an impact on the control of the reduction rate of the Ag ions and the aggregation process of Ag atoms during nanoparticle synthesis. Hence, hybrid nanomaterials based on PVP and AgNPs not only present antimicrobial activity against *E. coli, P. aeruginosa*, and *S. aureus*, but also have shown antifungal activity against *Candida albicans, Candida krusei, Candida tropicalis, Candida glabrata*, and *Aspergillus brasiliensis* [90–92].

Synthetic polymers open the possibility of mimicking the three dimensionality of naturally occurring nanostructures, such as micelles and vesicles, or even creating novel artificial nanostructured scaffolds, such as dendrimers (Fig. 6.5).

Polymeric micelles and vesicles are supramolecular structures that are easily obtained through the self-assembly of amphiphilic block copolymers into a monolayer, in the case of micelles, or a bilayer, in the case of vesicles. Polymeric micelles are formed by a hydrophobic core and a hydrophilic external layer solvated by water. Apart from the most known spherical shape in which micelles can assemble, other geometries such as cylinders and nanoplates can be adopted depending on the structure of the individual amphiphile and its packing configuration when assembled [93]. Independently of the shape, the hydrophobic core facilitates the encapsulation of hydrophobic drugs (*e.g.*, poor water-soluble antibiotics, antimicrobial peptides, and AgNPs), whereas the hydrophilic surface may be functionalized with ligands by both covalent and non-covalent interactions [94–96]. For example, in the field of photodynamic therapy, lipase-sensitive methoxy poly(ethylene glycol)-*block*-poly(ε-caprolactone) (mPEG-PCL) micelles have been developed to encapsulate and deliver the water-insoluble photosensitizer hyprocrellin A (HA) drug to treat *MRSA* infections. These micelles showed better anti-*MRSA* activity in periodontal infections in mice than the HA drug by itself. In addition, light irradiation of the HA-loaded mPEG-PCL micelles enhanced its antibacterial effect by triggering ROS production [97].

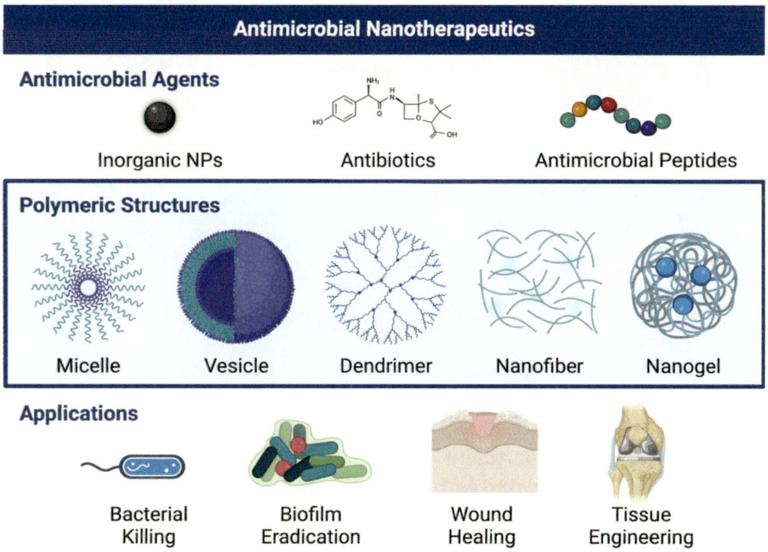

Figure 6.5 Polymeric nanomaterials can take the form of different 3D structures, such as micelles, vesicles, dendrimers, nanofibers, and nanogels. As such, they can incorporate a variety of antimicrobial agents, including metallic and metal oxide NPs, antibiotics, and antimicrobial peptides, finding applications in the eradication of planktonic and biofilm-forming bacteria, wound healing, and tissue engineering. Created with BioRender.com.

Polymeric vesicles or polymersomes are closed hollow nanostructures with a bilayer membrane composed of a hydrophobic portion separating two hydrophilic blocks that face the aqueous core or the outer aqueous phase (resembling a cell membrane). The molecular composition and length of the hydrophilic and hydrophobic blocks determine the vesicle properties, such as size, stability, rigidity, and biodegradability. As for micelles, polymersomes may adopt spherical or nonspherical structures [98] (Fig. 6.6).

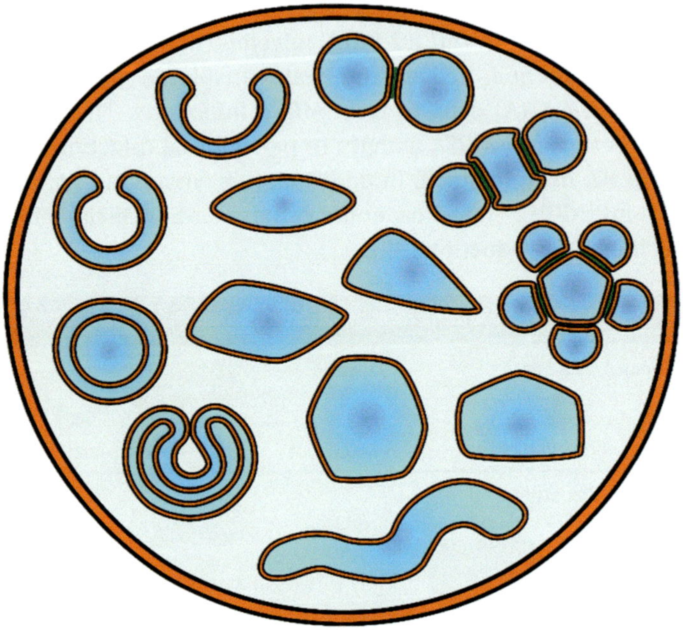

Figure 6.6 Schematic representation of nonspherical polymersomes engulfed within a larger spherical polymersome. The hydrophobic and hydrophilic domains are depicted in yellow and black, respectively. Water is represented in blue. The water compartment of the larger outer spherical polymersome is omitted for clarity. Reprinted with permission from Ref. [98].

As an additional benefit compared to micelles, polymersomes are capable of encapsulating both hydrophilic drugs in the lumen and hydrophobic drugs in the membrane. For example, poly (2-(methacryloyloxy) ethyl-phosphorylcholine)-*co*-poly(2-diisopropylamino)ethyl methacrylate) (PMPC-PDPA) vesicles have been successfully loaded with a large variety of antimicrobial agents, including proteins (*e.g.*, lysostaphin), small peptides (*e.g.*, vancomycin), glycols (*e.g.*, gentamicin), poorly water-soluble organic molecules (*e.g.*, rifampicin), and functionalized pyridines (*e.g.*, isoniazid), to treat intracellular infections caused by *S. aureus*, *Mycobacterium bovis*-attenuated Bacillus Calmette–Guérin (BCG), *Mycobacterium marinum*, and *M. tuberculosis* [99].

Dendrimers are nano-sized, monodisperse, and highly branched macromolecules with a central core surrounded by radially symmetric peripheral groups [100]. The repeated growth reactions during synthesis lead to a higher and tunable degree of branching (*i.e.*, higher surface polyvalency), which enables them to engage in multivalent interactions [100]. Dendrimers have the ability to control and engineer "critical nanoscale design parameters," overcoming limitations associated with other polymeric nanocarriers, such as high polydispersity, poorly controlled shape and/or surface chemistry, and inability to make systematic size modifications below the "critical micelle concentrations" (*i.e.*, 30–40 nm) [101]. The unique structure of dendrimers not only provides special opportunities for host–guest chemistry but also renders them promising scaffolds for drug loading within their nanocavities [94].

Dendrimers can be prepared by two different synthetic pathways: convergent and divergent. In convergent synthesis, dendrons, the branched repeating units, are first synthesized individually (up to the desired degree of branching) and then linked to a multifunctional core molecule, which will be at the center of the dendrimer. Conversely, in divergent synthesis, the dendrimer grows outward from the multifunctional core by the subsequential addition of new repeating units [102]. Many antimicrobial agents, including antimicrobial peptides, silver and metal oxide NPs, and commercial antibiotics have been successfully loaded within dendrimers [94]. For example, poly(amidoamine) (PAMAM) dendrimers loaded with sulfamethoxazole enhanced the aqueous solubility of the drug by a factor of 40, in addition to increasing its antibacterial activity against *E. coli* by a factor of 4 and 8 in dimethyl sulfoxide (DMSO) and 0.01M NaOH solutions, respectively [103]. On the other hand, platensimycin (PTM)-loaded PLGA and PAMAM dendrimers showed an enhanced antimicrobial activity against *MRSA*, as well as improved pharmacokinetics and reduced cytotoxicity when compared to the free antibiotic [104] (Fig. 6.7).

Finally, dendrimers can be, *per se*, effective antimicrobial agents. As for cationic antimicrobial peptides, dendrimers with positively charged surfaces usually have strong interactions with negatively charged bacterial cell membranes [94]. For instance, a cationic dendrimer designed by capping a mannose-functionalized PAMAM dendrimer with a quaternary ammonium compound, 1-hexadecyl-azaniabicyclo[2.2.2]octane (C_{16}-DABCO), showed to inhibit the growth of *Streptococcus oralis*, *S. aureus*, *Bacillus cereus*, *P. aeruginosa*, and *E. coli* at low micromolar MICs. Moreover, the activity of the dendrimer against these microorganisms is 10 times more potent than C_{16}-DABCO alone [105].

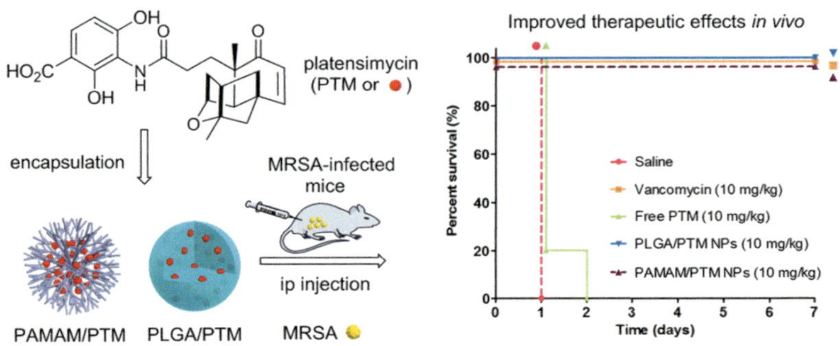

Figure 6.7 Schematic representation of PTM-loaded PLGA and PAMAM dendrimers and percentage of survival of mice infected with *MRSA* 7 days post-infection. Reprinted with permission from Ref. [104]. Copyright 2020 American Chemical Society.

Polymeric nanofibers are one-dimensional nanostructures whose diameter is less than 100 nm. This class of nanomaterials exhibits a high surface-to-weight ratio, which makes them suitable for a wide range of applications in the medical field, such as antimicrobial materials for wound healing [94, 106]. Polymeric nanofibers can be prepared through different techniques, including self-assembly [107], template synthesis [108], thermal-induced phase separation [109], melt-blown [110], and electrospinning [111]. The latter is a superior technique for preparing nanofibers from polymer solutions, with proper control over their chemical composition and diameter, in addition to its simplicity and versatility [112]. The simplest method to incorporate antimicrobial agents into nanofibers is by the solubilization/suspension of the target agents in the electrospinning precursor polymer solution [94]. Different blends of antibiotics (*e.g.*, ampicillin and ciprofloxacin) can be combined with different nanofiber mats to render antimicrobial activity against both gram-positive and gram-negative bacteria [113, 114]. However, one of the main drawbacks of these systems is the burst release of the loaded drugs, which limits their application when sustained mid- to long-term antimicrobial activity is desired. Polymers can also be combined with other nanomaterials, such as metallic NPs or carbon nanotubes, before electrospinning, opening new avenues to combat antimicrobial resistance [115].

Finally, nanogels or hydrogels are cross-linked, water-swellable, biocompatible polymeric networks that can swell or shrink upon external physical or chemical stimuli [116, 117]. Their tunable size, from nanometers to micrometers, their large surface area for multivalent bioconjugation, and their stimuli-responsive swelling and collapsing triggered by external factors (*e.g.*, pH, temperature, enzymes, or ionic strength) provide them unique advantages for on-demand delivery of antimicrobial agents [94, 116]. Nanogels can be synthesized by a myriad of methods depending on the raw materials adopted, yet their synthesis strategies roughly fall into three major groups: monomer polymerization, physical or chemical cross-linking, and template-assisted nanofabrication [118]. The obtained three-dimensional aqueous network structure of nanogels enables the incorporation of small bioactive molecules, as well as macromolecules such as DNA and proteins [119, 120]. However, nanogels loaded with small molecules have seen more significant development compared to their biomacromolecule-loaded counterparts, primarily because the latter exhibit greater instability due to their large molecular weight and complex structure [120–122]. At the target site, drug release occurs through three main mechanisms: (i) gel degradation; (ii) gel expansion and drug diffusion; and (iii) cleavage of drugs covalently linked to the nanogel matrix [123].

In line with this, the encapsulation of conventional antibiotics into nanogels has been widely studied aiming to protect drugs from degradation and enhance their targeted delivery to infected tissues at therapeutic concentrations [19, 124]. For example, several studies have used nanogels to deliver the last-resort antibiotic vancomycin in higher dosages as a tactic to overcome drug resistance [118, 125–127]. Xiong et al. developed a bacterial lipase–sensitive polymeric triple-layered nanogel (TLN) using the amphiphilic diblock copolymer mPEG-PCL to initiate the ring-opening polymerization of the difunctional monomer 3-oxapentane-1,5-diyl bis(ethylene phosphate) (DEGDP). In this manner, hydrophobic PCL segments collapsed and surrounded the cross-linked polyphosphoester core of DEGDP, leading to a compact and hydrophobic matrix that behaves as a molecular fence and prevents nonspecific drug leakage. Therefore, only when the TLN senses lipase-secreting bacteria will the PCL fence degrade to release loaded vancomycin [127]. The same research group reported a strategy to target macrophages infected with *MRSA* using mannosylated nanogels with a polyphosphoester cross-linked core and encapsulated vancomycin. The antibiotic is only released in the presence of bacterial phosphatase and phospholipases, which efficiently degrade the nanogel core (Fig. 6.8) [126].

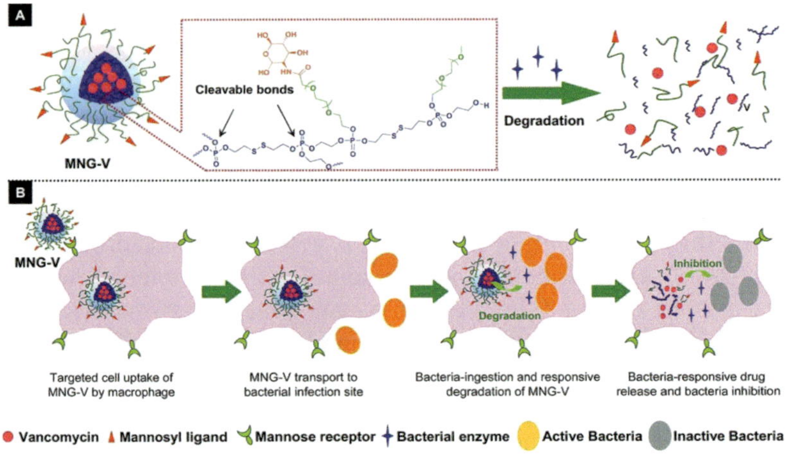

Figure 6.8 **(A)** Schematic representation of vancomycin-loaded mannosylated nanogels (MNG-V) and antibiotic release by the action of bacterial phosphatases and phospholipases. **(B)** Schematic illustration of nanogel-uptake by macrophages, intracellular trafficking, nanogel degradation, vancomycine release, and bacterial killing. Reprinted with permission from Ref. [126].

As for antibiotics, the incorporation of AgNPs in nanogels renders their controlled and sustained release while limiting nonspecific cellular uptake and possible cytotoxicity [118]. For example, bactericidal activity against *S. aureus* and *E. coli* has been achieved by the photoresponsive release of AgNPs from chitosan and aniline nanogels [128, 129] (Fig. 6.9).

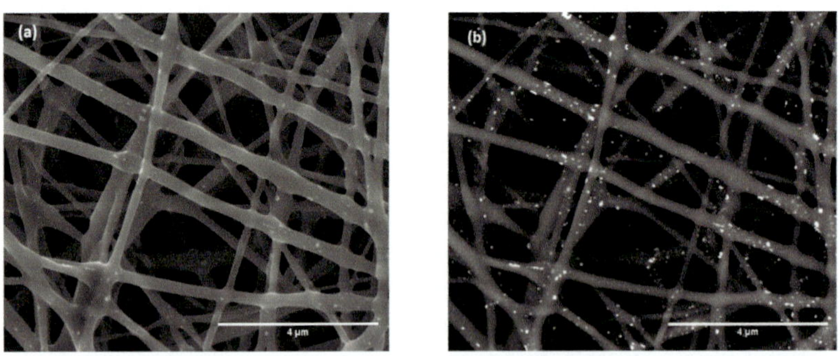

Figure 6.9 Field-emission scanning electron microscopy (FE-SEM) image of PCL nanofiber mats with AgNP-nanogels immobilized onto their surface. **(a)** secondary and **(b)** backscattering electrons show the difference between electronic density of the AgNP-nanogels and the PCL nanofibers. Reprinted with permission from Ref. [129].

Many efforts have also been made in the design of environmentally friendly hydrogel nanocomposites by using biodegradable synthetic polymers, such as poly(*N*-isopropylarclamide) (PNIPAAm), or naturally occurring polymers, such as polysaccharides [130–134]. For instance, agarose-based hydrogels containing tannic acid-Fe(III) NPs presented excellent photothermal conversion capability, being able to kill 99% of bacteria within 10 min of near-infrared radiation (NIR). The bactericidal effect was evaluated *in vitro* and *in vivo* using mice with *S. aureus*–infected wounds [132].

Alike other nanostructures, hydrogel applications go beyond antibiotic delivery. Hydrogels with intrinsic antimicrobial moieties, such as surface-functionalized with quaternary ammonium compounds or cationic guanidine groups, have been developed to physically damage the bacterial cell membrane (*i.e.*, generate pores and alter permeability) and, ultimately, cause bacterial cell death [118, 135–139].

6.2.3 Polymer–Antibiotic Conjugates

As illustrated by the many examples earlier, the physicochemical interactions between an antibiotic and its carrier offer a means to entrap these bioactive agents in polymeric nanomaterials. However, this approach has two main drawbacks: the loading capacity, often limited to ≤10% weight, and the occurrence of burst release, in which more than 50% of the drug can be released within 24 h [140].

On the other hand, the covalent ligation of small bioactive molecules to polymeric carriers (*i.e.*, drug–polymer conjugates) offers unique advantages, such as controlled drug release, active intracellular delivery, prolonged drug circulation in the bloodstream, reduced immunogenicity, enhanced stability, and the potential for targeted delivery through the incorporation of targeting moieties. In addition, the kinetics of drug release can be tailored based on the bonds that link the antibiotic to the polymer (*e.g.*, ester, amide, urethane), the type of formulation (*e.g.*, powder, hydrogel, coating, microsphere), and the polymer chemical composition. Beyond linker chemistry, numerous other physicochemical factors, including steric effects, polymer molecular weight, and nanocarrier composition, play a significant role in drug cleavage and release rates [140, 141].

A thorough assessment and careful selection of all system design parameters should be conducted to allow a proper spatiotemporal control over the drug release.

Norfloxacin (NOR) covalently linked to dextran with one of two tetrapeptide linkers (*i.e.*, Gly-Phe-Ala-Leu-NOR or Gly-Phe-Leu-Gly-NOR) is an example of an early study in which the bioactive agent was linked through an amide bond susceptible to lysosomal conditions. The use of cathepsin B at pH 5.5 during 7 h, however, did not promote the release of free drug in any of the two systems, but of the dimers Leu-NOR (80%) and Gly-NOR (25%) [142]. When the linker was replaced by a methoxy-terminated Gly-Phe-Gly-Gly linker, in which NOR was covalently bound to the α-carbon of Gly at the C-terminus, and mannosylated dextran polymer was used, NOR was released as a free drug—27% release at pH 7.4, 40% release at pH 5.5, and ≈65% release at pH 5.5 in the presence of cathepsin B, in a 24 h period [143].

In a different study, polymeric nanobiotics composed of an isoniazid (INH)-conjugated α-keto polyester and encapsulated clofazimine (CFZ) showed to efficiently reduce granuloma and bacterial burden in a zebrafish model of mycobacterial infection, outperforming the combination therapy with the free drugs at the same concentration [144] (Fig. 6.10).

Drug-conjugated hybrid nanomaterials have also been developed for drug delivery applications. For instance, azithromycin (AZM)-conjugated clustered NPs (AZM-DA NPs) have been developed as therapeutic agents for the treatment of *P. aeruginosa*–induced chronic lung infection. AZM-DA NPs are prepared by the electrostatic complexation between AZM-conjugated poly(amidoamine) dendrimer (PAMAM-AZM) and 2,3-dimethyl maleic anhydride (DA)-modified poly(ethylene glycol)-*block*-polylysine (PEG-*b*-PLys). In an acidic biofilm microenvironment (pH 6.0), AZM-DA NPs disassemble and promote the release of the secondary cationic PAMAM-AZM NPs. The small size and positive charge of PAMAM-AZM NPs enhanced the permeabilization of both inner and outer bacterial membranes and, consequently, increased the internalization of AZM as compared to the free AZM (Fig. 6.11) [145].

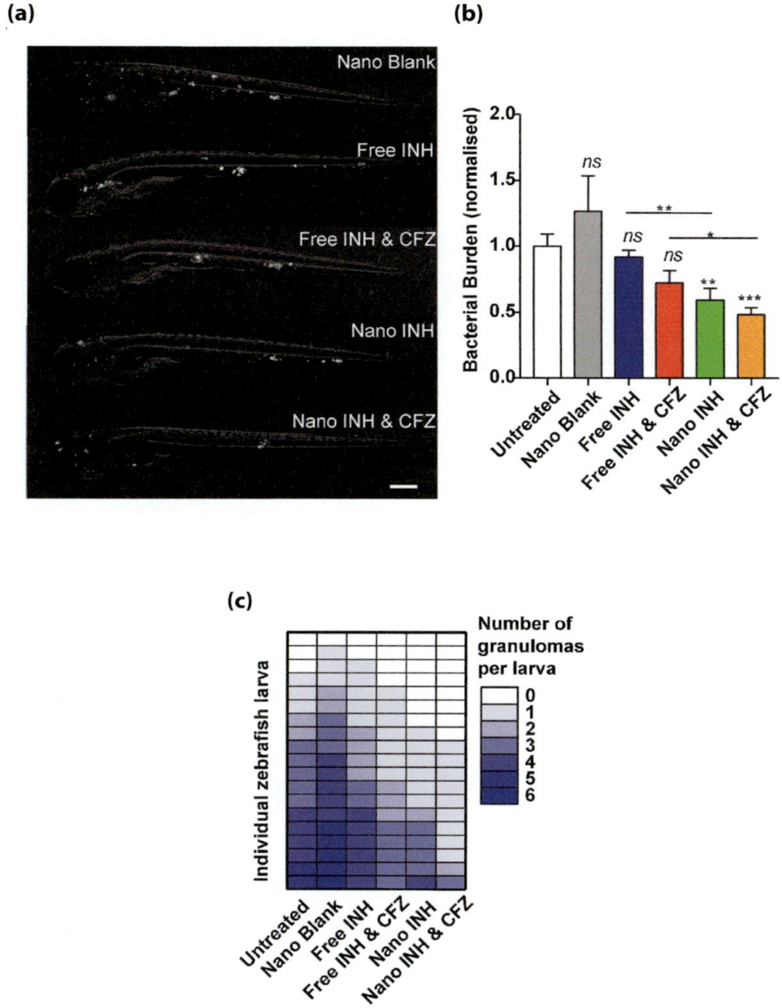

Figure 6.10 Effect of nanobiotics at 3 days post-infection (dpi) on zebrafish infected with fluorescently labeled *M. marinum*. **(a)** Representative images (scale bar, 200 μm). **(b)** Quantification of bacterial load (results plotted as mean ± SEM from two independent experiments; *n* = 21). **(c)** Quantification of granuloma number at 3 dpi. Results are plotted as mean ± SEM from two independent experiments (*n* = 19). Reprinted with permission from Ref. [144].

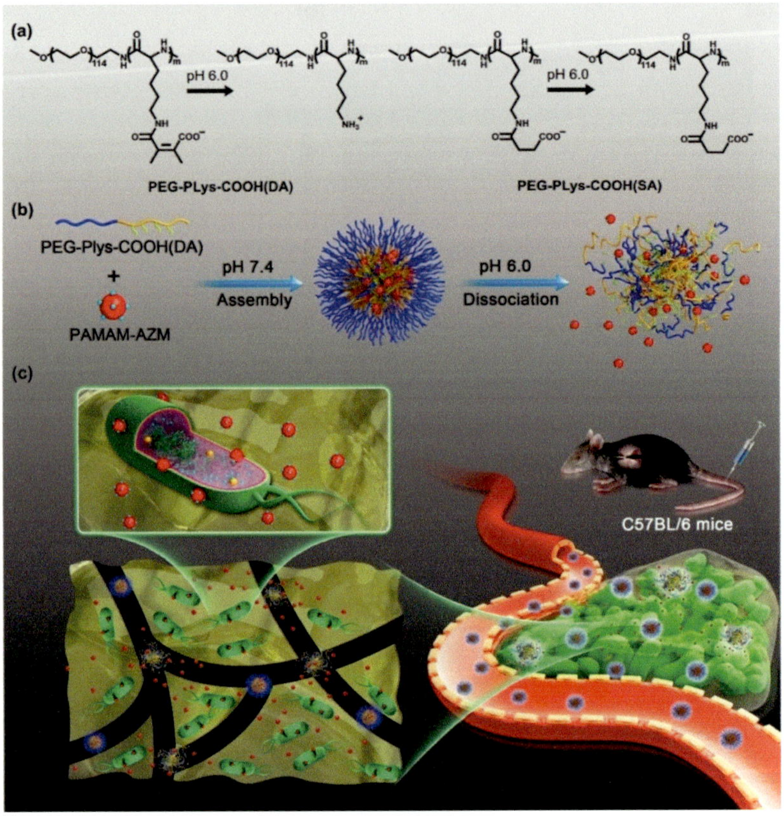

Figure 6.11 **(a)** Change in chemical structure of 2,3-dimethyl maleic anhydride (DA) and succinic acid (SA) (negative control) modified poly(ethylene glycol)-block-polylysine (PEG-b-PLys) in acidic pH. **(b)** Illustration of the self-assembly of azithromycin-conjugated clustered (AZM-DA) NPs at pH 7.4 and release of secondary AZM-conjugated poly(amidoamine) dendrimer (PAMAM-AZM) NPs in an acidic biofilm microenvironment. **(c)** Illustration of the accumulation of AZM-DA NPs in biofilms and subsequent release of PAMAM-AZM NPs for enhanced biofilm penetration, permeabilization of the bacterial membrane, and increased AZM internalization. Reprinted with permission from Ref. [145]. Copyright 2020 American Chemical Society.

6.3 Liposomes and Lipid-based Nanoparticles

Lipids are a wide group of small biomolecules that can be either hydrophobic or amphiphilic and play a vital role in physiological and pathophysiological events of living systems [146, 147]. As for amphiphilic block copolymers, amphiphilic lipids, such as phospholipids, spontaneously self-assemble into ordered lyotropic liquid-crystalline phases (*e.g.*, monolayers, lipid bilayers, micelles, liposomes, and tubules) in the presence of water [148]. Liposomes are spherical hollow nanostructures composed by at least one phospholipid bilayer with an inner aqueous compartment. Like polymersomes, liposomes can deliver both hydrophilic and hydrophobic drugs. Liposomes building blocks are charged or neutral lipids that can be natural, such as L-α-phosphatidylcholine (PC) and 1,2-dipalmitoyl-*sn*-glycero-3-phosphocholine (DPPC), or synthetic, such as 1,2-dioleoyl-*sn*-glycero-3-phospho-L-serine (DOPS) and 1,2-dioleoyl-3-trimethylammonium-propane (DOTAP) [149]. With recognized biocompatibility and biodegradability, liposomes have been approved for clinical use for almost 30 years now [150].

In a study by Sande et al., the anti-staphylococcal activity of vancomycin-loaded liposomes has been investigated through two different liposome formulations: (i) dicethylphosphate (DCP) liposomes, composed of 1,2-distearoyl-*sn*-glycero-3-phosphocholine (DSPC), DCP, and cholesterol; and (ii) dimyristoylphoshatidylglycerol (DMPG) liposomes, composed of DSPC, DMPG, and cholesterol. The DCP and DMPG liposomal vancomycin formulations showed a two-fold to four-fold decrease in MICs compared to the free drug, when tested against the hospital-associated *MRSA* strain NRS-35 and the community-associated *MRSA* strain LAC. The DPC-based liposome formulation was the most efficacious, showing bacteria-killing efficacy against seven *MRSA* strains. The authors suggested that the increased bacteria-killing efficacy of the liposomes might be related to the fusion of liposomal components with the bacterial cell wall, enabling an increased amount of antibiotic to reach the cytoplasm [151].

Mycobacterium abscessus nontuberculous infections have risen considerable concern since this human pathogen, responsible for a wide spectrum of soft tissue infections, is naturally resistant to many common antibiotics [152–153]. Rifampicin is a broad-spectrum antibiotic and a first line drug used for the treatment of drug-susceptible tuberculosis, with proven antibacterial activity against intracaseum bacilli [154]. However, rifampicin is not suitable for the clinical management of *M. abscessus* lung disease, since this bacterium possesses monooxygenase enzymes capable of inactivating this antibiotic [155, 156]. On the other hand, liposomal formulations composed of hydrogenated phosphatidylcholine (HSPC) and 1,2-dipalmitoyl-sn-glycero-3-phosphorylglycerol (DPPG) loaded with rifampicin significantly increased drug intramacrophage bioavailability and reduced intracellular mycobacterial viability, when compared to the free drug [154]. In a recent review article, Gosh and De presented a list of FDA-approved liposome-based antibiotic delivery systems [157].

Other types of lipid-based antimicrobial delivery systems, such as lipidic micelles [158], vitamin-lipid NPs [159], solid lipid NPs (SLNs) [160], nanostructured lipid carriers (NLCs) [161], and lipid-coated hybrid (LCH) NPs [162], have been developed for the eradication of a wide number of bacterial infections, including the ones caused by *Helicobacter pylori*, *MRSA*, *E. coli*, and *P. aeruginosa*, among others.

6.4 Concluding Remarks

The rapid emergence of antibiotic-resistant bacteria poses one of the greatest threats to public health as the effectiveness of conventional therapies and commercial antibiotics has dropped. In the race for the discovery of new strategies to prevent a scenario in which commonplace infections prove fatal, nanomaterials stand in the limelight due to their unique physicochemical properties that can be seized to overcome common resistance mechanisms.

NP-driven drug delivery has emerged as a beacon of hope, shielding antibiotics from enzymatic degradation, enhancing their targeted delivery to afflicted sites in therapeutically potent concentrations, and minimizing undesired side effects. Drugs can either be entrapped or chemically conjugated to NPs, with the latter offering a myriad of possibilities in orchestrating spatiotemporal controlled release of the therapeutic payload. Meanwhile, nanomaterials can also display intrinsic antimicrobial properties, either by direct disruption of bacterial cell membranes (*e.g.*, nanoparticles functionalized with cationic groups) or by instigating the generation of ROS (*e.g.*, metallic nanoparticles).

The clinical implementation of nanotherapeutics still faces considerable challenges, mainly related to their complex chemistry and polydispersity, which pose difficulties related to cost-effectiveness, scale-up, and Chemistry, Manufacturing, and Controls (CMC) management. Still, the development of computational approaches that allow a better understanding of nano-bio interactions and predictive biodistribution, pharmacokinetics, and toxicology, along with a harmonized international regulatory framework, is expected to facilitate clinical translation in the near future.

Funding

Íris Luz Batalha would like to acknowledge the support of a fellowship from the La Caixa Foundation (ID 100010434) and the European Union's Horizon 2020 research and innovation program under the Marie Skłodowska-Curie grant agreement No 847648. The fellowship code is LCF/BQ/PI21/11830004. Catia D. F. Lopes thanks funding from the European Union's Horizon Europe research and innovation program under the Marie Skłodowska-Curie grant agreement No 101066836. José Muñoz-López gratefully acknowledges the support of the Ministerio de Ciencia, Innovación y Universidades of Spain through the fellowship PRE2020-092717.

References

1. World Health Organization, *Antibiotic Resistance*, https://www.who.int/news-room/fact-sheets/detail/antibiotic-resistance (accessed in October 2023).
2. W. C. Reygaert and W. C. Reygaert, An Overview Of The Antimicrobial Resistance Mechanisms Of Bacteria, *AIMS Microbiol.* **4**, 482 (2018).
3. J. M. V. Makabenta, A. Nabawy, C.-H. Li, S. Schmidt-Malan, R. Patel, and V. M. Rotello, Nanomaterial-based therapeutics for antibiotic-resistant bacterial infections, *Nat. Rev. Microbiol.* **19**, 1 (2021).
4. A. Gupta, R. F. Landis, and V. M. Rotello, Nanoparticle-based antimicrobials: Surface functionality is critical, *F1000Research* **5**, 364 (2016).
5. L.-S. Wang, A. Gupta, and V. M. Rotello, Nanomaterials for the treatment of bacterial biofilms, *ACS Infect. Dis.* **2**, 3 (2016).

6. J. J. Aguilera-Correa, J. Esteban, and M. Vallet-Regí, Inorganic and polymeric nanoparticles for human viral and bacterial infections prevention and treatment, *Nanomaterials* **11**, 1 (2021).
7. E. Sánchez-López et al., Metal-based nanoparticles as antimicrobial agents: An overview, *Nanomaterials* **10**, 292 (2020).
8. L. Milkovic, A. Cipak Gasparovic, M. Cindric, P.-A. Mouthuy, and N. Zarkovic, Short overview of ROS as cell function regulators and their implications in therapy concepts, *Cells* **8**, 793 (2019).
9. E. Pedone, G. Fiorentino, S. Bartolucci, and D. Limauro, Enzymatic antioxidant signatures in hyperthermophilic archaea, *Antioxidants* **9**, 703 (2020).
10. P. Yuan, X. Ding, Y. Y. Yang, and Q.-H. Xu, Metal nanoparticles for diagnosis and therapy of bacterial infection, *Adv. Healthc. Mater.* **7**, 1701392 (2018).
11. S. Temizel-Sekeryan and A. L. Hicks, Global environmental impacts of silver nanoparticle production methods supported by life cycle assessment, *Resour. Conserv. Recycl.* **156**, 104676 (2020).
12. B. Schäfer, J. Tentschert, and A. Luch, Nanosilver in consumer products and human health: More information required! *Environ. Sci. Technol.* **45**, 7589 (2011).
13. Z. Li, J. Yang, and X. J. Loh, Polyhydroxyalkanoates: Opening doors for a sustainable future, *NPG Asia Mater.* **8**, 4 (2016).
14. Z. Zhang, O. Ortiz, R. Goyal, and J. Kohn, Biodegradable polymers, in *Principles of Tissue Engineering* (Fourth Edition), edited by R. Lanza, R. Langer, and J. Vacanti (Academic Press, Boston, 2014), pp. 441–473.
15. V. Kundrat, N. Cernekova, A. Kovalcik, V. Enev, and I. Marova, Drug release kinetics of electrospun PHB meshes, *Materials* **12**, 12 (2019).
16. Z.-C. Xing, W.-P. Chae, J.-Y. Baek, M.-J. Choi, Y. Jung, and I.-K. Kang, *In vitro* assessment of antibacterial activity and cytocompatibility of silver-containing PHBV nanofibrous scaffolds for tissue engineering, *Biomacromolecules* **11**, 1248 (2010).
17. J. L. Castro-Mayorga, A. Martínez-Abad, M. J. Fabra, C. Olivera, M. Reis, and J. M. Lagarón, Stabilization of antimicrobial silver nanoparticles by a polyhydroxyalkanoate obtained from mixed bacterial culture, *Int. J. Biol. Macromol.* **71**, 103 (2014).
18. D. Da Costa, C. Exbrayat-Héritier, B. Rambaud, S. Megy, R. Terreux, B. Verrier, and C. Primard, Surface charge modulation of rifampicin-loaded PLA nanoparticles to improve antibiotic delivery in *Staphylococcus aureus* biofilms, *J. Nanobiotechnol.* **19**, 12 (2021).

19. K. Zhou, X. Wang, D. Chen, Y. Yuan, S. Wang, C. Li, Y. Yan, Q. Liu, L. Shao, L. Huang, Z. Yuan, and S. Xie ., Enhanced treatment effects of tilmicosin against *Staphylococcus aureus* cow mastitis by self-assembly sodium alginate-chitosan nanogel, *Pharmaceutics* **11**, 10 (2019).

20. P. Courvalin, Vancomycin resistance in gram-positive cocci, *Clin. Infect. Dis.* **42**, S25 (2006).

21. T. Kruk, K. Szczepanowicz, J. Stefańska, R. P. Socha, and P. Warszyński, Synthesis and antimicrobial activity of monodisperse copper nanoparticles, *Colloids Surf. B Biointerfaces* **128**, 17 (2015).

22. B. Das and S. Patra, Antimicrobials: Meeting the challenges of antibiotic resistance through nanotechnology, in *Nanostructures for Antimicrobial Therapy*, edited by A. Ficai and A. M. Grumezescu (Elsevier, 2017), pp. 1–22.

23. M. S. Usman, M. E. E. Zowalaty, K. Shameli, N. Zainuddin, M. Salama, and N. A. Ibrahim, Synthesis, characterization, and antimicrobial properties of copper nanoparticles, *Int. J. Nanomed.* **8**, 4467 (2013).

24. D. Longano, N. Ditaranto, L. Sabbatini, L. Torsi, and N. Cioffi, Synthesis and antimicrobial activity of copper nanomaterials, in *Nano-Antimicrobials: Progress and Prospects*, edited by N. Cioffi and M. Rai (Springer, Berlin, Heidelberg, 2012), pp. 85–117.

25. F. Nishino, M. Jeem, L. Zhang, K. Okamoto, S. Okabe, and S. Watanabe, Formation of CuO nano-flowered surfaces via submerged photosynthesis of crystallites and their antimicrobial activity, *Sci. Rep.* **7**, 1 (2017).

26. U. Bogdanović, V. Vodnik, M. Mitrić, S. Dimitrijević, S. D. Škapin, V. Žunič, M. Budimir, and M. Stoiljković, Nanomaterial with high antimicrobial efficacy: Copper/polyaniline nanocomposite, *ACS Appl. Mater. Interfaces* **7**, 1955 (2015).

27. Y.-H. Hsueh, P.-H. Tsai, and K.-S. Lin, pH-dependent antimicrobial properties of copper oxide nanoparticles in *Staphylococcus aureus*, *Int. J. Mol. Sci.* **18**, 4 (2017).

28. U. Halder, R. K. Roy, R. Biswas, D. Khan, K. Mazumder, and R. Bandopadhyay, Synthesis of copper oxide nanoparticles using capsular polymeric substances produced by *Bacillus altitudinis* and investigation of its efficacy to kill pathogenic *Pseudomonas aeruginosa*, *Chem. Eng. J. Adv.* **11**, 100294 (2022).

29. J. Balcucho, D. M. Narváez, and J. L. Castro-Mayorga, Antimicrobial and biocompatible polycaprolactone and copper oxide nanoparticle

wound dressings against methicillin-resistant *Staphylococcus aureus*, *Nanomaterials* **10**, 9 (2020).

30. F. A. Bezza, S. M. Tichapondwa, and E. M. N. Chirwa, Fabrication of monodispersed copper oxide nanoparticles with potential application as antimicrobial agents, *Sci. Rep.* **10**, 1 (2020).
31. M. Okkeh, N. Bloise, E. Restivo, L. De Vita, P. Pallavicini, and L. Visai, Gold nanoparticles: Can they be the next magic bullet for multidrug-resistant bacteria? *Nanomaterials* **11**, 2 (2021).
32. S. Kalita, R. Kandimalla, A. C. Bhowal, J. Kotoky, and S. Kundu, Functionalization of β-lactam antibiotic on lysozyme capped gold nanoclusters retrogress MRSA and its persisters following awakening, *Sci. Rep.* **8**, 1 (2018).
33. G. A. Marcelo, M. P. Duarte, and E. Oliveira, Gold@mesoporous silica nanocarriers for the effective delivery of antibiotics and by-passing of β-lactam resistance, *SN Appl. Sci.* **2**, 1354 (2020).
34. S. V. Gudkov, D. E. Burmistrov, D. A. Serov, M. B. Rebezov, A. A. Semenova, and A. B. Lisitsyn, A mini review of antibacterial properties of ZnO nanoparticles, *Front. Phys.* **9**, (2021).
35. M. Divya, B. Vaseeharan, M. Abinaya, S. Vijayakumar, M. Govindarajan, N. S. Alharbi, S. Kadaikunnan, J. M. Khaled, and G. Benelli, Biopolymer gelatin-coated zinc oxide nanoparticles showed high antibacterial, antibiofilm and anti-angiogenic activity, *J. Photochem. Photobiol. B* **178**, 211 (2018).
36. K. Elumalai and S. Velmurugan, Green synthesis, characterization and antimicrobial activities of zinc oxide nanoparticles from the leaf extract of *Azadirachta indica* (L.), *Appl. Surf. Sci.* **345**, 329 (2015).
37. J. Han, D. Zhao, D. Li, X. Wang, Z. Jin, and K. Zhao, Polymer-based nanomaterials and applications for vaccines and drugs, *Polymers* **10**, 31 (2018).
38. H. S. Azevedo, T. C. Santos, and R. L. Reis, Controlling the degradation of natural polymers for biomedical applications, in *Natural-based Polymers for Biomedical Applications*, edited by R. L. Reis, N. M. Neves, J. F. Mano, M. E. Gomes, A. P. Marques, and H. S. Azevedo (Woodhead Publishing, 2008), pp. 106–128.
39. G. Chen, I. Roy, C. Yang, and P. N. Prasad, Nanochemistry and nanomedicine for nanoparticle-based diagnostics and therapy, *Chem. Rev.* **116**, 2826 (2016).
40. A. Qiu, Y. Wang, G. Zhang, and H. Wang, Natural polysaccharide-based nanodrug delivery systems for treatment of diabetes, *Polymers* **14**, 15 (2022).

41. A. Plucinski, Z. Lyu, and B. V. K. J. Schmidt, Polysaccharide nanoparticles: From fabrication to applications, *J. Mater. Chem. B* **9**, 7030 (2021).

42. S. Khattak, F. Wahid, L.-P. Liu, S.-R. Jia, L.-Q. Chu, Y.-Y. Xie, Z.-X. Li, and C. Zhong, Applications of cellulose and chitin/chitosan derivatives and composites as antibacterial materials: Current state and perspectives, *Appl. Microbiol. Biotechnol.* **103**, 1989 (2019).

43. M. G. Arafa, H. A. Mousa, and N. N. Afifi, Preparation of PLGA-chitosan based nanocarriers for enhancing antibacterial effect of ciprofloxacin in root canal infection, *Drug Deliv.* **27**, 26 (2019).

44. N. A. Kumar, N. S. Rejinold, P. Anjali, A. Balakrishnan, R. Biswas, and R. Jayakumar, Preparation of chitin nanogels containing nickel nanoparticles, *Carbohydr. Polym.* **97**, 469 (2013).

45. A. A. Oun and J.-W. Rhim, Effect of isolation methods of chitin nanocrystals on the properties of chitin-silver hybrid nanoparticles, *Carbohydr. Polym.* **197**, 349 (2018).

46. A. A. Oun and J.-W. Rhim, Preparation of multifunctional chitin nanowhiskers/ZnO-Ag NPs and their effect on the properties of carboxymethyl cellulose-based nanocomposite film, *Carbohydr. Polym.* **169**, 467 (2017).

47. D. Solairaj, P. Rameshthangam, P. Muthukumaran, and J. Wilson, Studies on electrochemical glucose sensing, antimicrobial activity and cytotoxicity of fabricated copper nanoparticle immobilized chitin nanostructure, *Int. J. Biol. Macromol.* **101**, 668 (2017).

48. A. M. Youssef, M. S. Abdel-Aziz, and S. M. El-Sayed, Chitosan nanocomposite films based on Ag-NP and Au-NP biosynthesis by bacillus subtilis as packaging materials, *Int. J. Biol. Macromol.* **69**, 185 (2014).

49. A. M. Youssef, H. Abou-Yousef, S. M. El-Sayed, and S. Kamel, Mechanical and antibacterial properties of novel high performance chitosan/nanocomposite films, *Int. J. Biol. Macromol.* **76**, 25 (2015).

50. M. R. P, K. Muraleedaran, and V. M. A. Mujeeb, Applications of chitosan powder with *in situ* synthesized nano ZnO particles as an antimicrobial agent, *Int. J. Biol. Macromol.* **77**, 266 (2015).

51. B. Li, Y. Zhang, Y. Yang, W. Qiu, X. Wang, B. Liu, Y. Wang, and G. Sun, Synthesis, characterization, and antibacterial activity of chitosan/TiO_2 nanocomposite against *Xanthomonas oryzae* Pv. Oryzae, *Carbohydr. Polym.* **152**, 825 (2016).

52. A. D. Baldwin and K. L. Kiick, Polysaccharide-modified synthetic polymeric biomaterials, *Biopolymers* **94**, 128 (2010).

53. G. Baier, A. Cavallaro, K. Vasilev, V. Mailänder, A. Musyanovych, and K. Landfester, Enzyme responsive hyaluronic acid nanocapsules containing polyhexanide and their exposure to bacteria to prevent infection, *Biomacromolecules* **14**, 1103 (2013).

54. I. S. Bayer, Hyaluronic acid and controlled release: A review, *Molecules* **25**, 2649 (2020).

55. A. Mero and M. Campisi, Hyaluronic acid bioconjugates for the delivery of bioactive molecules, *Polymers* **6**, 2 (2014).

56. D. R. Vogus, M. A. Evans, A. Pusuluri, A. Barajas, M. Zhang, V. Krishnan, M. Nowak, S. Menegatti, M. E. Helgeson, T. M. Squires, and S. Mitragotri, A hyaluronic acid conjugate engineered to synergistically and sequentially deliver gemcitabine and doxorubicin to treat triple negative breast cancer, *J. Controlled Release* **267**, 191 (2017).

57. T. A. Debele, L.-Y. Yu, C.-S. Yang, Y.-A. Shen, and C.-L. Lo, pH- and GSH-sensitive hyaluronic acid-MP conjugate micelles for intracellular delivery of doxorubicin to colon cancer cells and cancer stem cells, *Biomacromolecules* **19**, 3725 (2018).

58. E. Yoo, A. C. D. Salyer, M. J. H. Brush, Y. Li, K. L. Trautman, N. M. Shukla, A. De Beuckelaer, S. Lienenklaus, K. Deswarte, B. N. Lambrecht, B. G. De Geest, and S. A. David, Hyaluronic acid conjugates of TLR7/8 agonists for targeted delivery to secondary lymphoid tissue, *Bioconjug. Chem.* **29**, 2741 (2018).

59. Y. Luo, X. Wang, D. Du, and Y. Lin, Hyaluronic acid-conjugated apoferritin nanocages for lung cancer targeted drug delivery, *Biomater. Sci.* **3**, 1386 (2015).

60. F. Dosio, S. Arpicco, B. Stella, and E. Fattal, Hyaluronic acid for anticancer drug and nucleic acid delivery, *Adv. Drug Deliv. Rev.* **97**, 204 (2016).

61. R. A. Shapi'i, S. H. Othman, N. Nordin, R. Kadir Basha, and M. Nazli Naim, Antimicrobial properties of starch films incorporated with chitosan nanoparticles: *In vitro* and *in vivo* evaluation, *Carbohydr. Polym.* **230**, 115602 (2020).

62. M. Babaee, F. Garavand, A. Rehman, S. Jafarazadeh, E. Amini, and I. Cacciotti, Biodegradability, physical, mechanical and antimicrobial attributes of starch nanocomposites containing chitosan nanoparticles, *Int. J. Biol. Macromol.* **195**, 49 (2022).

63. H. Tian, J. Yan, A. V. Rajulu, A. Xiang, and X. Luo, Fabrication and properties of polyvinyl alcohol/starch blend films: Effect of composition and humidity, *Int. J. Biol. Macromol.* **96**, 518 (2017).

64. R. Yoksan and S. Chirachanchai, Silver nanoparticle-loaded chitosan–starch based films: Fabrication and evaluation of tensile, barrier and antimicrobial properties, *Mater. Sci. Eng. C* **30**, 891 (2010).

65. Z. Abdollahi, E. N. Zare, F. Salimi, I. Goudarzi, F. R. Tay, and P. Makvandi, Bioactive carboxymethyl starch-based hydrogels decorated with CuO nanoparticles: Antioxidant and antimicrobial properties and accelerated wound healing *in vivo*, *Int. J. Mol. Sci.* **22**, 5 (2021).

66. M. E. Villanueva, A. M. del R. Diez, J. A. González, C. J. Pérez, M. Orrego, L. Piehl, S. Teves, and G. J. Copello, Antimicrobial activity of starch hydrogel incorporated with copper nanoparticles, *ACS Appl. Mater. Interfaces* **8**, 16280 (2016).

67. N. Zainuddin, I. Ahmad, H. Kargarzadeh, and S. Ramli, Hydrophobic kenaf nanocrystalline cellulose for the binding of curcumin, *Carbohydr. Polym.* **163**, 261 (2017).

68. W. Y. Tong, A. Y. K. bin Abdullah, N. A. S. Binti Rozman, M. I. A. bin Wahid, Md. S. Hossain, L. C. Ring, Y. Lazim, and W.-N. Tan, Antimicrobial wound dressing film utilizing cellulose nanocrystal as drug delivery system for curcumin, *Cellulose* **25**, 631 (2018).

69. S. Das, B. Ghosh, and K. Sarkar, Nanocellulose as sustainable biomaterials for drug delivery, *Sens. Int.* **3**, 100135 (2022).

70. O. A. Madkhali, S. S. Moni, M. H. Sultan, H. A. Bukhary, M. Ghazwani, N. A. Alhakamy, A. M. Meraya, S. Alshahrani, S. S. Alqahtani, M. Ali Bakkari, M. I. Alam, and M. E. Elmobark, Formulation and evaluation of injectable dextran sulfate sodium nanoparticles as a potent antibacterial agent, *Sci. Rep.* **11**, 1 (2021).

71. N. Watcharadulyarat, M. Rattanatayarom, N. Ruangsawasdi, N. Hongdilokkul, and N. Patikarnmonthon, Dextran-based nanoparticles for encapsulation of ciprofloxacin, *J. Phys. Conf. Ser.* **2175**, 012006 (2022).

72. P. Prakash, W.-H. Lee, C.-Y. Loo, H. S. J. Wong, and T. Parumasivam, Advances in polyhydroxyalkanoate nanocarriers for effective drug delivery: An overview and challenges, *Nanomaterials* **12**, 1 (2022).

73. Y. B. Kim and R. W. Lenz, Polyesters from microorganisms, in *Biopolyesters*, edited by W. Babel and A. Steinbüchel (Springer, Berlin, Heidelberg, 2001), pp. 51–79.

74. B. H. A. Rehm, Bacterial polymers: Biosynthesis, modifications and applications, *Nat. Rev. Microbiol.* **8**, 578 (2010).

75. J. Chesterman, Z. Zhang, O. Ortiz, R. Goyal, and J. Kohn, Biodegradable polymers, in *Principles of Tissue Engineering* (Fifth Edition), edited by

R. Lanza, R. Langer, J. P. Vacanti, and A. Atala (Academic Press, 2020), pp. 317–342.

76. V. S. Giourieva, R. M. Papi, and A. A. Pantazaki, Polyhydroxyalkanoates applications in antimicrobial agents delivery and wound healing, in *Biotechnological Applications of Polyhydroxyalkanoates*, edited by V. C. Kalia (Springer, Singapore, 2019), pp. 49–76.

77. A. P. Desbois and V. J. Smith, Antibacterial free fatty acids: Activities, mechanisms of action and biotechnological potential, *Appl. Microbiol. Biotechnol.* **85**, 1629 (2010).

78. C. B. Huang, Y. Alimova, T. M. Myers, and J. L. Ebersole, Short- and medium-chain fatty acids exhibit antimicrobial activity for oral microorganisms, *Arch. Oral Biol.* **56**, 650 (2011).

79. N. Naveen, R. Kumar, S. Balaji, T. S. Uma, T. S. Natrajan, and P. K. Sehgal, Synthesis of nonwoven nanofibers by electrospinning: A promising biomaterial for tissue engineering and drug delivery, *Adv. Eng. Mater.* **12**, B380 (2010).

80. G. R. de Almeida Neto, M. V. Barcelos, M. E. A. Ribeiro, M. M. Folly, and R. J. S. Rodríguez, Formulation and characterization of a novel PHBV nanocomposite for bone defect filling and infection treatment, *Mater. Sci. Eng. C Mater. Biol. Appl.* **104**, 110004 (2019).

81. B. K. Lee, Y. Yun, and K. Park, PLA micro- and nano-particles, *Adv. Drug Deliv. Rev.* **107**, 176 (2016).

82. M. Zhang, Y. Liao, X. Tong, and F. Yan, Novel urea derivative-loaded PLGA nanoparticles to inhibit caries-associated *Streptococcus mutans*, *RSC Adv.* **12**, 4072 (2022).

83. Y. Savelyev, A. Gonchar, B. Movchan, A. Gornostay, S. Vozianov, A. Rudenko, R. Rozhnova, and T. Travinskaya, Antibacterial polyurethane materials with silver and copper nanoparticles, *Mater. Today Proc.* **4**, 87 (2017).

84. J.-T. Sun, J.-W. Li, C.-H. Tsou, J.-C. Pang, R.-J. Chung, and C.-W. Chiu, Polyurethane/nanosilver-doped halloysite nanocomposites: Thermal, mechanical properties, and antibacterial properties, *Polymers* **12**, 11 (2020).

85. C.-H. Wang, G.-G. Hou, Z.-Z. Du, W. Cong, J.-F. Sun, Y.-Y. Xu, and W.-S. Liu, Synthesis, characterization and antibacterial properties of polyurethane material functionalized with quaternary ammonium salt, *Polym. J.* **48**, 3 (2016).

86. V. H. Pino-Ramos, J. C. Audifred-Aguilar, R. Sánchez-Obregón, and E. Bucio, Antimicrobial polyurethane catheters synthesized by grafting-

radiation method doped with silver nanoparticles, *React. Funct. Polym.* **167**, 105006 (2021).

87. S. K. Desai, D. Mondal, and S. Bera, Polyurethane-functionalized starch nanocrystals as anti-tuberculosis drug carrier, *Sci. Rep.* **11**, 8331 (2021).

88. F. Kamali, K. Faghihi, and F. Mirhoseini, High antibacterial activity of new eco-friendly and biocompatible polyurethane nanocomposites based on Fe_3O_4/Ag and starch moieties, *Polym. Eng. Sci.* **62**, 1444 (2022).

89. H. R. Ashjari, M. S. S. Dorraji, V. Fakhrzadeh, H. Eslami, M. H. Rasoulifard, M. Rastgouy-Houjaghan, P. Gholizadeh, and H. S. Kafil, Starch-based polyurethane/CuO nanocomposite foam: Antibacterial effects for infection control, *Int. J. Biol. Macromol.* **111**, 1076 (2018).

90. R. Bryaskova, D. Pencheva, S. Nikolov, and T. Kantardjiev, Synthesis and comparative study on the antimicrobial activity of hybrid materials based on silver nanoparticles (AgNps) stabilized by polyvinylpyrrolidone (PVP), *J. Chem. Biol.* **4**, 185 (2011).

91. N. Sheikh, A. Akhavan, and M. Z. Kassaee, Synthesis of antibacterial silver nanoparticles by γ-irradiation, *Phys. E Low-Dimens. Syst. Nanostruct.* **42**, 132 (2009).

92. K.-H. Cho, J.-E. Park, T. Osaka, and S.-G. Park, The study of antimicrobial activity and preservative effects of nanosilver ingredient, *Electrochimica Acta* **51**, 956 (2005).

93. L. Rodríguez-Arco, A. Poma, L. Ruiz-Pérez, E. Scarpa, K. Ngamkham, and G. Battaglia, Molecular bionics: Engineering biomaterials at the molecular level using biological principles, *Biomaterials* **192**, 26 (2019).

94. Y. Wang and H. Sun, Polymeric nanomaterials for efficient delivery of antimicrobial agents, *Pharmaceutics* **13**, 2108 (2021).

95. W. Lin, K. Huang, Y. Li, Y. Qin, D. Xiong, J. Ling, G. Yi, Z. Tang, J. Lin, Y. Huang, C. Yang, and J. Wang, Facile *in situ* preparation and *in vitro* antibacterial activity of PDMAEMA-based silver-bearing copolymer micelles, *Nanoscale Res. Lett.* **14**, 256 (2019).

96. S. Rigo, D. Hürlimann, L. Marot, M. Malmsten, W. Meier, and C. G. Palivan, Decorating nanostructured surfaces with antimicrobial peptides to efficiently fight bacteria, *ACS Appl. Bio Mater.* **3**, 1533 (2020).

97. L.-Y. Guo, S.-Z. Yan, X. Tao, Q. Yang, Q. Li, T.-S. Wang, S.-Q. Yu, and S.-L. Chen, Evaluation of hypocrellin A-loaded lipase sensitive polymer micelles for intervening methicillin-resistant *Staphylococcus aureus*

antibiotic-resistant bacterial infection, *Mater. Sci. Eng. C* **106**, 110230 (2020).

98. C. K. Wong, M. H. Stenzel, and P. Thordarson, Non-spherical polymersomes: Formation and characterization, *Chem. Soc. Rev.* **48**, 4019 (2019).
99. F. Fenaroli, J. D. Robertson, E. Scarpa, V. M. Gouveia, C. Di Guglielmo, C. De Pace, P. M. Elks, A. Poma, D. Evangelopoulos, J. O. Canseco, T. K. Prajsnar, H. M. Marriott, D. H. Dockrell, S. J. Foster, T. D. McHugh, S. A. Renshaw, J. S. Martí, G. Battaglia, and L. Rizzello, Polymersomes eradicating intracellular bacteria, *ACS Nano* **14**, 8287 (2020).
100. E. Abbasi, S. F. Aval, A. Akbarzadeh, M. Milani, H. T. Nasrabadi, S. W. Joo, Y. Hanifehpour, K. Nejati-Koshki, and R. Pashaei-Asl, Dendrimers: Synthesis, applications, and properties, *Nanoscale Res. Lett.* **9**, 247 (2014).
101. R. M. Kannan, E. Nance, S. Kannan, and D. A. Tomalia, Emerging concepts in dendrimer-based nanomedicine: From design principles to clinical applications, *J. Intern. Med.* **276**, 579 (2014).
102. P. Patel, V. Patel, and P. M. Patel, Synthetic strategy of dendrimers: A review, *J. Indian Chem. Soc.* **99**, 100514 (2022).
103. M. Ma, Y. Cheng, Z. Xu, P. Xu, H. Qu, Y. Fang, T. Xu, and L. Wen, Evaluation of polyamidoamine (PAMAM) dendrimers as drug carriers of anti-bacterial drugs using sulfamethoxazole (SMZ) as a model drug, *Eur. J. Med. Chem.* **42**, 93 (2007).
104. X. Liu, Z. Wang, X. Feng, E. Bai, Y. Xiong, X. Zhu, B. Shen, Y. Duan, and Y. Huang, Platensimycin-encapsulated poly(lactic-co-glycolic acid) and poly(amidoamine) dendrimers nanoparticles with enhanced anti-staphylococcal activity *in vivo*, *Bioconjug. Chem.* **31**, 1425 (2020).
105. H. W. VanKoten, W. M. Dlakic, R. Engel, and M. J. Cloninger, Synthesis and biological activity of highly cationic dendrimer antibiotics, *Mol. Pharm.* **13**, 3827 (2016).
106. C. Pathak, F. U. Vaidya, and S. M. Pandey, Mechanism for development of nanobased drug delivery system, in *Applications of Targeted Nano Drugs and Delivery Systems*, edited by S. S. Mohapatra, S. Ranjan, N. Dasgupta, R. K. Mishra, and S. Thomas (Elsevier, 2019), pp. 35–67.
107. L. Schnaider, S. Brahmachari, N. W. Schmidt, B. Mensa, S. Shaham-Niv, D. Bychenko, L. Adler-Abramovich, L. J. W. Shimon, S. Kolusheva, W. F. DeGrado, and E. Gazit, Self-assembling dipeptide antibacterial nanostructures with membrane disrupting activity, *Nat. Commun.* **8**, 1 (2017).

108. L. Feng, S. Li, H. Li, J. Zhai, Y. Song, L. Jiang, and D. Zhu, Super-hydrophobic surface of aligned polyacrylonitrile nanofibers, *Angew. Chem. Int. Ed.* **41**, 1221 (2002).

109. P. X. Ma and R. Zhang, Synthetic nano-scale fibrous extracellular matrix, *J. Biomed. Mater. Res.* **46**, 60 (1999).

110. J. Drabek and M. Zatloukal, Meltblown technology for production of polymeric microfibers/nanofibers: A review, *Phys. Fluids* **31**, 091301 (2019).

111. M. Elbahri, S. Homaeigohar, R. Abdelaziz, T. Dai, R. Khalil, and A. U. Zillohu, Smart metal–polymer bionanocomposites as omnidirectional plasmonic black absorber formed by nanofluid filtration, *Adv. Funct. Mater.* **22**, 4771 (2012).

112. J. Xue, T. Wu, Y. Dai, and Y. Xia, Electrospinning and electrospun nanofibers: Methods, materials, and applications, *Chem. Rev.* **119**, 5298 (2019).

113. A. R. Unnithan, N. A. M. Barakat, P. B. Tirupathi Pichiah, G. Gnanasekaran, R. Nirmala, Y.-S. Cha, C.-H. Jung, M. El-Newehy, and H. Y. Kim, Wound-dressing materials with antibacterial activity from electrospun polyurethane–dextran nanofiber mats containing ciprofloxacin HCl, *Carbohydr. Polym.* **90**, 1786 (2012).

114. H. Liu, K. K. Leonas, and Y. Zhao, Antimicrobial properties and release profile of ampicillin from electrospun poly(ε-caprolactone) nanofiber yarns, *J. Eng. Fibers Fabr.* **5**, 10 (2010).

115. K. Song, Q. Wu, Y. Qi, and T. Kärki, Electrospun nanofibers with antimicrobial properties, in *Electrospun Nanofibers*, edited by M. Afshari (Woodhead Publishing, 2017), pp. 551–569.

116. J. K. Oh, R. Drumright, D. J. Siegwart, and K. Matyjaszewski, The development of microgels/nanogels for drug delivery applications, *Prog. Polym. Sci.* **33**, 448 (2008).

117. T. Kaewruethai, C. Laomeephol, Y. Pan, and J. A. Luckanagul, Multifunctional polymeric nanogels for biomedical applications, *Gels* **7**, 228 (2021).

118. D. Keskin, G. Zu, A. M. Forson, L. Tromp, J. Sjollema, and P. van Rijn, Nanogels: A novel approach in antimicrobial delivery systems and antimicrobial coatings, *Bioact. Mater.* **6**, 3634 (2021).

119. S. H. Gehrke, L. H. Uhden, and J. F. McBride, Enhanced loading and activity retention of bioactive proteins in hydrogel delivery systems, *J. Controlled Release* **55**, 21 (1998).

120. R. Kandil and O. M. Merkel, Recent progress of polymeric nanogels for gene delivery, *Curr. Opin. Colloid Interface Sci.* **39**, 11 (2019).

121. J. B. Thorne, G. J. Vine, and M. J. Snowden, Microgel applications and commercial considerations, *Colloid Polym. Sci.* **289**, 625 (2011).

122. R. Cheng, F. Feng, F. Meng, C. Deng, J. Feijen, and Z. Zhong, Glutathione-responsive nano-vehicles as a promising platform for targeted intracellular drug and gene delivery, *J. Controlled Release* **152**, 2 (2011).

123. C. Li, S. R. Obireddy, and W.-F. Lai, Preparation and use of nanogels as carriers of drugs, *Drug Deliv.* **28**, 1594 (2021).

124. S. N. Kłodzińska, F. Wan, H. Jumaa, C. Sternberg, T. Rades, and H. M. Nielsen, Utilizing nanoparticles for improving anti-biofilm effects of azithromycin: A head-to-head comparison of modified hyaluronic acid nanogels and coated poly (lactic-co-glycolic acid) nanoparticles, *J. Colloid Interface Sci.* **555**, 595 (2019).

125. P. C. Appelbaum and B. Bozdogan, Vancomycin resistance in *Staphylococcus aureus*, *Clin. Lab. Med.* **24**, 381 (2004).

126. M.-H. Xiong, Y.-J. Li, Y. Bao, X.-Z. Yang, B. Hu, and J. Wang, Bacteria-responsive multifunctional nanogel for targeted antibiotic delivery, *Adv. Mater.* **24**, 6175 (2012).

127. M.-H. Xiong, Y. Bao, X.-Z. Yang, Y.-C. Wang, B. Sun, and J. Wang, Lipase-sensitive polymeric triple-layered nanogel for "on-demand" drug delivery, *J. Am. Chem. Soc.* **134**, 4355 (2012).

128. C. A. S. Ballesteros, J. C. Bernardi, D. S. Correa, and V. Zucolotto, Controlled release of silver nanoparticles contained in photoresponsive nanogels, *ACS Appl. Bio Mater.* **2**, 644 (2019).

129. C. A. S. Ballesteros, D. S. Correa, and V. Zucolotto, Polycaprolactone nanofiber mats decorated with photoresponsive nanogels and silver nanoparticles: Slow release for antibacterial control, *Mater. Sci. Eng. C* **107**, 110334 (2020).

130. G. R. Bardajee, Z. Hooshyar, and H. Rezanezhad, A novel and green biomaterial based silver nanocomposite hydrogel: Synthesis, characterization and antibacterial effect, *J. Inorg. Biochem.* **117**, 367 (2012).

131. S. Ravindra, K. Varaprasad, V. Rajinikanth, A. F. Mulaba-Bafubiandi, and K. Venkata Surya Ramam, Studies on curcumin loaded poly(N-isopropylacrylamide) silver nanocomposite hydrogels for antibacterial and drug releasing applications, *J. Macromol. Sci. Part A* **50**, 1230 (2013).

132. H. Deng, Z. Yu, S. Chen, L. Fei, Q. Sha, N. Zhou, Z. Chen, and C. Xu, Facile and eco-friendly fabrication of polysaccharides-based nanocomposite hydrogel for photothermal treatment of wound infection, *Carbohydr. Polym.* **230**, 115565 (2020).

133. H. Wang, J. Li, N. Ding, X. Zeng, X. Tang, Y. Sun, T. Lei, and L. Lin, Eco-friendly polymer nanocomposite hydrogel enhanced by cellulose nanocrystal and graphitic-like carbon nitride nanosheet, *Chem. Eng. J.* **386**, 124021 (2020).

134. M. C. Coll Ferrer, S. Dastgheyb, N. J. Hickok, D. M. Eckmann, and R. J. Composto, Designing nanogel carriers for antibacterial applications, *Acta Biomater.* **10**, 2105 (2014).

135. S. Sagbas, N. Aktas, and N. Sahiner, Modified biofunctional p(tannic acid) microgels and their antimicrobial activity, *Appl. Surf. Sci.* **354**, 306 (2015).

136. K. E. Richardson, Z. Xue, Y. Huang, Y. Seo, and Y. Lapitsky, Physicochemical and antibacterial properties of surfactant mixtures with quaternized chitosan microgels, *Carbohydr. Polym.* **93**, 709 (2013).

137. C. Echeverría, A. Aragón-Gutiérrez, M. Fernández-García, A. Muñoz-Bonilla, and D. López, Thermoresponsive poly(N-isopropylacrylamide-co-dimethylaminoethyl methacrylate) microgel aqueous dispersions with potential antimicrobial properties, *Polymers* **11**, 4 (2019).

138. H. Han, J. Zhu, D.-Q. Wu, F.-X. Li, X.-L. Wang, J.-Y. Yu, and X.-H. Qin, Inherent guanidine nanogels with durable antibacterial and bacterially antiadhesive properties, *Adv. Funct. Mater.* **29**, 1806594 (2019).

139. L. Qian, H. Xiao, G. Zhao, and B. He, Synthesis of modified guanidine-based polymers and their antimicrobial activities revealed by AFM and CLSM, *ACS Appl. Mater. Interfaces* **3**, 1895 (2011).

140. I. Ekladious, Y. L. Colson, and M. W. Grinstaff, Polymer–drug conjugate therapeutics: Advances, insights and prospects, *Nat. Rev. Drug Discov.* **18**, 4 (2019).

141. N. D. Stebbins, M. A. Ouimet, and K. E. Uhrich, Antibiotic-containing polymers for localized, sustained drug delivery, *Adv. Drug Deliv. Rev.* **78**, 77 (2014).

142. V. Coessens, E. H. Schacht, and D. Domurado, Synthesis and *in vitro* stability of macromolecular prodrugs of norfloxacin, *J. Controlled Release* **47**, 283 (1997).

143. E. Roseeuw, V. Coessens, E. Schacht, B. Vrooman, D. Domurado, and G. Marchal, Polymeric prodrugs of antibiotics with improved efficiency, *J. Mater. Sci. Mater. Med.* **10**, 743 (1999).

144. I. L. Batalha, A. Bernut, M. Schiebler, M. M. Ouberai, C. Passemar, C. Klapholz, S. Kinna, S. Michel, K. Sader, P. Castro-Hartmann, S. A. Renshaw, M. E. Welland, and R. A. Floto, Polymeric nanobiotics as a novel treatment for mycobacterial infections, *J. Controlled Release* **314**, 116 (2019).
145. Y. Gao, J. Wang, M. Chai, X. Li, Y. Deng, Q. Jin, and J. Ji, Size and charge adaptive clustered nanoparticles targeting the biofilm microenvironment for chronic lung infection management, *ACS Nano* **14**, 5686 (2020).
146. S. Subramaniam, E. Fahy, S. Gupta, M. Sud, R. W. Byrnes, D. Cotter, A. R. Dinasarapu, and M. R. Maurya, Bioinformatics and systems biology of the lipidome, *Chem. Rev.* **111**, 6452 (2011).
147. E. Fahy, S. Subramaniam, R. C. Murphy, M. Nishijima, C. R. H. Raetz, T. Shimizu, F. Spener, G. Van Meer, M. J. O. Wakelam, and E. A. Dennis, Update of the LIPID MAPS comprehensive classification system for lipids, *J. Lipid Res.* **50**, S9 (2009).
148. N. Monteiro, A. Martins, R. L. Reis, and N. M. Neves, Liposomes in tissue engineering and regenerative medicine, *J. R. Soc. Interface* **11**, 20140459 (2014).
149. A. Gonzalez Gomez and Z. Hosseinidoust, Liposomes for antibiotic encapsulation and delivery, *ACS Infect. Dis.* **6**, 896 (2020).
150. Y. Barenholz, Doxil®: The first FDA-approved nano-drug: Lessons learned, *J. Control. Release* **160**, 117 (2012).
151. L. Sande, M. Sanchez, J. Montes, A. J. Wolf, M. A. Morgan, A. Omri, and G. Y. Liu, Liposomal encapsulation of vancomycin improves killing of methicillin-resistant *Staphylococcus aureus* in a murine infection model, *J. Antimicrob. Chemother.* **67**, 2191 (2012).
152. R. Nessar, E. Cambau, J. M. Reyrat, A. Murray, and B. Gicquel, *Mycobacterium abscessus*: A new antibiotic nightmare, *J. Antimicrob. Chemother.* **67**, 810 (2012).
153. J. P. Sarathy, L. E. Via, D. Weiner, L. Blanc, H. Boshoff, E. A. Eugenin, C. E. Barry, and V. A. Dartois, Extreme drug tolerance of *Mycobacterium tuberculosis* in caseum, *Antimicrob. Agents Chemother.* **62**, 1 (2018).
154. F. Rinaldi, P. N. Hanieh, S. Sennato, F. De Santis, J. Forte, M. Fraziano, S. Casciardi, C. Marianecci, F. Bordi, and M. Carafa, Rifampicin–liposomes for Mycobacterium abscessus infection treatment: Intracellular uptake and antibacterial activity evaluation, *Pharmaceutics* **13**, 7 (2021).
155. M. Gorzynski, T. Week, T. Jaramillo, E. Dzalamidze, and L. Danelishvili, Mycobacterium abscessus genetic determinants associated with the intrinsic resistance to antibiotics, *Microorganisms* **9**, 2527 (2021).

156. P. Rudra, K. Hurst-Hess, P. Lappierre, and P. Ghosh, High levels of intrinsic tetracycline resistance in Mycobacterium abscessus are conferred by a tetracycline-modifying monooxygenase, *Antimicrob. Agents Chemother* **62**, 10.1128/aac.00119 (2018).

157. R. Ghosh and M. De, Liposome-based antibacterial delivery: An emergent approach to combat bacterial infections, *ACS Omega* **8**, 35442 (2023).

158. D.-Y. Wang, H. C. van der Mei, Y. Ren, H. J. Busscher, and L. Shi, Lipid-based antimicrobial delivery-systems for the treatment of bacterial infections, *Front. Chem.* **7**, (2020).

159. X. Hou, X. Zhang, W. Zhao, C. Zeng, B. Deng, D. W. McComb, S. Du, C. Zhang, W. Li, and Y. Dong, Vitamin lipid nanoparticles enable adoptive macrophage transfer for the treatment of multidrug-resistant bacterial sepsis, *Nat. Nanotechnol.* **15**, 1 (2020).

160. A. Walduck, P. Sangwan, Q. Anh Vo, J. Ratcliffe, J. White, B. W. Muir, and N. Tran, Treatment of *Staphylococcus aureus* skin infection *in vivo* using rifampicin loaded lipid nanoparticles, *RSC Adv.* **10**, 33608 (2020).

161. C. L. Seabra, C. Nunes, M. Brás, M. Gomez-Lazaro, C. A. Reis, I. C. Gonçalves, S. Reis, and M. C. L. Martins, Lipid nanoparticles to counteract gastric infection without affecting gut microbiota, *Eur. J. Pharm. Biopharm.* **127**, 378 (2018).

162. L. Jiang, H. Wee Lee, and S. C. Joachim Loo, Therapeutic lipid-coated hybrid nanoparticles against bacterial infections, *RSC Adv.* **10**, 8497 (2020).

Chapter 7

Nanotechnology Applied to Tackle Malaria

Yunuen Avalos-Padilla[a,b] **and**
Xavier Fernàndez-Busquets[a,b,c]

[a]*Barcelona Institute for Global Health (ISGlobal, Hospital Clínic-University of Barcelona), Rosselló 149-153, 08036 Barcelona, Spain*
[b]*Nanomalaria Group, Institute for Bioengineering of Catalonia (IBEC), The Barcelona Institute of Science and Technology, Baldiri Reixac 10-12, 08028 Barcelona, Spain*
[c]*Nanoscience and Nanotechnology Institute (IN2UB), University of Barcelona, Martí i Franquès 1, 08028 Barcelona, Spain*
yavalos@ibecbarcelona.eu, xfernandez@ibecbarcelona.eu

With 608,000 estimated deaths in 2022 and ca. 249 million cases worldwide [1], malaria still represents an important global health problem, particularly in low-income countries. Despite the efforts made to control its spread, drug resistance as well as environmental and social concerns are the main challenges to eradicating this disease. Therefore, there is an extreme need for alternative therapeutical approaches that can help to overcome such problems.

Nanotherapeutics for Infectious Diseases
Edited by Íris Luz Batalha

ISBN 978-981-5129-36-6 (Hardcover), 978-1-003-56678-6 (eBook)
www.jennystanford.com

In the last years, nanotechnology has been explored as an interesting strategy against malaria, although the research conducted in this field remains scarce [2].

7.1 Biology of Malaria Parasites

Malaria is caused by protozoan parasites of the genus *Plasmodium* transmitted by several species of *Anopheles* mosquitoes and is considered one of the most serious and lethal diseases [3]. There are six different species of *Plasmodium* able to infect humans: *P. falciparum, P. malariae, P. vivax, P. ovale curtisi, P. ovale wallikeri,* and *P. knowlesi*. Among those, *P. falciparum* and *P. vivax* represent the major threat [4, 5].

Malaria affects millions of people mostly in tropical and subtropical areas in developing countries, causing the highest mortality in the African region [6]. The most common manifestations of this disease include headache, fever, sweating, vomiting, chills, and nausea [7]. However, when not treated properly, especially in pregnant women and children under the age of five, *P. falciparum* malaria can progress to life-threatening cases where it is possible to observe severe anemia, metabolic acidosis resulting in renal failure, or cerebral malaria, which eventually leads to encephalitis, shock, coma, and death [8].

Plasmodium, an obligate intracellular parasite of hepatocytes and erythrocytes, undergoes a complex life cycle, divided between a vertebrate (intermediate host) and the mosquito vector (final host) (Fig. 7.1). The different *Plasmodium* species are diverse in their morphology and life cycle, which begins when a female *Anopheles* mosquito takes a blood meal and injects sporozoites under the skin. The motile sporozoites enter the bloodstream, reach the liver, and enter hepatocytes [9]. With this, the parasite evades host clearance and escapes the immune response [10]. Inside the hepatocytes, *Plasmodium* establishes a parasitophorous vacuole and begins the first round of asexual replication [11]. Then the multinucleated exo-erythrocytic schizont (also called meront) is formed, a process that can last from two to several days, depending on the species [12]. At this stage, there are no clinical signs or symptoms in the host. In the case of *P. vivax* and *P. ovale*, the parasite can enter a period of latency forming a non-replicating hypnozoite, which allows long-term

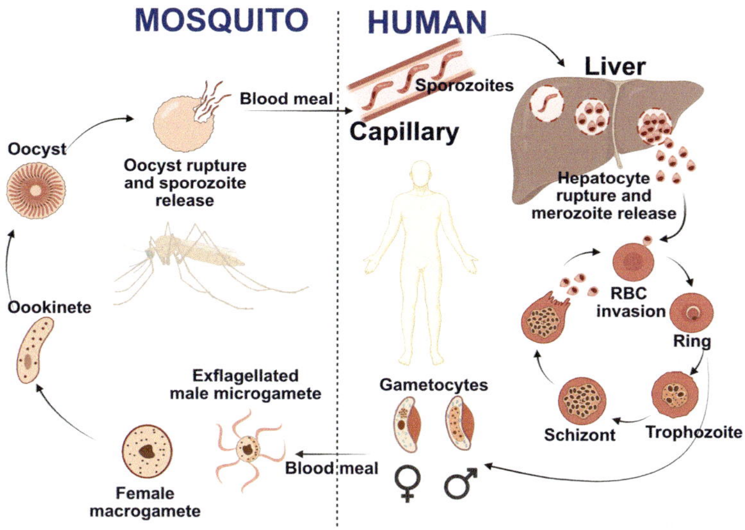

Figure 7.1 *Plasmodium* life cycle. During a blood meal, an infected female *Anopheles* mosquito injects sporozoites into the human host. Sporozoites travel through the blood stream and reach the liver, where they infect hepatocytes and mature into schizonts, which eventually rupture and release merozoites (exo-erythrocytic schizogony). Afterward, merozoites infect red blood cells (RBCs), where they undergo asexual reproduction (erythrocytic schizogony). Ring stages progress into trophozoites and schizonts, which finally rupture, releasing daughter merozoites. Some parasites can differentiate into sexual erythrocytic stages (gametocytes), which are ingested by a mosquito during its blood meal. Inside the mosquito stomach, microgametes fuse with macrogametes generating zygotes, which in turn become motile and elongated ookinetes. The ookinetes invade the mosquito midgut wall and develop into oocysts that grow and rupture, releasing sporozoites. Finally, sporozoites reach the salivary glands of the mosquito, which is now able to inoculate parasites into a new human host, perpetuating the malaria life cycle. Created with BioRender.com.

survival of the pathogen and is associated to malaria relapses [13]. The cycle continues when thousands of daughter merozoites are released from the hepatocyte back into the bloodstream via the liver sinusoids [14]. The merozoites invade red blood cells (RBCs), where the second asexual schizogony occurs. During this asexual replication, up to 32 merozoites are produced inside a parasitophorous vacuole over the course of 24 to 72 h [15]. Parasitized RBCs (pRBCs) rupture in synchrony, provoking increases in cytokine production that leads to the clinical symptoms of malaria, with disease severity correlating with parasite load [16]. After several rounds of invasion and growth, an acute, and ultimately chronic infection, is established. Some species, such as *P. vivax*, are restricted to reticulocytes, which limits the total parasitemia [17]. However, in the case of *P. falciparum*, there is no such restriction, and the pathogen can infect a high number of RBCs leading to an extraordinary parasite burden [12]. Finally, a subpopulation of asexual stages commits to gametocytogenesis and produces male and female sexual gametocytes [18]. Mature gametocytes can then circulate in the blood for several days, maximizing the chances of transmission to mosquitoes, which take the parasites with a blood meal. Ingested gametocytes differentiate into eight microgametes (male) or one macrogamete (female), which fuse to produce the zygote. In here, homologous recombination occurs and yields significant genetic variability of progeny [19]. The zygote then transforms into a motile ookinete, crosses the epithelial layer of the midgut wall, and forms an oocyst. At this stage, parasites undergo the third cycle of asexual replication to produce thousands of sporozoites that are released into the insect's hemolymph. Sporozoites finally reach the salivary glands of the mosquito, where they remain until transmitted to a new host through a mosquito bite, to start the cycle again [12, 20].

7.2 Current Preventive, Diagnostic, and Therapeutic Strategies Against Malaria

The five major intervention strategies for malaria reduction are vector control, chemopreventive measures, early diagnosis, proper chemotherapy for infected individuals, and vaccination approaches [21].

7.2.1 Vector Control

Vector control has a significant role in preventing and reducing the transmission of malaria [22]. The different approaches to control mosquito populations can be divided into five classes: source reduction (environmental control), trapping (mechanical control), insecticides (chemical control), genetic control, and biological control [23].

Environmental control refers to all methods based on, *e.g.*, removing impermanent water tanks, management of wetlands, and sealing permanent water containers. However, this is not enough to eliminate vector populations due to unseen or inaccessible reservoirs [24]. To lure and remove mosquitoes, the most common traps used for *Anopheles* species are bed net-, entry/exit-, sentinel-, capture/kill-, light-, and chemical-based traps [25], which are commonly treated with insecticides and combinations of attractants [26, 27]. The insecticides used to kill *Anopheles* mosquitoes include natural (plant-based [28]) or synthetic chemicals that allow instant mosquito population reduction when the previously mentioned methods are not sufficient to reduce mosquito numbers below certain thresholds [23]. The first widely deployed antimalarial insecticide was dichlorodiphenyltrichloroethane (DDT), first synthesized in the 19th century [29], which initially decreased overall malaria cases. Nevertheless, due to risks for human health and the environment, the use of insecticides, including DDT, has been reduced over the years [30]. There are four types of insecticides currently employed in the control of malaria vectors: organochlorines, organophosphates, carbamates, and pyrethroids; 12 insecticides classified within these four types are recommended for indoor residual spraying, whereas only pyrethroids are recommended for use in long-lasting insecticidal nets [31]. However, resistance to insecticides is widespread in *Anopheles* mosquitoes, especially to pyrethroids, jeopardizing the success of malaria control [32].

For the past few years, bio-pesticides have gained interest as an alternative control approach that targets female mosquitoes either at the larval stages or at mature adult stages [30]. These biological control agents include bacteria, fungi, viruses, protozoa, fish, and natural predators [33]. Among them, spore-forming bacteria such as *Bacillus thuringiensis* and *Bacillus sphaericus* had been investigated against *Anopheles* populations [34]. However, due to their limited availability and high cost, these agents are still under study [35]. Finally, genetic control is used as an alternative to conventional methods for vector control. This approach is characterized by gene-drive technologies that rapidly spread desirable traits through the use of genetic constructs, which alter mosquito inheritance and spread fast through vector populations, acting as biocontrol agents [36]. The intended outcomes of such strategies are population suppression, population replacement, and other key aspects of genetic element persistence [37].

7.2.2 Chemoprevention

In 2012, the World Health Organization (WHO) recommended seasonal malaria chemoprevention (SMC), which, since then, has been implemented in several low-income countries with high transmission rates [38]. SMC targets children from 5 months to 5 years during the intense malaria transmission period, just after the rainy season. Children receive monthly doses of sulfadoxine/pyrimethamine and amodiaquine (Fig. 7.2) for up to 4 months. This scheme combines an antifolate (sulfadoxine/pyrimethamine) that

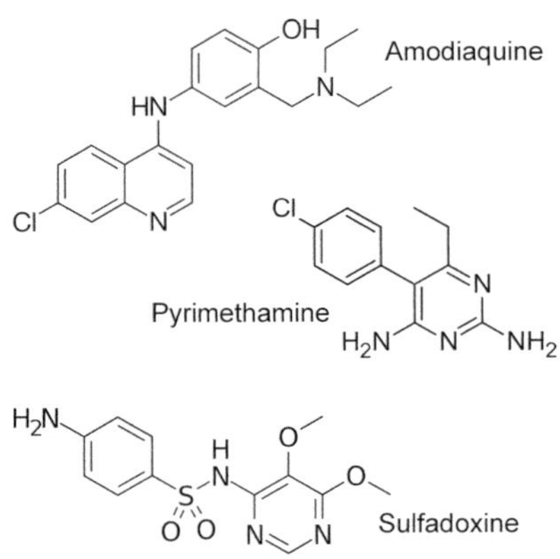

Figure 7.2 Chemical structure of antimalarial drugs currently used in chemoprevention.

inhibits either dihydrofolate reductase or dihydropteroate synthase, which are key enzymes in de novo folate biosynthesis, an important pathway for the synthesis of nitrogen-containing bases and some amino acids [39]. The other drug used in chemoprevention is a quinoline derivative, which operates through the inhibition of the hemoglobin degradation pathway preventing heme polymerization, thus resulting in the accumulation of drug–heme complexes, which results in the lysis of the membrane and parasite death [40]. Each month, amodiaquine is given in daily doses for 3 days, and a single dose of sulfadoxine/pyrimethamine is given in the first day of the preventive treatment [41]. SMC has been effective in preventing morbidity and mortality from malaria [42], and some studies have tested its efficacy in pregnant women [43], whereas adverse reactions have been rarely reported [44]. However, despite the success achieved in the last years, one of the main drawbacks of this strategy is that some of the target zones are hard-to-reach areas because of poor infrastructure and political instability [41].

7.2.3 Diagnostic Approaches

A prompt and accurate diagnosis is essential for both effective disease management and malaria surveillance. Clinical diagnosis, based on observation of the patient's symptoms, is the least expensive method. However, due to overtreatment and frequent misdiagnosis, the WHO has recommended since 2010 that, before treatment, all patients with suspected malaria should be confirmed by microscopy or a rapid diagnostic test (RDT) identifying parasite antigens in blood [45].

Most routine diagnostic laboratories are able to detect infection at parasitemia levels of 0.01% of the total RBC count (approximately 500 parasites/μL) [46]. The current "gold-standard" method for malaria diagnosis is the visual inspection of blood smears using light microscopy, which requires highly trained and experienced personnel and cannot provide an early diagnosis of the disease [47, 48]. The quantitative buffy coat (QBC) method, designed to enhance microscopy detection, involves the fluorescent staining of the parasite's nuclei in micro-hematocrit tubes and its subsequent detection by epi-fluorescence microscopy [49]. Although QBC is simple and reliable, it requires specialized equipment, cannot determine species, and is more expensive than conventional microscopy [45]. On the other hand, RDTs employ antibody–antigen binding on test sticks, but they only provide a qualitative measure of infection [46]. Currently, there are around 200 different RDTs available on the market, with variable sensitivity and specificity; some of them detect individual *Plasmodium* species, while others detect multiple species [5]. RDTs have a sensitivity of ca. 100 parasites/μL, but they frequently lead to false positives and false negatives [50]. In addition, RDTs are not considered very reliable for non-*falciparum* infections [51].

The drawbacks of current malaria diagnostic techniques have spurred the development of molecular approaches for *Plasmodium* detection that provide high sensitivity and specificity toward different species. Among these, polymerase chain reaction (PCR)-based detection is recommended by the WHO but only for epidemiological research and surveys mapping sub-microscopic infections [47]. PCR can detect as few as 1–5 parasites/μL of blood, although it is a complex methodology with high cost, needs trained technicians, and is difficult to perform in rural settings [45].

7.2.4 Chemotherapeutic Strategies

Malaria has been traditionally treated with natural products found in barks, roots, or leaves of plants and trees [52, 53]. However, only in the last century the active ingredients of these natural compounds have been identified and isolated as drugs [54]. Among these, the most successful was quinine, a quinolone-containing alkaloid isolated from the bark of Peruvian cinchona trees that has been used as an antimalarial since the 17th century [55]. Quinine set the basis for the development of other drug analogs, including quinacrine [56] and chloroquine [57] (Fig. 7.3). In the 1950s, chloroquine was the treatment of choice for malaria, and its use continued globally for decades due to its efficacy, bioavailability, low toxicity, and affordability [58]. Chloroquine and DDT were the frontline in the WHO's global malaria eradication campaign [59]. However, in the 1960s, chloroquine resistance emerged and the need for new drugs became urgent.

As malaria cases increased over the globe, high-throughput screening for drug discovery and development work, most notably by the U.S. military, led to the synthesis of atovaquone (proguanil), pyrimethamine, primaquine, and amodiaquine (the last two, chloroquine analogs), as well as the arylaminoalcohols mefloquine and halofantrine [56] (Fig. 7.4).

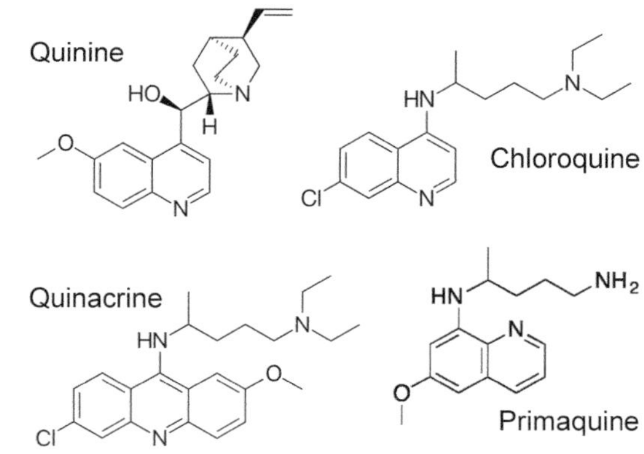

Figure 7.3 Chemical structures of some quinoline derivative antimalarial drugs.

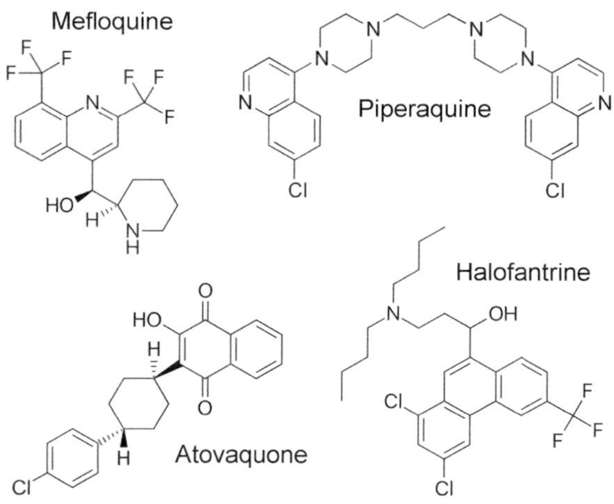

Figure 7.4 Chemical structures of some antimalarial drugs.

Unfortunately, parasites also evolved rapid resistance to these drugs, reducing their efficacy from the late 1980s. Piperaquine, a *bis* 4-aminoquinoline chloroquine derivative synthesized simultaneously by the Chinese National Malaria Elimination program and Rhône-Poulenc in France [60], was able to overcome chloroquine resistance and has been recently used in combination therapies in Southeast Asia [61]. Finally, artemisinin, originally extracted from the Chinese sweet wormwood, *Artemisia annua*, and its derivatives artemether, artesunate, and dihydroartemisinin (Fig. 7.5) are fast-acting compounds widely used for treating malaria [62]. They have a strong activity against asexual parasites, with the ability to kill all blood stages [63]. Furthermore, these compounds are also able to eliminate *P. falciparum* gametocytes [64].

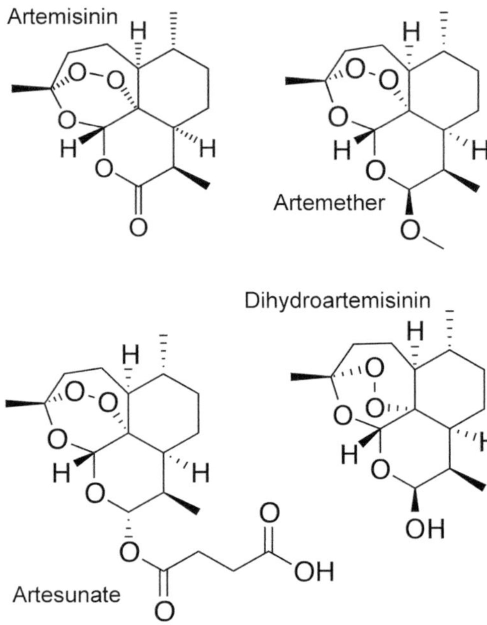

Figure 7.5 Chemical structures of artemisinin and some of its derivatives.

Currently, the WHO recommends for malaria treatment the use of two or more compounds with different modes of action [5]. In the last years, artemisinin-based combination therapies (ACTs) have been successful in controlling malaria, with the global burden showing more than 30% reduction from 2000 to 2015 [65]. For the treatment of uncomplicated *P. falciparum* malaria, the recommended ACTs are based on the combination of artemisinin or one of its derivatives with a partner drug that has a longer half-life, including lumefantrine, piperaquine, mefloquine, amodiaquine, sulfadoxine, and more recently pyronaridine [5, 66]. Among other drugs with potential as antimalarials are the plant-derived products cryptolepine [66], curcumin [67], fosmidomycin [68], artemisone [69], and artemiside [70] (Fig. 7.6), to cite some examples.

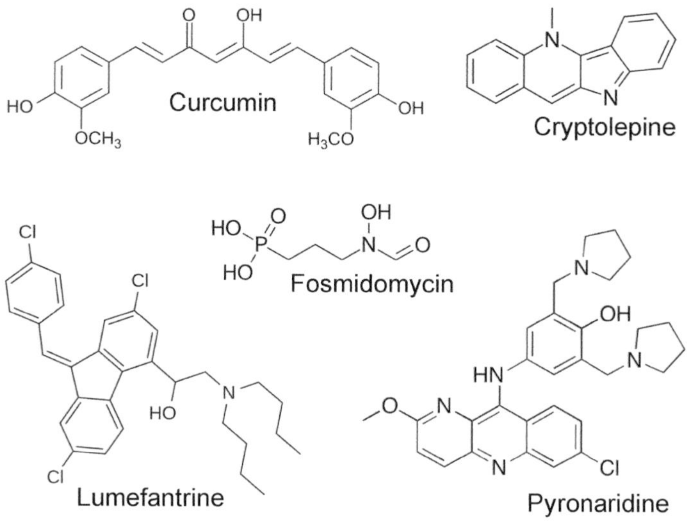

Figure 7.6 Chemical structures of some antimalarial drugs.

7.2.5 Vaccines

Malaria vaccine development has been delayed by the complexity of the *Plasmodium* life cycle [71], its widespread antigenic variation [72], a poor understanding of the interaction between the parasite and the host immune system [73, 74], and the relative scarcity of research funding [75]. Current malaria vaccine candidates are directed against human (pre-erythrocytic and blood-stage vaccines) and mosquito stages (transmission-blocking vaccines). However, very few antigens have been studied for potential vaccine development [76]. An ideal malaria vaccine would completely inhibit the first parasite forms following an infected mosquito bite, blocking further stages and preventing transmission [77].

RTS,S/AS01 (Mosquirix^{TM}) (Fig. 7.7), the first licensed malaria vaccine, has shown partial protection against the disease in young children. Work on this vaccine started in 1987, and pilot implementation began in Ghana, Kenya, and Malawi in 2019 among children aged 5 to 17 months in a four-dose scheme [77]. RTS,S/AS01 was engineered using the pre-erythrocytic circumsporozoite protein (CSP) and a viral envelope protein of the hepatitis B virus to which a chemical adjuvant was added to increase immune system response [78, 79].

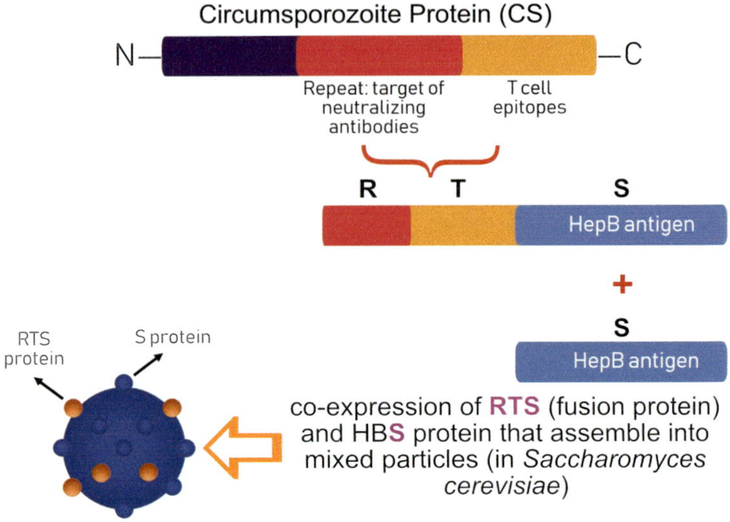

Figure 7.7 RTS,S recombinant protein virus-like particle. The RTS,S antigen consists of two proteins, RTS (fusion protein of a portion of the circumsporozoite protein from *P. falciparum* and the amino terminal end of the hepatitis B virus S protein) and the same hepatitis B surface antigen (S), which are co-expressed in *Saccharomyces cerevisiae* and assemble intracellularly and spontaneously into mixed polymeric particulate structures. Adapted from Ref. [79].

No other malaria vaccine has advanced to Phase III clinical testing. However, other candidate vaccines against *P. falciparum* are under development: PfSPZ is a radiation-attenuated sporozoite vaccine, and R21 is a virus-like particle-based CSP vaccine, similar to RTS,S. In both cases, an enhanced immune response has been observed when compared to the placebo [80–82].

7.3 Development of Resistance to Antimalarial Drugs

Drug resistance is defined as the ability of a pathogen to survive or multiply despite the properly administered and dosed medication [83]. The factors that contribute to resistance evolution in malaria parasites include the intrinsic disadvantages of antimalarial drugs, such as low bioavailability (poor aqueous solubility, permeability, and bio-stability), the wide spectrum of side effects [84], and the delayed parasite's clearance [85]. Unfortunately, resistance has emerged to all classes of antimalarial drugs deployed so far (Fig. 7.8). Antimalarial drug misuse (poor patient compliance) has put an enormous drug-selection pressure on *Plasmodium* to evolve resistance mechanisms [52]. Moreover, antimalarial drug resistance usually appears outside the African region in low mixed-strain transmission settings, before expanding to high-transmission settings in Sub-Saharan Africa [86]. For instance, resistance to chloroquine, mefloquine, and sulfadoxine-pyrimethamine initially began in Southeast Asia and latter spread to the more vulnerable African region [87–89].

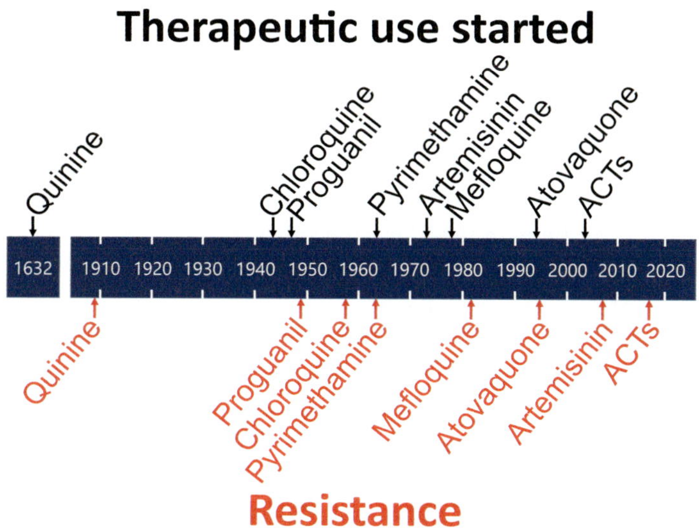

Figure 7.8 Timeline of antimalarial resistance evolution. The chronogram shows the years when main antimalarials were first applied therapeutically and when resistance to them emerged. Adapted from Ref. [89].

The ability of *Plasmodium* to develop resistance is primarily due to a combination of the high burden of parasites in the bloodstream of an infected person during the asexual blood stage of infection, and the variant expression of the parasite's proteome [83, 86]. High-grade resistance occurs when the methods to clear parasitemia fail; hence, an important marker of resistance is delayed parasite clearance time. In the era of antimalarial combination therapies, a major challenge is that therapy failure may not be observed even when the parasites are resistant to one of the partner drugs because the other one remains active [83]. Moreover, artemisinin resistance was detected in early 2009 in Southeast Asia [90], and more recently in Africa [91] and South America [92], threatening the recent progress made in malaria reduction and highlighting the need for new interventions.

7.4 Nanotechnology-based Approaches Against Malaria

Nowadays, therapeutic approaches for malaria are challenged by unwanted side effects, the potential for development of drug resistance, and the lack of specific targeting to intracellular parasites, resulting in high dose requirements. To overcome these problems, nanotechnology-based strategies have been developed toward the generation of new alternatives to diagnose, prevent, and treat malaria, as outlined in the following subsections.

7.4.1 Diagnostic Applications

Disease diagnostics have benefited from nanotechnology primarily by exploiting the unique characteristics of nanomaterials such as their optical, fluorescent, and magnetic properties to impart new device capabilities [93].

Aptamers are single-stranded nucleic acids (DNA or RNA) or peptides that have a defined three-dimensional structural conformation and a specific property of "induced fit binding," which gives them high sensitivity and selectivity toward a target molecule [94]. These nanomaterials have been incorporated into colorimetric, electrochemical, and fluorescence-based biosensors for malaria diagnosis. For instance, Lee et al. identified DNA aptamers against *P. falciparum* lactate dehydrogenase (*Pf*LDH), which bind to the target protein with high sensitivity (K_d = 16.8–49.6 nM) [95, 96]. *Pf*LDH aptamers have also been combined with a DNA origami scaffold (produced by the arbitrary folding of DNA to create two- and three-dimensional shapes [97]) and atomic force microscopy as a diagnostic tool [98]. On the other hand, using pan-LDH as target, aptamers were incorporated along cationic polymers in a gold nanoparticle–based colorimetric sensor (Fig. 7.9A). This method allowed the detection of less than 100 parasites/μL in 40 min [99]. Other colorimetric gold nanoparticle–based assays incorporating aptamers include the studies by Dirkzwager et al. and Jain et al., which allowed a qualitative correlation for infection using the naked eye [100, 101].

Quantum dots (QDs) are fluorescent semiconductor nanoparticles, previously used for cellular imaging and diagnosis and the study of intracellular processes at the single-molecule level [102]. These materials have been explored for malaria diagnosis by labeling pRBCs, where poly(ethylene) glycol (PEG)-coated positively charged QDs offered a faster result and higher sensitivity than the classical LDH assay [103]. More recently, nanosheets, a nanomaterial with better properties than QDs in terms of stability, have been used. For instance, a *Pf*LDH-specific aptamer functionalized with 6-carboxyfluorescein was adsorbed on the surface of MoS_2 nanosheets, to which energy was transferred through fluorescence resonance energy transfer (FRET). In the case of positive samples, the aptamer desorbed from the nanosheet and bound to the sample; as a consequence, FRET did not occur and an emission peak proportional to the concentration of the protein in the sample was observed [104, 105].

An alternative immunochromatographic assay employed nano-gold labeling techniques and a combination of monoclonal antibodies against a *P. falciparum*-specific protein and a pan-*Plasmodium*-specific LDH. By using this technique, the authors were able to detect malaria with a sensitivity of 96.46% for *P. falciparum* and 96.83% for non-*falciparum* species (Fig. 7.9B) [106].

Naturally occurring nanomaterials have been explored by Garrett et al., where nanostructures of the purple and blue regions on the wings of a butterfly combined with surface-enhanced Raman scattering were able to detect hemozoin in lysates from malaria-infected blood samples containing 0.0005% pRBCs [46].

Nanocarriers have also been used to improve signals of redox reactions in electrochemical sensors. For example, a copper-doped ZnO nanofiber-based electrochemical immunosensor was used for the sensitive detection of the *P. falciparum* histidine-rich protein 2 (*Pf*HRP2) [107]. More recently, a highly sensitive aptamer-based

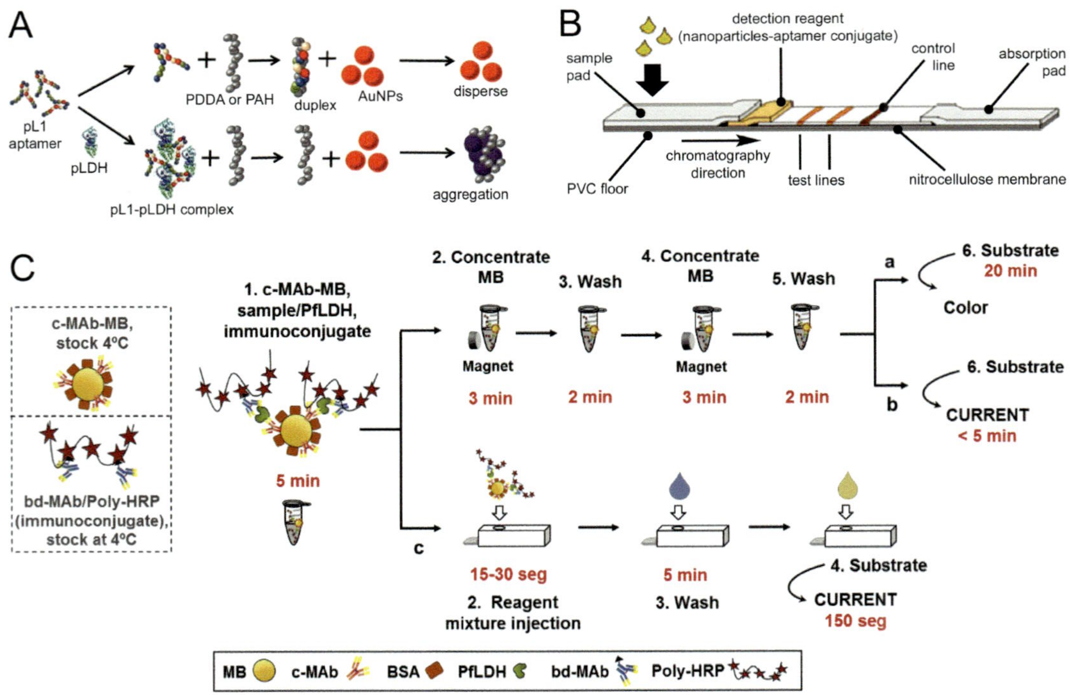

Figure 7.9 Examples of nanotechnology-based malaria diagnostic applications. **(A)** Schematic illustration of a colorimetric aptasensor for the diagnosis of malaria based on cationic polymers and gold nanoparticles for pan-LDH detection. pLDH: pan-LDH; PDDA: poly(diallyldimethylammonium chloride); PAH: poly(allylamine hydrochloride). Reprinted with permission from Ref. [99]. **(B)** Schematic of a nano-gold immunochromatographic assay for malaria detection. Adapted from Ref. [106]. **(C)** Different formats of a single-step magneto-immunoassay for malaria quantitative diagnosis in whole blood using *Pf* LDH, magnetic beads, poly-horseradish peroxidase (poly-HRP) and microfluidic paper electrodes. **(a)** Spectrophotometric detection. **(b)** Electrochemical detection using a screen-printed carbon electrode (SPCE). **(c)** Electrochemical detection on a microfluidic paper, double-sided SPCE. MB: magnetic beads; c-Mab: capture antibody; BSA: bovine serum albumin; bd-Mab: detection antibody. Reprinted with permission from Ref. [112].

electrochemical impedance spectroscopy sensor was reported to have a detection limit of 0.84 pM *Pf*LDH in tenfold-diluted serum samples with a good reusability of the electrode for multiple analysis [108]. On the other hand, a multiwalled carbon nanotube coupled to a ZnO nanofiber–based chemiresistive biosensor has been developed for the detection of HRP2 with a detection limit of 0.97 fg/mL [109].

The use of nanotechnology can also improve the sensitivity of the classical immunosensor based on enzyme-linked immunosorbent assay (ELISA). Here, the approach taken by Hemben et al. for the detection of *Pf*HRP2 consisted in specific monoclonal antibodies adsorbed onto a screen-printed gold electrode surface, which led to a detection limit of 2.95 ng/mL in serum samples [110]. Similarly, magnetic nanoparticles have also been used to improve the sensitivity of *Pf*HRP2 and *Pf*LDH detection, where antibodies were functionalized on a magnetic nanoparticle surface reaching detection limits of 0.36 and 2.47 ng/mL for the respective proteins [111, 112]. The *Pf*LDH-based sensor (Fig. 7.9C) detected parasitemias as low as 0.006% in *P. falciparum* cultures [111].

Finally, paper-based sensors are an ideal platform for malaria diagnosis because they allow the detection of malaria *in situ* and are cheap and affordable for low-income countries. An example of these sensors is a gold nanoparticle–amplified immunoassay in the form of a two-dimensional paper network–based lateral flow test with a detection limit of 2.9 ng/mL. Here, a murine antibody against *Pf*HRP2 was used as the capture reagent and the secondary antibody was tagged with gold nanoparticles allowing the visualization of the binding reaction, improved with a gold enhancement reagent [113]. Another example is the study by Xu et al., who developed an "easy-to-handle" test based on the paper folding technique of origami to enable the sequential steps of DNA extraction through loop-mediated isothermal amplification (a low-cost alternative for DNA amplification in a single tube), which finally allowed fluorescence detection of positive malaria samples using a simple flashlight [114].

7.4.2 Passive and Active Targeted Drug Delivery

7.4.2.1 Lipid-based nanoparticles

The term "liposomes" refers to synthetic assemblies at the nanoscale comprising one or several phospholipid bilayers enclosing an aqueous core. In this type of nanoparticles, both hydrophilic and hydrophobic molecules can be captured either in the aqueous lumen or in the lipophilic layers, respectively [115] (Fig. 7.10). For malaria therapy, these particles allow the encapsulation of drugs and can help to adjust their bioavailability and dose within the body, which can lessen toxic effects and the risk of drug resistance evolution [116, 117]. In general, liposomal formulation treatments are superior relative to the free drug, improving bioavailability, targeted delivery, and controlled release [30]. Liposomes are often used for specific targeting, which can be passive or active. Passive targeting is achieved using conventional liposomes and long-circulating liposomes (usually PEGylated). On the contrary, active targeting is promoted by the surface activation of the liposome through the inclusion of ligands such as glycolipids, peptides, antibodies, or other proteins [118].

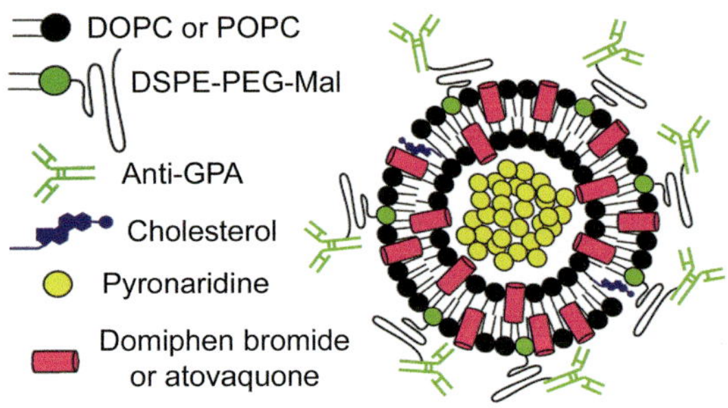

Figure 7.10 Scheme of an immunoliposome designed for targeted antimalarial combination therapy. DOPC: 1,2-dioleoyl-*sn*-glycero-3-phosphocholine; POPC: 1-palmitoyl-2-oleoyl-sn-glycero-3-phosphocholine; DSPE: 1,2-distearoyl-*sn*-glycero-3-phosphorylethanolamine; Mal: maleimide; GPA: glycophorin A. Reprinted with permission from Ref. [131].

Liposomes are an excellent tool for targeting the hepatic forms of malaria parasites, mainly due to the high accumulation levels (between 70–80%) of this type of nanocarrier in the liver Kupffer cells after oral or parenteral delivery [119]. Indeed, liposomized primaquine, an effective drug against intrahepatic forms of all *Plasmodium* species, showed 100% antimalarial activity in a murine model [120, 121], while unspecific toxicity was substantially reduced compared with the free drug. Moreover, primaquine accumulation in liver and spleen was observed together with a significant increase in the circulation time of the encapsulated drug [122]. Liposome composition can also be modulated to achieve selective targeting. As an example of this, Arica et al. evaluated the influence of the liposomal membrane rigidity on primaquine tissue distribution and found that gel-state liposomes accumulated in the liver and spleen more than liquid-state liposomes [123]. Another study showed that the size of liposomes also affected their distribution: Hepatocytes were more efficiently reached by smaller liposomes, although, due to their small size, the amount of encapsulated primaquine was low [124].

Both passive and active targeting have been explored to arrest the growth of intraerythrocytic forms of malaria parasites. In these studies, different lipid compositions and liposome physicochemical properties have been tested to achieve the maximum encapsulation rates and antimalarial activity. For example, Crommelin et al. evaluated the antimalarial efficacy of encapsulated chloroquine versus the free drug and found that the liposomal formulation had a higher ability to prevent and treat *Plasmodium berghei* (*P. berghei*) infection in a murine model [116], in some cases reaching 100% protection [125]. Moreover, they found that the different administration routes and rigidities of the liposomal membrane led to different pharmacological activities and toxicities [125, 126]. It is important to mention that, in liposomes, the overall lipid composition and charge has a strong influence on encapsulation rates. This has to be considered together with the charge of the drug cargo, as in most cases the loading method is based on transmembrane pH gradients. Following this line, Moles et al. designed a mathematical model to predict drug encapsulation as a function of pH and phospholipid compositions, focusing on liposomal RBC delivery [127] (Fig. 7.11). Moreover, the authors demonstrated that their model fitted experimental data, providing an excellent tool for liposomal formulations in the field of malaria.

Active targeting has been achieved *in vitro* and *in vivo* with several formulations. For instance, the administration of chloroquine encapsulated in liposomes targeted to pRBCs through a covalently linked antibody against surface antigens in drug-resistant *P. berghei* reduced parasitemia between 75% and 90% in infected mice [128]. On the other hand, phosphatidylcholine liposomes decorated with a short peptide from the CSP of *P. berghei* (responsible for glycosaminoglycan recognition in the liver stage) were successfully used to treat infected mice and showed high accumulation in hepatocytes and nonparenchymal cells [129]. Using a similar approach, Moles et al. designed immunoliposomes incorporating monoclonal antibodies directed against the erythrocyte surface protein glycophorin A and observed that these particles bound to both RBCs and pRBCs with 100% efficiency. Moreover, when loaded with chloroquine, these liposomes dramatically improved parasite clearance in *P. falciparum*-infected mice [130]. In a follow-up study by the same group, liposomes were designed to deliver a combination therapy to pRBCs. In this case, either domiphen bromide or atovaquone was encapsulated in the membrane, whereas pyronaridine was enclosed in the lumen of the liposome. In all tested formulations, the activities of the drugs were significantly higher than their free forms in *in vitro P. falciparum*-infected RBC cultures [131].

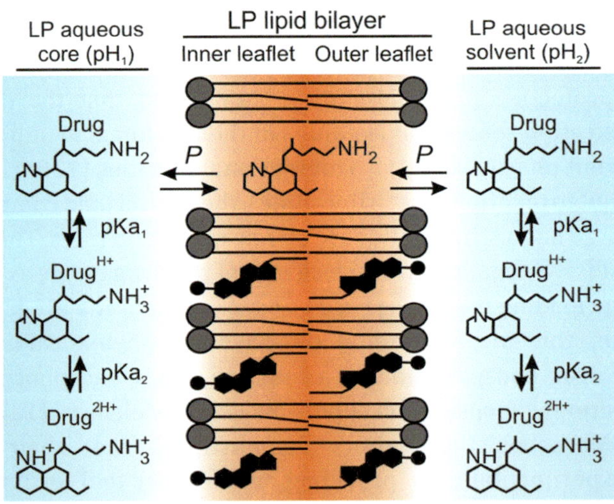

Figure 7.11 Distribution model adapted for amphiphilic diprotic basic drugs in liposomal systems. The illustration presents primaquine as an example. LP: liposome; *P*: partition coefficient. Reprinted with permission from Ref. [127].

Additionally, other lipid-based nanoparticles have also been tested for antimalarial delivery. Nanostructured lipid carriers (NLCs, also named deformable liposomes) are composed of both solid and liquid biocompatible lipids and co-surfactants, which are able to transport hydrophilic and lipophilic drugs [132] (Fig. 7.12A). These second-generation nanomaterials were tested by Parashar et al., who developed NLCs co-loaded with artemether and lumefantrine and tested them in *P. berghei*–infected mice, where they showed an improved activity when compared with the free drugs [133]. Another study of an NLC loaded with artemether reported an amelioration of therapeutic activity in cerebral malaria-induced mice [134]. Other drugs, including arteether [135], miconazole nitrate [136], and curcumin [137], have also been encapsulated within NLCs with promising results. More recently, Dong et al. designed heparin-decorated NLCs to co-encapsulate artemether with protoporphyrin IX and iron-loaded transferrin [138]. The nanoformulation showed specific and efficient targeting into intraerythrocytic parasites as well as an increased inhibitory activity against *P. falciparum*–infected RBCs and *P. berghei*–infected mice, thus postulating itself as an interesting antimalarial strategy [138].

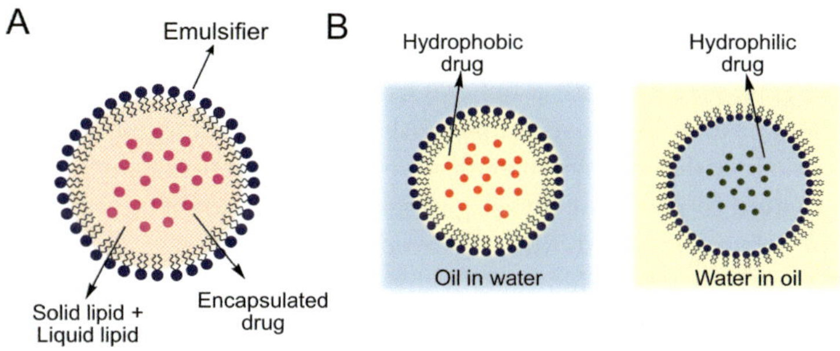

Figure 7.12 Lipid-based nanoparticles. **(A)** Typical composition of a nanostructured lipid carrier. **(B)** Sub-classes of nanoemulsions.

Lipid nanoemulsions are oil-in-water or water-in-oil dispersions widely used for the encapsulation and active delivery of bioactive lipophilic compounds [139] (Fig. 7.12B). Primaquine (the only drug used to treat the relapsing form of malaria) incorporated in oral lipid nanoemulsions had activity against *P. berghei*–infected mice and improved oral bioavailability [140]. Curcuminoid-loaded nanoemulsions have also been proved to efficiently increase antimalarial activity when compared to free curcuminoids [141]; similarly, dihydroartemisinin-loaded nanoemulsions exhibited good clearance of parasitemia in infected mice [142]. Finally, these nanomaterials have also been used to carry antisense oligonucleotides against plasmodial topoisomerase II, imparting them higher stability and antimalarial activity [143].

7.4.2.2 Nucleic acid–based nanoparticles

Nucleic acids are also used to functionalize nanoparticles for the treatment of malaria. Specific silencing of essential genes by antisense oligomers has been proposed as a strategy to block malaria development and minimize drug resistance [144]. For instance, phosphorothioate antisense oligodeoxynucleotides (ODNs) [145] (Fig. 7.13A) complexed with the biodegradable polymer chitosan to form solid nanoparticles were tested in *in vitro P. falciparum*–infected RBCs, showing a stronger inhibitory effect (~87% parasite growth inhibition) compared with free ODNs (~68% inhibition). Moreover, the nanoparticles were found to protect ODNs from nuclease degradation, thus extending their half-life [146]. Chitosan nanoparticles have been further explored to deliver double-stranded RNA for targeting the DNA topoisomerase II gene from *P. falciparum* in pRBC cultures, which resulted in a 71% growth inhibition [147].

Peptide nucleic acids (PNAs) are metabolically stable DNA mimic molecules that efficiently hybridize to complementary RNA [148] (Fig. 7.13B). With a neutral backbone, PNAs should not have delivery issues as those found with negatively charged antisense oligonucleotides. Kolevzon et al., who used cell-penetrating peptides conjugated to PNAs for investigating their gene-silencing activity in *P. falciparum*–infected RBCs, proved this hypothesis. They found that these nanoparticles specifically downregulated the expression of the targeted endogenous *Pfsec13* gene, involved in vesicular transport, which resulted in a significantly reduced parasite viability in cultured pRBCs [149].

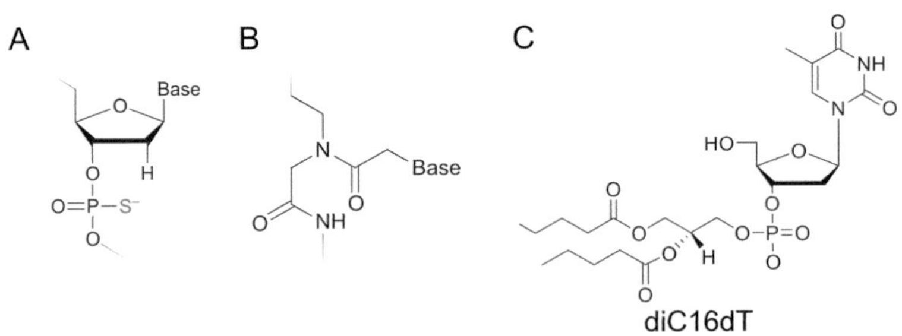

Figure 7.13 Nucleic acid–based nanoparticles. Chemical structure of **(A)** phosphorothioate antisense oligodeoxynucleotide, **(B)** peptide nucleic acids, and **(C)** example of a nucleolipid. Reprinted with permission from Refs. [145] and [153].

Nucleolipid supramolecular assemblies are also promising drug delivery systems, particularly for nucleic acids. They are made of lipid chains and base–base interactions [150, 151] (Fig. 7.13C) and have been identified in eukaryotic and prokaryotic cells [152]. The antimalarial drug methylene blue has been encapsulated in this type of carriers using the anionic nucleotide-lipid diC16dT (thymidine 3′-[1,2-dipalmitoyl-*sn*-glycero-3-phosphate]), and the cationic nucleoside-lipid DOTAU (2′,3′-dioleyl-5′-deoxy-5′-trimethyl-ammonium-uridine). In both cases, the nano-assemblies prevented early drug degradation but failed in increasing its protective effect against *P. falciparum* [153].

7.4.2.3 Polymer-based nanoparticles

Polymer-based nanoparticles are biodegradable and biocompatible solid-colloidal structures ranging from 1 to 1000 nm in size. One of the most common nanoparticles within this class are made of chitin-derived chitosan, although other natural polymers are also used for their production, including alginate, hydroxyapatite, and hyaluronic acid [154]. Chitosan-tripolyphosphate nanoparticles encapsulating chloroquine were tested in *P. berghei*–infected mice, where they showed therapeutic effect at a lower dose relative to the free drug [155]. In a follow-up study, both chloroquine and nanoconjugated chloroquine administered to *P. berghei*–infected mice increased antioxidant levels in liver and spleen while reducing stress markers and pro-inflammatory cytokines, but the nanoparticle formulation was more efficient in carrying out these processes, indicating its potential use in the treatment of malaria [156].

Curcumin is a good antimalarial but has poor bioavailability and stability; to overcome these limitations, it was encapsulated in chitosan nanoparticles, which resulted in higher stability and drug uptake by mouse RBCs, as compared to free curcumin. Moreover, this nanoformulation improved the drug bioavailability in plasma when delivered orally and showed a partial protective effect in infected mice [157]. In a different approach, chitosan and chondroitin sulfate were employed to generate artemether-loaded nanoparticles for transdermal delivery [158]. Although antimalarial activity was not tested, penetration through skin in *ex vivo* studies and the release of artemether from the nanoparticles were reported.

Dendrimers are nano-sized synthetic polymers, radially symmetric with well-defined, homogeneous, and monodisperse structures consisting of tree-like branching units built around a small molecule or a linear polymer core [159]. These macromolecules consist in an exterior layer containing functional groups, useful for the conjugation of targeting moieties, and an inner core where most of the drug cargo is contained [160]. Dendrimers based on 2,2-bis (hydroxymethyl) propionic acid (bis-MPA) and Pluronic polymers (Fig. 7.14) containing chloroquine and primaquine were tested for their targeting ability and antimalarial activity against *Plasmodium*-infected RBCs. The encapsulated compounds showed

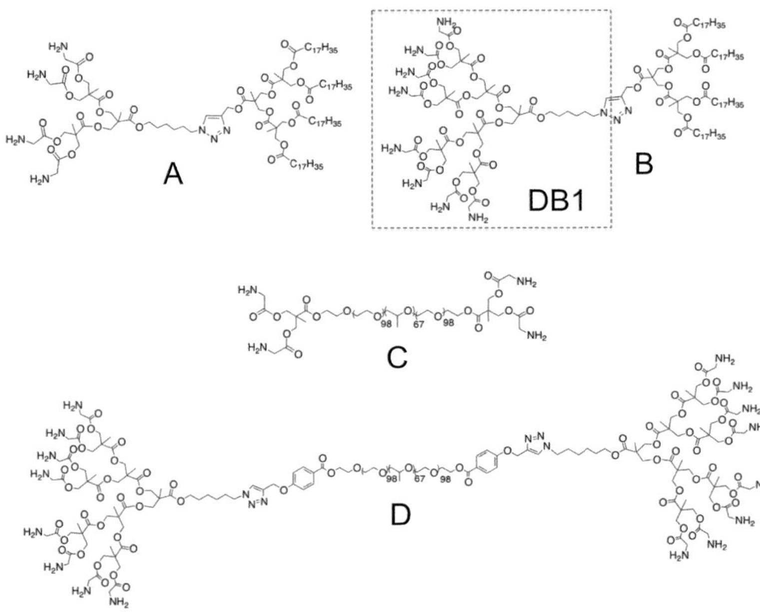

Figure 7.14 Molecular structures of dendrimers used for antimalarial drug encapsulation. **(A)** and **(B)** represent Janus dendrimers, DB1 a dendron, and **(C)** and **(D)** two hybrid dendritic-linear-dendritic block copolymers. Reprinted with permission from Ref. [161].

specific targeting mechanisms to pRBCs compared to naïve RBCs, and a slightly improved *in vivo* therapeutic efficacy as demonstrated in a *Plasmodium yoelii* murine malaria model. The authors hypothesized that the specific targeting observed was due to the chemical structure of the dendrimers; in particular, the chemical interplay between copolymer and drug, the size, and the shape of the nanoparticles [161]. On the other hand, the improved antimalarial activity of the dendritic derivatives C-primaquine and D-chloroquine could be directly related to structures C and D as they are poloxamer derivatives that have been previously reported to form micelles (Fig. 7.15) [162] and to interact with cell membranes [163], which facilitates the internalization of their cargos. Martí Coma-Cros et al. explored the use of two more dendritic nanocarriers for antimalarial drugs: a hybrid dendritic-linear-dendritic block copolymer, and an amino-terminated dendronized hyperbranched polymer that formed unimolecular micelles. Both macromolecules were able to enter pRBCs and, when loaded with chloroquine, inhibited *Plasmodium* growth *in vitro* and *in vivo* [164]. Another reported approach was the generation of poly-L-lysine dendrimers having PEG-1000 as core, loaded with chloroquine. These nanocarriers offered a better protection of the drug and a more controlled delivery compared to its free formulation both *in vitro* and *in vivo* [165].

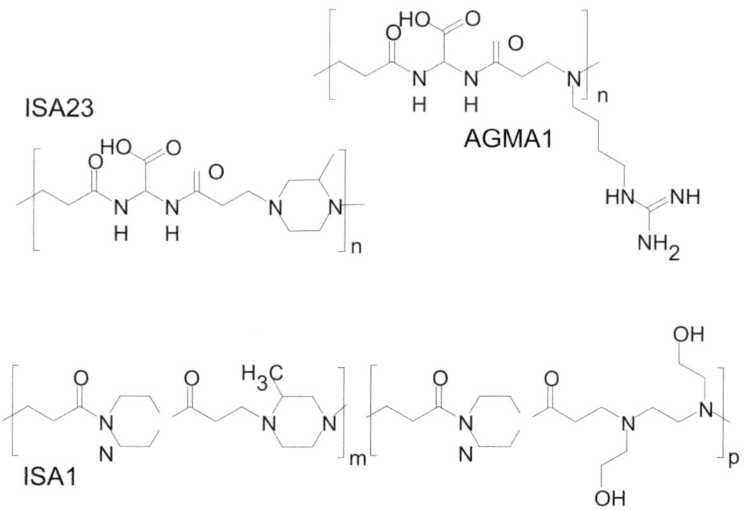

Figure 7.15 Molecular structures of the polyamidoamines AGMA1, ISA23, and ISA1. Reprinted with permission from Ref. [168].

Polyamidoamines (PAA) are a sub-class of dendrimers composed of repetitively branched subunits of amide and amine moieties [166]. Due to the long blood circulation time of some PAAs, these polymers have been used for drug delivery [167]. In the case of malaria research, the PAA structures AGMA1, ISA1, and ISA23 (Fig. 7.15) have been tested for the specific delivery of chloroquine and primaquine to pRBCs [168]. The polyelectrolyte nature and the presence of amine groups in PAAs aid in the specific internalization of the nanoparticles in pRBCs through the new permeation pathways [169]. Moreover, *in vivo* assays in *P. yoelii*–infected mice showed that animals treated with chloroquine-loaded PAAs were cured from infection, demonstrating the potential use of this type of polymers for antimalarial therapy [168].

Last, but not least, polymers have also been extensively used to generate nanotechnological strategies for the improvement of malaria vaccines as will be discussed later.

7.4.2.4 Protein-based nanoparticles

Protein nanoparticles represent an attractive alternative to the synthetic polymers commonly used in nanotechnology as they have several advantages, including higher biocompatibility and biodegradability. These nanoparticles can be generated using proteins such as fibroin, albumin, gelatin, gliadine, keratin, silk, collagen, elastin, and ferritin, among others [170, 171]. There is a growing number of studies dealing with the production of protein-based nanoparticles carrying antimalarial drugs. Gelatin is a widely used protein in nanotechnology [172], being also employed as an adjuvant and as a plasma extender system [173]. Bajpai et al. designed gelatin nanoparticles loaded with chloroquine and demonstrated release of the encapsulated drug, although its activity was not tested in parasitized cells [174]. Nanoparticles based on human serum albumin were employed to deliver artemether into pRBCs, with results showing selective internalization into these cells and a 50% reduction in the drug dose required to achieve IC_{50} levels of inhibition in *P. berghei*–infected mice [175].

Self-assembling protein nanoparticles (SAPNs) are designed using coiled-coil protein-folding motifs and have been used to develop vaccines for different infectious diseases [176]. Kaba et al. produced different SAPN formulations containing epitopes from the *P. falciparum* CSP (*Pf*CSP) and portions of the Toll-like receptor 5 agonist flagellin as an intrinsic adjuvant (Fig. 7.16), designated as FMP014. They found an optimal ratio of the two proteins that showed the highest protective effect against a challenge with transgenic *P. berghei* sporozoites expressing *Pf*CSP, as indicated by antibody titers [177].

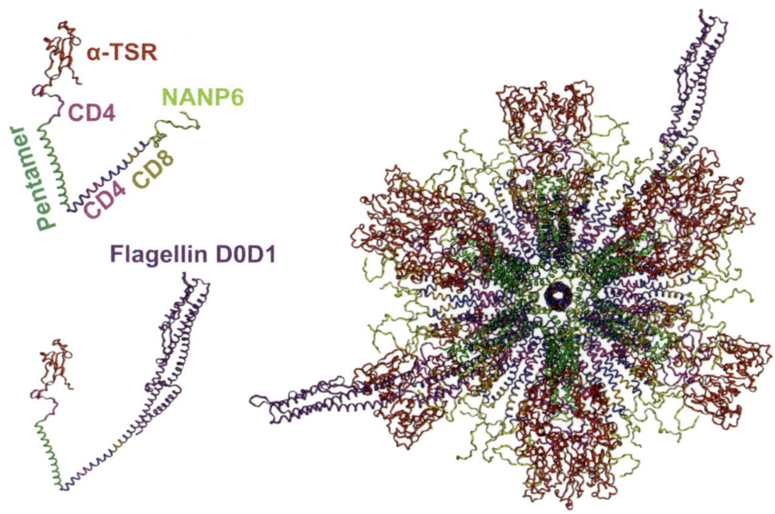

Figure 7.16 Computer model of SA-FMP014$_{58:2}$ SAPN co-assembly. On the left, an FMP014 monomer showing the relative positions of the NANP repeats, the αTSR region, the two CD4 epitopes and a flagellin D0D1 monomer. On the right, model of an assembled SA-FMP014 containing 58 FMP014 monomers and 2 flagellin D0D1 monomers. NANP$_6$: NANP repeat peptide. αTSR: carboxy-terminal region of the PfCSP. Adapted from Ref. [177].

7.4.2.5 Green nanoparticles

Metal-derived nanoparticles (mainly made of zinc, platinum, gold, silver, and palladium) are extensively used in a myriad of everyday products, including soaps, detergents, cosmetics, and pharmaceutical carriers [178]. Their production is driven by different methods, including lithography, UV irradiation, aerosol techniques, ultrasonic fields, and photochemical reduction [179]. Metal nanoparticles have been investigated for biomedical applications not only due to their non-specific bacterial toxicity but also for their effectiveness against pathogens listed as priority [180], including malaria parasites. Their antimicrobial activity is associated with different mechanisms, including (i) electrostatic-mediated attachment of the nanoparticles to membranes, which leads to disruption of cell walls and an increase in their permeability [181]; (ii) rise in the production of reactive oxygen species (ROS), leading to the oxidation of glutathione, thus suppressing the antioxidant mechanisms of bacteria against ROS [182]; (iii) binding of metal ions to cellular structures (*e.g.*, proteins or DNA), disrupting cell functions [183], and (iv) inhibition of ATP production/recycling [184], among others. However, despite their broad range of applications, metal-derived nanoparticles are often toxic for the host as they can promote oxidative stress and ROS production similar to the effect observed in the pathogenic agent.

Due to the wide usage of this type of nanoparticles, there is a growing need to develop processes for synthesizing them in an environmentally sustainable way. In this regard, biological methods using microorganisms, enzymes, and plant extracts have been suggested as possible ecologically friendly alternatives to generate them [185–187]. Biogenic reduction is a process similar to chemical reduction where a reducing agent is replaced by an extract of natural products with inherent stabilizing, growth-terminating, and capping properties [188]. In addition, green synthesized nanoparticles are advantageous because they are cost-effective as compared to conventional methods [189]. The most commonly used approach to biosynthesizing metal nanoparticles is using polyphenols, which are widely distributed compounds in plants [190].

Different publications have reported the preparation of silver nanoparticles using green synthesis from plant extracts, whose antimalarial activity has been tested. For instance, the antiplasmodial activity of silver nanoparticles synthesized using *Andrographis paniculata* leaves as capping agent was found to be around 83% by using 100 μg/mL of the generated nanoparticles [191]. Other examples of plants used for synthesizing silver nanoparticles with antimalarial activity include extracts of *Catharanthus roseus* [192], Ashoka and Neem [193], *Euphorbia hirta* [194], the seaweed *Ulva lactuca* [195], *Pteridium aquilinum* [196], *Azadirachta indica* [197], and *Ocimum sanctum* [198]. Other biological sources for green synthesis of silver nanoparticles with antimalarial applications have also been explored, such as the spongeweed *Codium tomentosum* [199] and *Eudrilus eugeniae* earthworms [200].

Gold nanoparticles biosynthesized using a cheap flower, *Couroupita guianensis,* were successfully tested against a chloroquine-resistant (CQ-r) *P. falciparum* strain [201]. Bark and leaf extracts of *Syzygium jambos* were employed by Dutta et al. to produce gold and silver nanoparticles, and their antiplasmodial potential in CQ-r and wild-type strains was compared with chemically synthesized nanoparticles. Results showed that biosynthesized nanoparticles had higher antimalarial activity and biocompatibility as compared to the chemically produced nanomaterials, thus suggesting that the antiparasitic activity is largely dependent on the nanoparticle microenvironment [202]. On the other hand, palladium nanoparticles synthesized using *Eclipta prostrata* extracts [203] and TiO_2 nanoparticles produced using *Momordica charantia* leaf extracts have also displayed protective antiplasmodial activity [204].

Finally, magnetic nanoparticles naturally produced by *Magnetospirillum gryphiswaldense*, a gram-negative bacteria found in water and sediments, were effective against CQ-r *P. falciparum* in *in vitro* cultures. However, the *in vivo* efficacy of these nanoparticles was lower than that of chloroquine-based treatments [205].

7.4.3 Vector Control Applications

Mosquito vectors benefit from new scenarios brought about by climate change and the increasing resistance to insecticides. Therefore, some current strategies focus on finding new ways of removing mosquito populations [23]. Nanotechnology-based methods have revolutionized a wide variety of research fields, including entomology.

Chitosan nanoparticles have many advantages over other nanomaterials, including greater stability, biodegradability, biocompatibility and water solubility, controlled drug release, lower toxicity, non-immunogenicity, and ease of fabrication, thus making them ideal candidates for mosquito control. Indeed, chitin is a major component of all insects, including mosquitoes [206]. Its biosynthesis depends on the chitin synthases, which catalyze the transfer of sugars from activated carbohydrate donors to acceptors present in mosquitoes [207]. Indeed, chitin synthase genes *CHS-A* and *CHS-B* are good target candidates as chitin is not synthesized in vertebrates [208]. Along this research line, chitosan/dsRNA-based nanoparticles were fed to *Anopheles gambiae* larvae to induce post-transcriptional gene silencing of *CHS-A* and *-B* [209]. Although the generated nanoparticles did not lead to larval mortality even at high concentrations, the authors successfully demonstrated the potential use of RNAi for mosquito control, perhaps through its incorporation into a pest management programme to increase the efficacy of insecticides [209].

Nanoemulsions from the essential oils of *Mentha* spp. and *Eucalyptus* spp. have shown greater repellent activity compared to N,N-diethyl-*m*-toluamide, the standard chemical compound against *Anopheles stephensi* mosquitoes, an Asian malaria vector [210]. Other essential oils such as estragon [211], dill [212], castor oil [213], cinnamon [214], and rose-scented geranium [215] have been used to produce nanoemulsions with larvicidal activity.

The misuse of chemical insecticides has led to environmental toxicity hazards and ecological problems; hence, green synthesized nanoparticles provide a safer alternative. For instance, silver nanoparticles produced using *Dicranopteris linearis, Chenopodium ambrosioides, Aristolochia indica, Gracilaria edulis, C. guianensis,* and *Phyllanthus niruri* plant extracts were used against *Anopheles* mosquitoes, showing significant larvicide activity [216, 217].

7.4.4 Vaccines

As mentioned earlier, the generation of an effective malaria vaccine has remained elusive mainly due to the complexity of the parasite life cycle and to its widespread antigenic variation. Advances in nanotechnology during the recent years have generated important contributions to vaccine development tools, enabling the precise and independent control of material parameters that have an effect on the immunological process [218].

Superparamagnetic nanoparticles are based on a core consisting of iron oxides that are responsive to external magnets and, therefore, can be targeted to specific disease sites. The core can be further functionalized with small drugs, proteins, or plasmids [219]. These particles were used by Al-Deen et al. to deliver via magnetofection a malaria DNA vaccine encoding a *P. yoelii* merozoite surface protein. This study showed the ability of the nanostructure to protect DNA molecules, with a low unspecific cytotoxicity as well as a high *in vitro* efficiency of the vaccine, as indicated by protein expression analysis [220].

SAPNs can overcome problems associated with linear peptide antigens that cannot generate protective immune responses and degrade in a rapid manner. Additionally, these particles allow the insertion of multiple epitope targets, each of them stimulating a different targeted immune response [221]. Using a wide screen for alpha-helical coiled coil motifs, Kulangara et al. identified the trophozoite export protein 1, which contains two potential vaccine candidates: the unstructured region P27A and the coiled-coil domain P27 [222]. These resulting particles triggered high levels of immunogenicity in a murine model and induced immune responses similar to those observed in immunized individuals from malaria-endemic regions [223]. Finally, a different construct that integrated *P. falciparum* antigens from different life cycle stages was tested in mice and rabbits and showed different immunogenicity depending on antigen presentation [224]. This indicates the importance of a correct folding and orientation, which can be stabilized by SAPNs, and their influence in vaccine formulations.

Recently, biodegradable pullulan-coated iron oxide nanoparticles in the viral particle size range were synthesized and tested as malaria vaccine carriers. The nanoparticles had very low toxicity and led to the induction of specific antibodies by using as antigen a combination of the merozoite surface proteins 4 and 5 in a murine model, a response similar to that observed with the use of classical adjuvants [225].

Huang et al. reported a new malaria vaccine technology using spontaneous nanoliposome antigen presentation through the mixing of preformed liposomes containing cobalt porphyrin-phospholipid and the poly(histidine)-tagged *Pfs*25 *P. falciparum* sexual stage protein, thus creating a simple system to assemble into liposomes any antigen modified with His-tags. With this approach, the vaccine was able to generate long-lasting functional antibodies against malaria in mice and rabbits [226].

7.5 Limitations in the Use of Nanotechnology Approaches in Malaria

As a general rule, the nanotechnological approaches used for malaria must be affordable and robust; they should ideally be self-propagating and capacity-building in order to engage local communities at the beginning of the engineering process, not at its end [64]. For instance, nanotechnologies designed as a set of protocols, modular experiments, and ready-to-run tests could be more easily integrated into the existing practice of low- and middle-income countries [64]. Unfortunately, most of the published nanotechnology literature in this field does not contemplate issues associated with local supply chains, and neither accounts for possibilities of domestic invention and fabrication [93]. Therefore, it is important to adapt protocols to the realities of local resources.

An additional limitation in the use of nanotechnology in malaria is related to the explored nanovectors for targeted delivery. It is well known that most antimalarial drugs need to cross several membranes: host RBC, parasitophorous vacuole, and parasite plasma membranes and, in some cases, the food vacuole or an organelle membrane, depending on the site of action of the drug. Nanovectors often fail to reach their target sites due to the low endocytic activity of mature RBCs. This is especially true in complex environments, such as *in vivo* systems, where different endocytic cells are present, often resulting in the failure of nanoparticles to reach their target even when their activity had been already demonstrated in *in vitro* cultures.

Diagnostic devices for global health should follow the ASSURED (affordable, sensitive, specific, user-friendly, rapid and robust, equipment-free, and deliverable to end user) criteria issued by the WHO to ensure their robustness in low resource settings. However, apart from paper-based technologies, the nanotechnological approaches proposed until now for malaria diagnosis fail to comply with these rules.

Finally, a barrier that malaria researchers often find is that the funding to translate their basic research into an applied technology is significantly inferior to the funding allocated for other infectious diseases prevalent in high-income regions.

7.6 Conclusion

As already proven for other diseases, nano-based approaches might improve the treatment and prevention of malaria (see Table 7.1). In the therapeutic field, the drawbacks limiting the efficacy of drugs against this disease, such as evolution of resistance, poor bioavailability, and long administration regimens, can be tackled by nanotechnology, which offers different approaches to address these issues. Unfortunately, these strategies have been, so far, limited to preclinical demonstrations mainly due to economic reasons.

Most malaria infections and deaths (94% of the whole) occur in Sub-Saharan Africa, one of the poorest regions on the planet. In general, malaria is a disease that imposes significant costs to both individuals and governments and, in most cases, treatments and prevention tools against the disease rely on the aids from public and private institutions from other countries. In the past few years, the decrease in malaria burden has stalled due to several reasons, including increased mosquito resistance to insecticides, evolution of parasite resistance to drugs, climate changes, and intensive population growth [203]. It is, therefore, necessary to design alternative tools that are not only highly effective but also affordable for low-income countries. In this sense, nanotechnology offers a wide spectrum of alternative approaches that have just begun being explored, such as nanovectors for targeted delivery and vaccine antigen presentation, nanodevices for disease diagnostics, or nanocarriers for new insecticides, to mention some.

In order to make a real impact on malaria, nano-based approaches must incorporate ethical, environmental, economic, legal, social, and cultural (E^3LSC) considerations, as defined by Salamanca-Buentello and Daar [227]. This translates into an equitable distribution of the innovations, waiving of intellectual properties (making nanotechnologies more affordable), and development of nanotechnology-based medical platforms in malaria high-risk countries.

Appendix

Table 7.1 Nanoparticles used for the treatment of malaria

Type	Antimalarial Agent	Target Species	Main Findings	Ref.
1. Lipid-based nanoparticles				
Liposome	Primaquine	*P. berghei*	Less drug-derived toxicity compared to the free form and 100% experimental antimalarial activity in mice with a single dose. The liposomes accumulated in the Kupffer cells after oral/parenteral delivery.	[117–120]
Liposome	Chloroquine	*P. berghei*	The liposomal formulation allowed the use of higher doses, which were efficient against CQ-resistant *P. berghei*, resulting in 100% protection.	[112, 122]
Liposome	Chloroquine	*P. berghei*	The inclusion of a targeting element (*Plasmodium*-infected erythrocyte-specific antibody) increased the specificity toward *P. berghei*-infected cells and reduced parasitemia levels between 75 and 90% in infected mice.	[124]
Liposome	—	*P. berghei*	The incorporation of a *Plasmodium* amino acid sequence proved to be effective for the specific delivery of agents into infected cells.	[125]

Type	Antimalarial Agent	Target Species	Main Findings	Ref.
Liposome	Chloroquine, primaquine, domiphen bromide, atovaquone	*P. falciparum* *P. berghei*	Anti-GPA carrying liposomes exhibited specific targeting to erythrocytes. Inmunoliposomes also showed a substantial increase in antimalarial activity in *in vitro* *Plasmodium* cultures and in the murine malaria model.	[126, 127]
NLC	Artemether and lumefantrine	*P. berghei*	Improved antimalarial activity compared to the free drug.	[129]
NLC	Artemether	*P. falciparum*	Better parasite clearance and higher survival in comparison to the free drug was observed in the infected mice.	[130]
NLC	Arteether	*P. berghei*	A complete eradication of the parasite from the blood was observed following administration of the nanoparticles to infected mice.	[131]
NLC	Miconazole nitrate	*P. berghei*	The nanoparticles showed good antimalarial activity in a murine malaria model.	[132]
NLC	Curcumin	*P. berghei*	The carriers showed improved anti-plasmodial activity, bioavailability, and chemical stability compared to free curcumin.	[133]
NLC	Artemether	*P. falciparum* *P. berghei*	Specific and efficient targeting to intraerythrocytic stages of *Plasmodium* as well as an increased inhibitory activity compared to the free drug.	[134]

(Continued)

Table 7.1 (*Continued*)

Type	Antimalarial Agent	Target Species	Main Findings	Ref.
Nanoemulsion	Primaquine	*P. berghei*	Design of an oral nanoemulsion that improved bioavailability and showed a better antimalarial activity.	[136]
Nanoemulsion	Dihydroarte-misinin	*P. berghei*	Improved antimalarial activity.	[138]
Nanoemulsion	Antisense oligonucleotide	*P. falciparum*	Efficient delivery to pRBCs and good antimalarial activity.	[139]
2. Nucleic acid–based nanoparticles				
Solid nanoparticles	ODNs	*P. falciparum*	Higher stability and stronger inhibitory effect compared with free ODNs.	[142]
Chitosan nanoparticles	dsRNA	*P. falciparum*	Good inhibitory activity and extended half-life or the dsRNA.	[143]
PNAs-CPP	PNAs	*P. falciparum*	Demonstration of the potential use of antisense PNA molecules to downregulate gene expression in the intraerythrocytic stages of *Plasmodium.*	[145]
Nucleolipid supramolecular assembly	Methylene blue	*P. falciparum*	Prevention of early drug inhibition.	[149]
3. Polymer-based nanoparticles				
Chitosan nanoparticles	Chloroquine	*P. berghei*	Inhibitory effect with a lower dose relative to the free drug.	[151, 152]
Chitosan nanoparticles	Curcumin	*P. berghei*	Higher stability and drug uptake compared to free curcumin.	[153]

Type	Antimalarial Agent	Target Species	Main Findings	Ref.
Chitosan-chondroitin sulfate nanoparticles	Artemether	—	Despite antimalarial activity was not tested, good penetration through the skin and release of the drug was reported *ex vivo.*	[154]
Dendrimers: bis-MPA and Pluronic	Chloroquine and primaquine	*P. falciparum* *P. yoelii*	Specific targeting toward pRBCs in cultures. Slightly improved antimalarial activity in a murine malaria model.	[157]
Dendrimers: HDLDBC-bGMPA and DHP-bMPA	Chloroquine, primaquine, and quinacrine	*P. falciparum* *P. yoelii*	High internalization rates into pRBCs and inhibition of *Plasmodium* growth *in vitro* and *in vivo.*	[160]
Dendrimers: poly-L-lysine and PEG-1000	Chloroquine	*P. falciparum*	Galactose-coated dendrimers could be used for controlled delivery of chloroquine more safely compared to its uncoated formulation both *in vitro* and *in vivo.*	[161]
Polyamidoamines: AGMA1, ISA1, and ISA23	Chloroquine and primaquine	*P. falciparum* *P. yoelii*	Specific internalization into pRBCs and 100% inhibition of *Plasmodium* growth *in vivo.*	[164]
4. Protein-based nanoparticles				
Gelatin nanoparticles	Chloroquine	—	High release of the encapsulated drug.	[169]
Human serum albumin nanoparticles	Artemether	*P. berghei*	Selective internalization into pRBCs and 50% reduction in the dose required to achieve the IC_{50}.	[170]
SAPN	*Pf*CSP epitopes	*P. falciparum*	The self-adjuvanted SAPNs were highly immunogenic and protective in the model system, making them an attractive new vaccine platform.	[172]

(*Continued*)

Table 7.1 (*Continued*)

Type	Antimalarial Agent	Target Species	Main Findings	Ref.
5. Green nanoparticles				
Silver nanoparticles	—	*Plasmodium*	Protective antiplasmodial activity.	[186–195]
Gold nanoparticles	—	*P. falciparum*	Good antimalarial activity against chloroquine-resistant strains.	[196, 197]
Palladium nanoparticles	—	*Plasmodium*	Protective antiplasmodial activity.	[198]
TiO_2 nanoparticles	—	*Plasmodium*	Protective antiplasmodial activity.	[199]
Magnetic nanoparticles	—	*P. falciparum*	Effective against chloroquine-resistant strains *in vitro*.	[200]

Acknowledgments

ISGlobal and IBEC are members of the CERCA Programme, Generalitat de Catalunya. We acknowledge support from the Spanish Ministry of Science, Innovation and Universities through the "Centro de Excelencia Severo Ochoa 2019-2023" Program (CEX2018-000806-S). This research is part of ISGlobal's Program on the Molecular Mechanisms of Malaria, which is partially supported by the Fundación Ramón Areces.

Funding

This work was funded by (i) Ministerio de Ciencia e Innovación/Agencia Estatal de Investigación (MCIN/AEI/10.13039/501100011033), grant numbers PCIN-2017-100 and PID2021-128325OB-I00, which included FEDER funds; (ii) ERA-NET Cofund EURONANOMED (http://euronanomed.net/), grant number 2017-178; and (iii) Generalitat de Catalunya, Spain (http://agaur.gencat.cat/), grant number 2021-SGR-00635.

References

1. World Health Organization (WHO), *World Malaria Report 2023*. Geneva: World Health Organization; 2023. License: CC BY-NC-SA 3.0 IGO.
2. B. B. Panda and R. K. Hazra, Nanotechnology and its role in malaria treatment, in *Green Nanoparticles: Synthesis and Biomedical Applications*, edited by J. K. Patra, L. F. Fraceto, G. Das, and E. V. Ramos Campos (Springer Nature, 2020), pp. 347–358.
3. M. E. Sinka et al., A global map of dominant malaria vectors, *Parasites Vectors* **5**, 69 (2012).
4. R. E. Howes, et al., Global epidemiology of *Plasmodium vivax. Am. J. Trop. Med. Hyg.* **95**(6 Suppl), 15 (2016).
5. World Health Organization (WHO), *World Malaria Report 2019*. Geneva: World Health Organization; 2019. License: CC BY-NC-SA 3.0 IGO.
6. World Health Organization (WHO), *World Malaria Report 2021*. Geneva: World Health Organization; 2021. License: CC BY-NC-SA 3.0 IGO.
7. G. R. López del Prado, et al., Malaria in developing countries. *J. Infect. Dev. Countries* **8**(1), 1 (2014).
8. I. Kleinschmidt, et al., Marked increase in child survival after four years of intensive malaria control. *Am. J. Trop. Med. Hyg.* **80**(6), 882 (2009).
9. R. Amino, et al., Quantitative imaging of *Plasmodium* transmission from mosquito to mammal. *Nat. Med.* **12**(2), 220 (2006).
10. J. Tavares, et al., Role of host cell traversal by the malaria sporozoite during liver infection. *J. Exp. Med.* **210**(5), 905 (2013).
11. M. Prudêncio, A. Rodriguez, and M. M. Mota, The silent path to thousands of merozoites: The *Plasmodium* liver stage. *Nat. Rev. Microbiol.* **4**(11), 849 (2006).
12. K. Venugopal, et al., *Plasmodium* asexual growth and sexual development in the haematopoietic niche of the host. *Nat. Rev. Microbiol.* **18**(3), 177 (2020).
13. W. A. Krotoski, The hypnozoite and malarial relapse. *Prog. Clin. Parasitol.* **1**, 1 (1989).
14. A. Sturm, et al., Manipulation of host hepatocytes by the malaria parasite for delivery into liver sinusoids. *Science* **313**(5791), 1287 (2006).

15. J. M. Matz, J. R. Beck, and M. J. Blackman, The parasitophorous vacuole of the blood-stage malaria parasite. *Nat. Rev. Microbiol.* **18**(7), 379 (2020).

16. J. W. Field, Blood examination and prognosis in acute falciparum malaria. *Trans. R. Soc. Trop. Med. Hyg.* **43**(1), 33 (1949).

17. J. C. Rayner, et al., Dramatic difference in diversity between *Plasmodium falciparum* and *Plasmodium vivax* reticulocyte binding-like genes. *Am. J. Trop. Med. Hyg.* **72**(6), 666 (2005).

18. M. Dyer and K.P. Day, Commitment to gametocytogenesis in *Plasmodium falciparum. Parasitol. Today* **16**(3), 102 (2000).

19. A. H. Lee, L. S. Symington, and D. A. Fidock, DNA repair mechanisms and their biological roles in the malaria parasite *Plasmodium falciparum. Microbiol. Mol. Biol. Rev.* **78**(3), 469 (2014).

20. D. Vlachou, et al., The developmental migration of *Plasmodium* in mosquitoes. *Curr. Opin. Genet. Dev.* **16**(4), 384 (2006).

21. S. Bhatt, et al., The effect of malaria control on *Plasmodium falciparum* in Africa between 2000 and 2015. *Nature* **526**(7572), 207 (2015).

22. J. O. Yukich, et al., Costs and consequences of large-scale vector control for malaria. *Malar. J.* **7**, 258 (2008).

23. T. Ahmed, et al., Climatic conditions: Conventional and nanotechnology based methods for the control of mosquito vectors causing human health issues. *Int. J. Environ. Res. Public Health* **16**(17), 3165 (2019).

24. D. Petrić, et al., Monitoring population and environmental parameters of invasive mosquito species in Europe. *Parasites Vectors* **7**, 187 (2014).

25. J. B. Lima, et al., Is there an efficient trap or collection method for sampling *Anopheles darlingi* and other malaria vectors that can describe the essential parameters affecting transmission dynamics as effectively as human landing catches? A review. *Mem. Inst. Oswaldo Cruz* **109**(5), 685 (2014).

26. D. L. Kline, Traps and trapping techniques for adult mosquito control. *J. Am. Mosq. Control Assoc.* **22**(3), 490 (2006).

27. P. A. Phillips-Howard, et al., Efficacy of permethrin-treated bed nets in the prevention of mortality in young children in an area of high perennial malaria transmission in western Kenya. *Am. J. Trop. Med. Hyg.* **68**(4 Suppl), 23 (2003).

28. R. Pavela, Essential oils for the development of eco-friendly mosquito larvicides: A review. *Ind. Crops Prod.* **76**, 174 (2015).

29. K. A. Hasall, *The Chemistry of Pesticides: Their Metabolism, Mode of Action and Uses in Crop Protection* (Macmillan, 1982).
30. K. Rahman, et al., Nano-biotechnology: A new approach to treat and prevent malaria. *Int. J. Nanomed.* **14**, 1401 (2019).
31. Malaria Consortium, https://www.malariaconsortium.org (Accessed 02/09/2021).
32. A. Djènontin, et al., Managing insecticide resistance in malaria vectors by combining carbamate-treated plastic wall sheeting and pyrethroidtreated bed nets. *Malar. J.* **8**, 233 (2009).
33. H. C. Chapman, Biological control of mosquito larvae. *Annu. Rev. Entomol.* **19**, 33 (1974).
34. F. Mahmood, Laboratory bioassay to compare susceptibilities of *Aedes aegypti* and *Anopheles albimanus* to *Bacillus thuringiensis* var. israelensis as affected by their feeding rates. *J. Am. Mosq. Control Assoc.* **14**(1), 69 (1998).
35. C. M. Ingabire, et al., Community-based biological control of malaria mosquitoes using *Bacillus thuringiensis* var. israelensis (Bti) in Rwanda: Community awareness, acceptance and participation. *Malar. J.* **16**(1), 399 (2017).
36. A. R. North, A. Burt, and H. C. J. Godfray, Modelling the potential of genetic control of malaria mosquitoes at national scale. *BMC Biol.* **17**(1), 26 (2019).
37. L. Alphey, Genetic control of mosquitoes. *Annu. Rev. Entomol.* **59**, 205 (2014).
38. M. Cairns, et al., Estimating the potential public health impact of seasonal malaria chemoprevention in African children. *Nat. Commun.* **3**, 881 (2012).
39. I. B. Müller and J. E. Hyde, Antimalarial drugs: Modes of action and mechanisms of parasite resistance. *Future Microbiol.* **5**(12), 1857 (2010).
40. A. M. Daskum, et al., Antimalarial chemotherapy, mechanisms of action and resistance to major antimalarial drugs in clinical use: A review. *Microbes Infect. Dis.* **2**, 130 (2020).
41. M. E. Coldiron, L. Von Seidlein, and R. F. Grais, Seasonal malaria chemoprevention: Successes and missed opportunities. *Malar. J.* **16**(1), 481 (2017).
42. ACCESS-SMC Partnership, Effectiveness of seasonal malaria chemoprevention at scale in west and central Africa: An observational study. *Lancet* **396**(10265), 1829 (2020).

43. N. Patson, et al., Systematic review of statistical methods for safety data in malaria chemoprevention in pregnancy trials. *Malar. J.* **19**(1), 119 (2020).

44. O. Attaher, et al., Effect of seasonal malaria chemoprevention on immune markers of exhaustion and regulation. *J. Infect. Dis.* 221(1), 138 (2020).

45. N. Tangpukdee, et al., Malaria diagnosis: A brief review. *Korean J. Parasitol.* **47**(2), 93 (2009).

46. N. L. Garrett, et al., Bio-sensing with butterfly wings: Naturally occurring nano-structures for SERS-based malaria parasite detection. *Phys. Chem. Chem. Phys.* **17**(33), 21164 (2015).

47. P. Berzosa, et al., Comparison of three diagnostic methods (microscopy, RDT, and PCR) for the detection of malaria parasites in representative samples from Equatorial Guinea. *Malar. J.* **17**(1), 333 (2018).

48. L. K. Erdman and K. C. Kain, Molecular diagnostic and surveillance tools for global malaria control. *Travel Med. Infect. Dis.* **6**(1–2), 82 (2008).

49. T. E. Clendennen 3rd, G. W. Long, and J. K. Baird, QBC and Giemsa-stained thick blood films: Diagnostic performance of laboratory technologists. *Trans. R. Soc. Trop. Med. Hyg.* **89**(2), 183 (1995).

50. A. Berhane, et al., Rapid diagnostic tests failing to detect *Plasmodium falciparum* infections in Eritrea: An investigation of reported false negative RDT results. *Malar. J.* **16**(1), 105 (2017).

51. B. A. Mathison and B. S. Pritt, Update on malaria diagnostics and test utilization. *J. Clin. Microbiol.* **55**(7), 2009 (2017).

52. K. J. Wicht, S. Mok, and D.A. Fidock, Molecular mechanisms of drug resistance in *Plasmodium falciparum* malaria. *Annu. Rev. Microbiol.* **74**, 431 (2020).

53. G. Covell, A brief review of the history and development of the more important antimalarial drugs. *Indian J. Malariol.* **1**(2), 231 (1947).

54. M. Schlitzer, Malaria chemotherapeutics part I: History of antimalarial drug development, currently used therapeutics, and drugs in clinical development. *ChemMedChem* **2**(7), 944 (2007).

55. J. Achan, et al., Quinine, an old anti-malarial drug in a modern world: Role in the treatment of malaria. *Malar. J.* **10**, 144 (2011).

56. D. Greenwood, Conflicts of interest: The genesis of synthetic antimalarial agents in peace and war. *J. Antimicrob. Chemother.* **36**(5), 857 (1995).

57. H. Andersag, S. Breitner, and H. Jung, *Quinoline Compound and Process of Making the Same*. US patent US2233970A (1941).

58. G. R. Coatney, Pitfalls in a discovery: The chronicle of chloroquine. *Am. J. Trop. Med. Hyg.* **12**, 121 (1963).

59. S. Litsios, The World Health Organization's changing goals and expectations concerning malaria, 1948–2019. *Hist. Cienc. Saude Manguinhos* **27**(suppl 1), 145 (2020).

60. T. M. Davis, et al., Piperaquine: A resurgent antimalarial drug. *Drugs* **65**(1), 75 (2005).

61. H. Y. Myint, et al., Efficacy and safety of dihydroartemisinin-piperaquine. *Trans. R. Soc. Trop. Med. Hyg.* **101**(9), 858 (2007).

62. G. A. Balint, Artemisinin and its derivatives: An important new class of antimalarial agents. *Pharmacol. Ther.* **90**(2–3), 261 (2001).

63. P. B. Memvanga and C. I. Nkanga, Liposomes for malaria management: The evolution from 1980 to 2020. *Malar. J.* **20**(1), 327 (2021).

64. S. Pukrittayakamee, et al., Activities of artesunate and primaquine against asexual- and sexual-stage parasites in falciparum malaria. *Antimicrob. Agents Chemother.* **48**(4), 1329 (2004).

65. R. T. Eastman and D. A. Fidock, Artemisinin-based combination therapies: A vital tool in efforts to eliminate malaria. *Nat. Rev. Microbiol.* **7**(12), 864 (2009).

66. C. W. Wright, et al., Synthesis and evaluation of cryptolepine analogues for their potential as new antimalarial agents. *J. Med. Chem.* **44**(19) 3187 (2001).

67. P. B. Memvanga, R. Coco, and V. Préat, An oral malaria therapy: Curcumin-loaded lipid-based drug delivery systems combined with beta-arteether. *J. Controlled Release* **172**(3), 904 (2013)

68. J. F. Fernandes, et al., Fosmidomycin as an antimalarial drug: A metaanalysis of clinical trials. *Future Microbiol.* **10**(8), 1375 (2015).

69. R. K. Haynes, et al., Artemisone: A highly active antimalarial drug of the artemisinin class. *Angew. Chem. Int. Ed.* **45**(13), 2082 (2006).

70. D. Coertzen, et al., Artemisone and artemiside are potent panreactive antimalarial agents that also synergize redox imbalance in *Plasmodium falciparum* transmissible gametocyte stages. *Antimicrob. Agents Chemother.* **62**(8), e02214 (2018).

71. N. Hall, et al., A comprehensive survey of the *Plasmodium* life cycle by genomic, transcriptomic, and proteomic analyses. *Science* **307**(5706), 82 (2005).

72. A. Scherf, J. J. Lopez-Rubio, and L. Riviere, Antigenic variation in *Plasmodium falciparum*. *Annu. Rev. Microbiol.* **62**, 445 (2008).

73. T. W. Ademolue and G. A. Awandare, Evaluating antidisease immunity to malaria and implications for vaccine design. *Immunology* **153**(4), 423 (2018).

74. J. Langhorne, et al., Immunity to malaria: More questions than answers. *Nat. Immunol.* **9**(7), 725 (2008).

75. A. Haakenstad, et al., Tracking spending on malaria by source in 106 countries, 2000-16: An economic modelling study. *Lancet Infect. Dis.* **19** (7), 703 (2019).

76. P. D. Crompton, S. K. Pierce, and L. H. Miller, Advances and challenges in malaria vaccine development. *J. Clin. Invest.* **120**(12), 4168 (2010).

77. M. B. Laurens, RTS,S/AS01 vaccine (Mosquirix): An overview. *Hum. Vaccines Immunother.* **16**(3), 480 (2020).

78. S. Casares, T. D. Brumeanu, and T. L. Richie, The RTS,S malaria vaccine. *Vaccine* **28**(31), 4880 (2010).

79. European Medicines Agency, Assessment report Mosquirix™, EMA/CHMP/439337/2015, European Medicines Agency: United Kingdom. p. 175 (2015).

80. K. A. Collins, et al., Enhancing protective immunity to malaria with a highly immunogenic virus-like particle vaccine. *Sci. Rep.* **7**, 46621 (2017).

81. A. S. Ishizuka, et al., Protection against malaria at 1 year and immune correlates following PfSPZ vaccination. *Nat. Med.* **22**(6), 614 (2016).

82. J. E. Epstein, et al., Live attenuated malaria vaccine designed to protect through hepatic CD8(+) T cell immunity. *Science* **334**(6055), 475 (2011).

83. A. N. Cowell and E. A. Winzeler, The genomic architecture of antimalarial drug resistance. *Brief Funct. Genomics* **18**(5), 314 (2019).

84. N. S. Santos-Magalhães and V. C. F. Mosqueira, Nanotechnology applied to the treatment of malaria. *Adv. Drug Deliv. Rev.* **62**(4–5), 560 (2010).

85. K. Thriemer, et al., Delayed parasite clearance after treatment with dihydroartemisinin-piperaquine in *Plasmodium falciparum* malaria patients in central Vietnam. *Antimicrob. Agents Chemother.* **58**(12), 7049 (2014).

86. N. J. White, Antimalarial drug resistance. *J. Clin. Invest.* **113**(8), 1084 (2004).

87. D. Menard and A. Dondorp, Antimalarial drug resistance: A threat to malaria elimination. *Cold Spring Harbor Perspect. Med.* **7**(7), a025619 (2017).

88. C. Roper, et al., Intercontinental spread of pyrimethamine-resistant malaria. *Science* **305**(5687), 1124 (2004).

89. F. Calderón, D. M. Wilson, and F. J. Gamo, Antimalarial drug discovery: Recent progress and future directions. *Prog. Med. Chem.* **52**, 97 (2013).

90. U. Samarasekera, Countries race to contain resistance to key antimalarial, *Lancet* **374**, 277 (2009).

91. A. Uwimana et al., Emergence and clonal expansion of *in vitro* artemisinin-resistant *Plasmodium falciparum kelch13* R561H mutant parasites in Rwanda, *Nat. Med.* **26**, 1602 (2020).

92. L. C. Mathieu et al., Local emergence in Amazonia of *Plasmodium falciparum k13* C580Y mutants associated with *in vitro* artemisinin resistance, *Elife* **9** (2020).

93. J. Gomez-Marquez and K. Hamad-Schifferli, Local development of nanotechnology-based diagnostics. *Nat. Nanotechnol.* **16**(5), 484 (2021).

94. K. V. Ragavan, et al., Advances in biosensors and optical assays for diagnosis and detection of malaria. *Biosens. Bioelectron.* **105**, 188 (2018).

95. S. Lee, et al., A highly sensitive aptasensor towards *Plasmodium* lactate dehydrogenase for the diagnosis of malaria. *Biosens. Bioelectron.* **35**(1), 291 (2012).

96. Y. W. Cheung, et al., Structural basis for discriminatory recognition of *Plasmodium* lactate dehydrogenase by a DNA aptamer. *Proc. Natl. Acad. Sci. U. S. A.* **110**(40), 15967 (2013).

97. J. Wang, et al., Application of DNA origami in nanobiomedicine. *Nan Fang Yi Ke Da Xue Xue Bao* **41**(6), 960 (2021).

98. M. Godonoga, et al., A DNA aptamer recognising a malaria protein biomarker can function as part of a DNA origami assembly. *Sci. Rep.* **6**, 21266 (2016).

99. W. Jeon, et al., A colorimetric aptasensor for the diagnosis of malaria based on cationic polymers and gold nanoparticles. *Anal. Biochem.* **439**(1), 11 (2013).

100. P. Jain, et al., Aromatic surfactant as aggregating agent for aptamer-gold nanoparticle-based detection of *Plasmodium* lactate dehydrogenase. *Mol. Biotechnol.* **58**(7), 497 (2016).

101. R. Dirkzwager, S. Liang, and J. Tanner, Development of aptamer-based point-of-care diagnostic devices for malaria using three-dimensional printing rapid prototyping. *ACS Sens.* **4**, 420 (2016).
102. T. Jamieson, et al., Biological applications of quantum dots. *Biomaterials* **28**(31), 4717 (2007).
103. M. J. Ku, et al., Quantum dots: A new tool for anti-malarial drug assays. *Malar. J.* **10**, 118 (2011).
104. Kenry, et al., Single-layer ternary chalcogenide nanosheet as a fluorescence-based "capture-release" biomolecular nanosensor. *Small* **13**(5), 1601925 (2017).
105. A. Geldert, et al., Enhancing the sensing specificity of a MoS2 nanosheetbased FRET aptasensor using a surface blocking strategy. *Analyst* **142**(14), 2570 (2017).
106. J. Wu, et al., Evaluation of wondfo rapid diagnostic kit (Pf-HRP2/PAN-pLDH) for diagnosis of malaria by using nano-gold immunochromatographic assay. *Acta Parasitol.* **59**(2), 267 (2014).
107. K. B. Paul, et al., A highly sensitive self-assembled monolayer modified copper doped zinc oxide nanofiber interface for detection of *Plasmodium falciparum* histidine-rich protein-2: Targeted towards rapid, early diagnosis of malaria. *Biosens. Bioelectron.* **80**, 39 (2016).
108. G. Figueroa-Miranda, et al., Aptamer-based electrochemical biosensor for highly sensitive and selective malaria detection with adjustable dynamic response range and reusability. *Sens. Actuators, B* **255**, 235 (2018).
109. K. B. Paul, et al., A multi-walled carbon nanotube-zinc oxide nanofiber based flexible chemiresistive biosensor for malaria biomarker detection. *Analyst* **142**(12), 2128 (2017).
110. A. Hemben, J. Ashley, and I. E. Tothill, Development of an immunosensor for PfHRP 2 as a biomarker for malaria detection. *Biosensors* (Basel) **7**(3), 28 (2017).
111. M. de Souza Castilho, et al., Magneto immunoassays for *Plasmodium falciparum* histidine-rich protein 2 related to malaria based on magnetic nanoparticles. *Anal. Chem.* **83**(14), 5570 (2011).
112. G. Ruiz-Vega, et al., Electrochemical POC device for fast malaria quantitative diagnosis in whole blood by using magnetic beads, Poly-HRP and microfluidic paper electrodes. *Biosens. Bioelectron.* **150**, 111925 (2020).
113. E. Fu, et al., Two-dimensional paper network format that enables simple multistep assays for use in low-resource settings in the context of malaria antigen detection. *Anal. Chem.* **84**(10), 4574 (2012).

114. G. Xu, et al., Paper-origami-based multiplexed malaria diagnostics from whole blood. *Angew. Chem. Int. Ed.* **55**(49), 15250 (2016).

115. G. Gregoriadis, Overview of liposomes. *J. Antimicrob. Chemother.* **28**(Suppl B), 39 (1991).

116. P. A. Peeters, et al., Chloroquine containing liposomes in the chemotherapy of murine malaria. *Parasitology* **98 Pt** 3, 381 (1989).

117. A. K. Agrawal, A. Singhal, and C. M. Gupta, Functional drug targeting to erythrocytes *in vivo* using antibody bearing liposomes as drug vehicles. *Biochem. Biophys. Res. Commun.* **148**(1), 357 (1987).

118. V. P. Torchilin, Recent advances with liposomes as pharmaceutical carriers. *Nat. Rev. Drug Discov.* **4**(2), 145 (2005).

119. E. Samuelsson, et al., Contribution of Kupffer cells to liposome accumulation in the liver. *Colloids Surf. B* **158**, 356 (2017).

120. P. Pirson, et al., Liposomes in the chemotherapy of experimental murine malaria. *Trans. R. Soc. Trop. Med. Hyg.* **73**(3), 347 (1979).

121. P. Pirson, et al., Primaquine liposomes in the chemotherapy of experimental murine malaria. *Ann. Trop. Med. Parasitol.* **74**(4), 383 (1980).

122. P. Pirson, R. Steiger, and A. Trouet, The disposition of free and liposomally encapsulated antimalarial primaquine in mice. *Biochem. Pharmacol.* **31**(21), 3501 (1982).

123. B. Arica, et al., Characterization, *in vitro* and *in vivo* studies on primaquine diphosphate liposomes. *J. Microencapsul.* **12**(5), 469 (1995).

124. G. Stensrud, et al., Formulation and characterisation of primaquine loaded liposomes prepared by a pH gradient using experimental design. *Int. J. Pharm.* **198**(2), 213 (2000).

125. D. J. A. Crommelin, et al., Liposomes and immunoliposomes for controlled release or site-specific delivery of anti-parasitic drugs and cytostatics. *J. Controlled Release* **16**, 147 (1991).

126. P. A. Peeters, et al., Chloroquine blood levels after administration of the liposome-encapsulated drug in relation to therapy of murine malaria. *Pharm. Res.* **6**(9), 787 (1989).

127. E. Moles, M. Kavallaris, and X. Fernàndez-Busquets, Modeling the distribution of diprotic basic drugs in liposomal systems: Perspectives on malaria nanotherapy. *Front. Pharm.* **10**, 1064 (2019).

128. M. Owais, et al., Chloroquine encapsulated in malaria-infected erythrocyte-specific antibody-bearing liposomes effectively controls

chloroquine-resistant *Plasmodium berghei* infections in mice. *Antimicrob. Agents Chemother.* **39**(1), 180 (1995).

129. K. J. Longmuir, et al., Effective targeting of liposomes to liver and hepatocytes *in vivo* by incorporation of a *Plasmodium* amino acid sequence. *Pharm. Res.* **23**(4), 759 (2006).

130. E. Moles, et al., Immunoliposome-mediated drug delivery to *Plasmodium*-infected and non-infected red blood cells as a dual therapeutic/prophylactic antimalarial strategy. *J. Controlled Release* **210**, 217 (2015).

131. A. Biosca, et al., An immunoPEGliposome for targeted antimalarial combination therapy at the nanoscale. *Pharmaceutics* **11**(7), 341 (2019).

132. I. Chauhan, et al., Nanostructured lipid carriers: A groundbreaking approach for transdermal drug delivery. *Adv. Pharm. Bull.* **10**(2), 150 (2020).

133. D. Parashar, N. P. Aditya, and R. S. Murthy, Development of artemether and lumefantrine co-loaded nanostructured lipid carriers: Physicochemical characterization and *in vivo* antimalarial activity. *Drug. Deliv.* **23**(1), 123 (2016).

134. R. Vanka, et al., Ameliorating the *in vivo* antimalarial efficacy of artemether using nanostructured lipid carriers. *J. Microencapsul.* **35**(2), 121 (2018).

135. Z. Ali, N. Mishra, and A. Baldi, Development and characterization of arteether-loaded nanostructured lipid carriers for the treatment of malaria. *Artif. Cells Nanomed. Biotechnol.* **44**(2), 545 (2016).

136. C. E. Umeyor, et al., Repositioning miconazole nitrate for malaria: Formulation of sustained release nanostructured lipid carriers, structure characterization and *in vivo* antimalarial evaluation. *J. Drug Deliv. Sci. Technol.* **61**, 102 (2021).

137. H. Rashidzadeh, et al., *In vivo* antiplasmodial activity of curcumin-loaded nanostructured lipid carriers. *Curr. Drug Deliv.* **16**(10), 923 (2019).

138. X. Dong, et al., Heparin-decorated nanostructured lipid carriers of artemether-protoporphyrin IX-transferrin combination for therapy of malaria. *Int. J. Pharm.* **605**, 120813 (2021).

139. Y. Tomii, Lipid formulation as a drug carrier for drug delivery. *Curr. Pharm. Des.* **8**(6), 467 (2002).

140. K. K. Singh and S. K. Vingkar, Formulation, antimalarial activity and biodistribution of oral lipid nanoemulsion of primaquine. *Int. J. Pharm.* 347(1-2), 136 (2008).

141. A. P. Nayak, et al., Curcuminoids-loaded lipid nanoparticles: Novel approach towards malaria treatment. *Colloids Surf. B*, **81**(1), 263 (2010).

142. C. E. Umeyor, et al., Development insights of surface modified lipid nanoemulsions of dihydroartemisinin for malaria chemotherapy: Characterization, and *in vivo* antimalarial evaluation. *Recent Pat. Biotechnol.* **13**(2), 149 (2019).

143. F. Bruxel, et al., Cationic nanoemulsion as a delivery system for oligonucleotides targeting malarial topoisomerase II. *Int. J. Pharm.* **416**(2), 402 (2011).

144. E. Ullu, C. Tschudi, and T. Chakraborty, RNA interference in protozoan parasites. *Cell Microbiol.* **6**(6), 509 (2004).

145. C. Rinaldi and M. J. A. Wood, Antisense oligonucleotides: The next frontier for treatment of neurological disorders. *Nat. Rev. Neurol.* **14**(1), 9 (2018).

146. F. Föger, et al., Inhibition of malarial topoisomerase II in *Plasmodium falciparum* by antisense nanoparticles. *Int. J. Pharm.* **319**(1–2), 139 (2006).

147. P. Attasart, et al., Inhibition of *Plasmodium falciparum* proliferation *in vitro* by double-stranded RNA nanoparticle against malaria topoisomerase II. *Exp. Parasitol.* **164**, 84 (2016).

148. P. E. Nielsen, et al., Sequence-selective recognition of DNA by strand displacement with a thymine-substituted polyamide. *Science* **254**(5037), 1497 (1991).

149. N. Kolevzon, et al., Use of peptide nucleic acids to manipulate gene expression in the malaria parasite *Plasmodium falciparum*. *PLoS One* **9**(1), e86802 (2014).

150. A. Nuthanakanti, et al., Self-assemblies of nucleolipid supramolecular synthons show unique self-sorting and cooperative assembling process. *Nanoscale* **11**(24), 11956 (2019).

151. P. V. Nguyen, et al., Oligonucleotide solid nucleolipid nanoparticles against antibiotic resistance of ESBL-producing bacteria. *Pharmaceutics* **14**(2): 299 (2022).

152. V. Allain, C. Bourgaux, and P. Couvreur, Self-assembled nucleolipids: From supramolecular structure to soft nucleic acid and drug delivery devices. *Nucleic Acids Res.* **40**(5), 1891 (2012).

153. K. Kowouvi, et al., Nucleoside–lipid-based nanocarriers for methylene blue delivery: Potential application as anti-malarial drug. *RSC Adv.* **9**, 18844 (2019).

154. H. Zhao, et al., Polymer-based nanoparticles for protein delivery: Design, strategies and applications. *J. Mater. Chem. B* **4**(23), 4060 (2016).

155. S. Tripathy, et al., Synthesis, characterization of chitosan–tripolyphosphate conjugated chloroquine nanoparticle and its *in vivo* anti-malarial efficacy against rodent parasite: A dose and duration dependent approach. *Int. J. Pharm.* **434**(1–2), 292 (2012).

156. S S. Tripathy, et al., A prospective strategy to restore the tissue damage in malaria infection: Approach with chitosan-trypolyphosphate conjugated nanochloroquine in Swiss mice. *Eur. J. Pharmacol.* **737**, 11 (2014).

157. F. Akhtar, M. M. Rizvi, and S. K. Kar, Oral delivery of curcumin bound to chitosan nanoparticles cured *Plasmodium yoelii* infected mice. *Biotechnol. Adv.* **30**(1), 310 (2012).

158. S. Talib, et al., Chitosan–chondroitin based artemether loaded nanoparticles for transdermal drug delivery system. *J. Drug Deliv. Sci. Technol.* **61**, 102281 (2021).

159. E. Abbasi, et al., Dendrimers: Synthesis, applications, and properties. *Nanoscale Res. Lett.* **9**(1), 247 (2014).

160. Z. Mhlwatika and B. A. Aderibigbe, Application of dendrimers for the treatment of infectious diseases. *Molecules* **23**(9), 2205 (2018).

161. J. Movellan, et al., Amphiphilic dendritic derivatives as nanocarriers for the targeted delivery of antimalarial drugs. *Biomaterials* **35**(27), 7940 (2014).

162. E. S. Lee, et al., Binary mixing of micelles using Pluronics for a nanosized drug delivery system. *Colloids Surf. B* **82**(1), 190 (2011)

163. G. Pembouong, et al., A comprehensive study in triblock copolymer membrane interaction. *J. Controlled Release* **151**(1), 57 (2011).

164. E. Martí Coma-Cros, et al., Micelle carriers based on dendritic macromolecules containing bis-MPA and glycine for antimalarial drug delivery. *Biomater. Sci.* **7**(4), 1661 (2019).

165. P. Agrawal, U. Gupta, and N. K. Jain, Glycoconjugated peptide dendrimers-based nanoparticulate system for the delivery of chloroquine phosphate. *Biomaterials* **28**(22), 3349 (2007).

166. P. Ferruti, Poly(amidoamine)s: Past, present, and perspectives. *J. Polym. Sci. Part A: Polym. Chem.* **51**(11), 2319 (2013).

167. S. Richardson, P. Ferruti, and R. Duncan, Poly(amidoamine)s as potential endosomolytic polymers: Evaluation *in vitro* and body distribution in normal and tumour-bearing animals. *J. Drug Target* **6**(6), 391 (1999).

168. P. Urbán, et al., Use of poly(amidoamine) drug conjugates for the delivery of antimalarials to *Plasmodium. J. Controlled Release* **177**, 84 (2014).

169. G. Bouyer et al., *Plasmodium falciparum* sexual parasites regulate infected erythrocyte permeability. *Commun. Biol.* **3**, 726 (2020).

170. S. Hong, et al., Protein-based nanoparticles as drug delivery systems. *Pharmaceutics* **12**(7), 604 (2020).

171. K. DeFrates, et al., Protein polymer-based nanoparticles: Fabrication and medical applications. *Int. J. Mol. Sci.* **19**(6), 1717 (2018).

172. M. Foox and M. Zilberman, Drug delivery from gelatin-based systems. *Expert Opin. Drug Delivery* **12**(9), 1547.

173. R. Nakaoka, Y. Tabata, and Y. Ikada, Potentiality of gelatin microsphere as immunological adjuvant. *Vaccine* **13**(7), 653 (1995).

174. A. K. Bajpai and J. Choubey, Design of gelatin nanoparticles as swelling controlled delivery system for chloroquine phosphate. *J. Mater. Sci.: Mater. Med.* **17**(4), 345 (2006).

175. A. A. Sidhaye, et al., Bio-inspired artemether-loaded human serum albumin nanoparticles for effective control of malaria-infected erythrocytes. *Nanomedicine* **11**(21), 2809 (2016).

176. C. P. Karch, et al., Self-assembling protein nanoparticles: Implications for HIV-1 vaccine development. *Nanomedicine* **13**(17), 2121 (2018).

177. S. A. Kaba, et al., Self-assembling protein nanoparticles with builtin flagellin domains increases protective efficacy of a *Plasmodium falciparum* based vaccine. *Vaccine* **36**(6), 906 (2018).

178. R. Bhattacharya and P. Mukherjee, Biological properties of "naked" metal nanoparticles. *Adv. Drug Deliv. Rev.* **60**(11), 1289 (2008).

179. P. Velusamy, et al., Bio-inspired green nanoparticles: Synthesis, mechanism, and antibacterial application. *Toxicol. Res.* **32**(2), 95 (2016).

180. Y. N. Slavin, et al., Metal nanoparticles: Understanding the mechanisms behind antibacterial activity. *J. Nanobiotechnol.* **15**(1), 65 (2017).

181. S. Käosaar, et al., Profiling of the toxicity mechanisms of coated and uncoated silver nanoparticles to yeast *Saccharomyces cerevisiae* BY4741 using a set of its 9 single-gene deletion mutants defective in oxidative stress response, cell wall or membrane integrity and endocytosis. *Toxicol. In Vitro* **35**, 149 (2016).

182. A. Abdal Dayem, et al., The role of reactive oxygen species (ROS) in the biological activities of metallic nanoparticles. *Int. J. Mol. Sci.* **18**(1) (2017)

183. W. Jiang, et al., Interaction between oxide nanoparticles and biomolecules of the bacterial cell envelope as examined by infrared spectroscopy. *Langmuir* **26**(23), 18071 (2010).

184. M. Yamanaka, K. Hara, and J. Kudo, Bactericidal actions of a silver ion solution on *Escherichia coli*, studied by energy-filtering transmission electron microscopy and proteomic analysis. *Appl. Environ. Microbiol.* **71**(11), 7589 (2005).

185. K. B. Narayanan and N. Sakthivel, Biological synthesis of metal nanoparticles by microbes. *Adv. Colloid Interface Sci.* **156**(1–2), 1 (2010).

186. P. Singh, et al., Biological synthesis of nanoparticles from plants and microorganisms. *Trends Biotechnol.* **34**(7), 588 (2016).

187. T. L. Riddin, M. Gericke, and C. G. Whiteley, Analysis of the interand extracellular formation of platinum nanoparticles by *Fusarium oxysporum* f. sp. lycopersici using response surface methodology. *Nanotechnology* **17**(14), 3482 (2006).

188. I. Hussain, et al., Green synthesis of nanoparticles and its potential application. *Biotechnol. Lett.* **38**(4), 545 (2016).

189. D. S. Goodsell, *Bionanotechnology: Lessons from Nature* (John Wiley & Sons, 2004).

190. P. Sathishkumar, et al., Flavonoids mediated "green" nanomaterials: A novel nanomedicine system to treat various diseases current trends and future perspective. *Mater. Lett.* **210**, 26 (2018).

191. C. Panneerselvam, S. Ponarulselvam, and K. Murugan, Potential anti-plasmodial activity of synthesized silver nanoparticle using *Andrographis paniculata* Nees (Acanthaceae). *Arch. Appl. Sci. Res.* **3**, 208 (2011).

192. S. Ponarulselvam, et al., Synthesis of silver nanoparticles using leaves of *Catharanthus roseus* Linn. G. Don and their antiplasmodial activities. *Asian Pac. J. Trop. Biomed.* **2**, 574 (2021).

193. A. Mishra, et al., Evaluation of antiplasmodial activity of green synthesized silver nanoparticles. *Colloids Surf. B* **111**, 713 (2013).

194. C. Panneerselvam, K. Murugan, and D. Amerasan, Biosynthesis of silver nanoparticles using plant extract and its anti-plasmodial property. *Adv. Mater. Res.* **1086**, 11 (2015).

195. K. Murugan, et al., Seaweed-synthesized silver nanoparticles: An ecofriendly tool in the fight against *Plasmodium falciparum* and its vector *Anopheles stephensi*? *Parasitol. Res.* **114**(11), 4087 (2015).

196. K. Murugan, et al., *In vivo* and *in vitro* effectiveness of *Azadirachta indica*–synthesized silver nanocrystals against *Plasmodium berghei* and *Plasmodium falciparum*, and their potential against malaria mosquitoes. *Res. Vet. Sci.* **106**, 14 (2016).

197. C. Panneerselvam, et al., Fern-synthesized nanoparticles in the fight against malaria: LC/MS analysis of *Pteridium aquilinum* leaf extract and biosynthesis of silver nanoparticles with high mosquitocidal and antiplasmodial activity. *Parasitol. Res.* **115**(3), 997 (2016).

198. M. Sardana, et al., Antiplasmodial activity of silver nanoparticles: A novel green synthesis approach. *Asian Pac. J. Trop. Biomed.* **8**, 268 (2018).

199. K. Murugan, et al., Eco-friendly drugs from the marine environment: Spongeweed-synthesized silver nanoparticles are highly effective on *Plasmodium falciparum* and its vector *Anopheles stephensi*, with little non-target effects on predatory copepods. *Environ. Sci. Pollut. Res. Int.* **23**(16), 16671 (2016).

200. A. Jaganathan, et al., Earthworm-mediated synthesis of silver nanoparticles: A potent tool against hepatocellular carcinoma, *Plasmodium falciparum* parasites and malaria mosquitoes. *Parasitol. Int.* **65**(3), 276 (2016).

201. J. Subramaniam, et al., Multipurpose effectiveness of *Couroupita guianensis*–synthesized gold nanoparticles: High antiplasmodial potential, field efficacy against malaria vectors and synergy with *Aplocheilus lineatus* predators. *Environ. Sci. Pollut. Res. Int.* **23**(8), 7543 (2016).

202. P. P. Dutta, et al., Antimalarial silver and gold nanoparticles: Green synthesis, characterization and *in vitro* study. *Biomed. Pharmacother.* **91**, 567 (2017).

203. G. Rajakumar, et al., Antiplasmodial activity of eco-friendly synthesized palladium nanoparticles using *Eclipta prostrata* extract against *Plasmodium berghei* in Swiss albino mice. *Parasitol. Res.* **114**(4), 1397 (2015).

204. A. O. Christy, et al., The antimalaria effect of *Momordica charantia* L. and *Mirabilis jalapa* leaf extracts using animal model. *J. Med. Plant Res.* **10**, 344 (2016).

205. K. Murugan, et al., Magnetic nanoparticles are highly toxic to chloroquine-resistant *Plasmodium falciparum*, dengue virus (DEN-2), and their mosquito vectors. *Parasitol. Res.* **116**(2), 495 (2017).

206. K. Nagpal, S. K. Singh, and D. N. Mishra, Chitosan nanoparticles: A promising system in novel drug delivery. *Chem. Pharm. Bull.* (Tokyo) **58**(11), 1423 (2010).

207. K. Y. Zhu, et al., Comparative studies on effects of three chitin synthesis inhibitors on common malaria mosquito (Diptera: Culicidae). *J. Med. Entomol.* **44**(6), 1047 (2007).

208. X. Zhang, et al., Identification and characterization of two chitin synthase genes in African malaria mosquito, *Anopheles gambiae*. *Insect Biochem. Mol. Biol.* **42**(9), 674 (2012).

209. X. Zhang, J. Zhang, and K. Y. Zhu, Chitosan/double-stranded RNA nanoparticle-mediated RNA interference to silence chitin synthase genes through larval feeding in the African malaria mosquito (*Anopheles gambiae*). *Insect Mol. Biol.* **19**(5), 683 (2010).

210. R. Mohammadi, et al., Nanoemulsified *Mentha piperita* and *Eucalyptus globulus* oils exhibit enhanced repellent activities against *Anopheles stephensi*. *Asian Pac. J. Trop. Biomed.* **12**(11), 520 (2019).

211. M. Osanloo, et al., Preparation and optimization nanoemulsion of Tarragon (*Artemisia dracunculus*) essential oil as effective herbal larvicide against *Anopheles stephensi*. *Ind. Crops Prod.* **109**, 214 (2017).

212. M. Osanloo, et al., Nanoemulsion of Dill essential oil as a green and potent larvicide against *Anopheles stephensi*. *Environ. Sci. Pollut. Res. Int.* **25**(7), 6466 (2018).

213. N. Sogan, et al., Larvicidal activity of *Ricinus communis* extract against mosquitoes. *J. Vector Borne Dis.* **55**(4), 282 (2018).

214. S. Firooziyan, et al., Preparation of nanoemulsion of *Cinnamomum zeylanicum* oil and evaluation of its larvicidal activity against a main malaria vector *Anopheles stephensi*. *J. Environ. Health Sci. Eng.* **19**(1), 1025 (2021).

215. M. Dehghankar, et al., Bioactivities of rose-scented geranium nanoemulsions against the larvae of *Anopheles stephensi* and their gut bacteria. *PLoS One* **16**(2), e0246470 (2021).

216. G. Benelli, Plant-mediated biosynthesis of nanoparticles as an emerging tool against mosquitoes of medical and veterinary importance: A review. *Parasitol. Res.* **115**(1), 23 (2016).

217. G. Benelli, et al., Mosquito control with green nanopesticides: Towards the One Health approach? A review of non-target effects. *Environ. Sci. Pollut. Res. Int.* **25**(11), 10184 (2018).

218. P. Lung, J. Yang, and Q. Li, Nanoparticle formulated vaccines: Opportunities and challenges. *Nanoscale* **12**(10), 5746 (2020).

219. T. Neuberger, et al., Superparamagnetic nanoparticles for biomedical applications: Possibilities and limitations of a new drug delivery system. *J. Magn. Magn. Mater.* **239**(1), 483 (2005).

220. F. N. Al-Deen, et al., Superparamagnetic nanoparticles for effective delivery of malaria DNA vaccine. *Langmuir* **27**(7), 3703 (2011).

221. M M. B. Laurens, Novel malaria vaccines. *Hum. Vaccines Immunother.* **17**(11), 4549 (2021).

222. C. Kulangara, et al., Cell biological characterization of the malaria vaccine candidate trophozoite exported protein 1. *PLoS One* **7**(10), e46112 (2012).

223. C. P. Karch, et al., The use of a *P. falciparum* specific coiled-coil domain to construct a self-assembling protein nanoparticle vaccine to prevent malaria. *J. Nanobiotechnol.* **15**(1), 62 (2017).

224. C. G. Schneider, et al., Orientation of antigen display on self-assembling protein nanoparticles influences immunogenicity. *Vaccines* (Basel) **9**(2), 103 (2021).

225. L. Powles, et al., Pullulan-coated iron oxide nanoparticles for bloodstage malaria vaccine delivery. *Vaccines* (Basel) **8**(4), 651 (2020).

226. W. C. Huang, et al., A malaria vaccine adjuvant based on recombinant antigen binding to liposomes. *Nat. Nanotechnol.* **13**(12), 1174 (2018).

227. F. Salamanca-Buentello and A. S. Daar, Nanotechnology, equity and global health. *Nat. Nanotechnol.* **16**, 358 (2021).

Chapter 8

Overview of Nanotherapeutics for Fungal Infections

Filipa Sousa,[a] Domingos Ferreira,[a] Salette Reis,[b] and Paulo Costa[a]
[a]*UCIBIO—Applied Molecular Biosciences Unit, MedTech-Laboratory of Pharmaceutical Technology, Faculty of Pharmacy, University of Porto, 4050-313 Porto, Portugal*
[b]*LAQV-REQUIMTE, Department of Chemical Sciences, Faculty of Pharmacy, University of Porto, 4050-313 Porto, Portugal*
filipambtsousa@gmail.com

The increasing incidence of fungal infections has become a worrisome public health matter, which has been intensified by a rise in host predisposition factors. Notwithstanding all the antifungal drugs available for the treatment of these infections, their efficacy is

---

This chapter, except for Section 8.5, is based on a review article on a similar topic from the same authors [1]
F. Sousa, D. Ferreira, S. Reis, and P. Costa, *Pharmaceuticals* **13**, 248 (2020).

Nanotherapeutics for Infectious Diseases
Edited by Íris Luz Batalha

ISBN 978-981-5129-36-6 (Hardcover), 978-1-003-56678-6 (eBook)
www.jennystanford.com

debatable, and their side effects cannot be overlooked. Bearing that in mind, it is of upmost importance to synthetize new and innovative carriers for these medicines, not only to fight emerging fungal infections but also to obviate the uprising of drug-resistant strains. Albeit this is considered a challenging task, novel nanocarriers are now under thorough study to deliver compounds with antifungal potential to many different (and even new) cellular targets. This chapter summarizes the state-of-the-art strategies that are being investigated so that antifungal therapy is upgraded and adverse effects of conventional drugs are diminished. The bidirectional relationship between "mycology" and "nanotechnology" will also be approached. Additionally, the chapter emphasizes on *in vivo* and *in vitro* models, which have been employed to study fungal infections and the efficacy of antifungal drug candidates.

8.1 Introduction

There is a wide range of fungal infections, from superficial (*i.e.*, affecting skin) to systemic, with invasion of internal organs [3]. Fungal infections affect millions of people every year worldwide. Of these, 1.5 million are invasive fungal infections, therefore requiring advanced treatment and hospitalization. Most of these disseminated infections are caused by Candida, Cryptococcus, Aspergillus, and Pneumocystis species, which cause cryptococcosis, candidiasis, aspergillosis, and pneumocystis pneumonia, respectively [4].

Superficial fungal infections are rather common and, despite rarely being life threatening, can spread to other skin regions and even become widespread. Furthermore, they can be transmitted to other people and may cause secondary bacterial skin infections, harming the quality of a person's life. Skin mycoses are classified according to the causative fungal agents into dermatophytosis, yeast infections, and mold infections [3].

Invasive fungal infections represent a significant burden on healthcare systems, having high morbidity and mortality rates. These rates are most worrisome among immunocompromised patients that are more prone to opportunistic infections, such as patients with acquired immune deficiency syndrome (AIDS), transplant patients whose immune systems are suppressed to prevent organ rejection, patients with cancer who are taking immunosuppressive chemotherapy, or autoimmune patients undergoing immunosuppressive therapy [4, 5].

The major currently available agents to treat invasive fungal infections can be grouped into four main classes according to their mechanism of action: polyenes, azoles, allylamines, and echinocandins (Table 8.1) [6]. They all present drawbacks when it comes to spectrum of activity, drug–drug interactions, pharmacokinetics and pharmacodynamics, resistance mechanisms, and the toxicity of the compounds themselves. Furthermore, there are some limitations in terms of clinical efficacy and efficiency, mainly because of their physicochemical properties such as their hydrophobic character, which leads to a low solubility in water, and also selectivity problems deriving from the similarities between fungi and human cells [5, 7].

Nevertheless, an emerging necessity is the design and development of new drug delivery systems or even new antifungals, owing to the following facts [10]:

- There are 20–40% mortality rates with invasive mycoses, so these figures need to be reduced.
- The increase in patients undergoing prolonged antifungal therapies reflects the need to develop better fungicidal drugs and thus reduce the length of the treatments and the associated costs.
- There is still space for improvement in pharmacokinetics and pharmacodynamics, in order to reduce the frequency of drug use.
- More attention needs to be given to the host toxicities and drug–drug interactions of current therapies so that their effects can be eliminated or, at least, minimized.
- New chemical class therapeutics with different mechanisms of action are needed; this way, these new drugs may synergize with the current ones to allow better responses.
- There is an alarming growth in antifungal resistance in all therapeutic groups [10].

As an emerging field of science, nanotechnology has shown an undeniable versatility and has ushered in a revolution in medical treatments, quicker diagnosis, cellular regeneration, and drug delivery [11, 12]. The materials used to produce nanoparticles can be roughly divided into three main groups: polymers, lipids, and metals, each one giving rise to nanoparticles with different properties [13]. New drug delivery systems can also be classified according to composition and type, as shown in Table 8.2. The main representatives of each of these three different groups of nanoparticles are mentioned in Fig. 8.1.

Table 8.1 Targets of each chemical class of antifungals

Class	Target (Mechanism of Action)	Antifungal	
Azoles	Ergosterol (inhibition of lanosterol 14-α-demethylase)	Imidazoles	Miconazole
			Econazole
			Ketoconazole
			Clotrimazole
		Triazoles	Itraconazole
			Fluconazole
			Voriconazole
Allylamines	Ergosterol (inhibition of squalene epoxidase)	Terbinafine	
		Naftifine	
		Butenafine	
Polyenes	Cell membrane (production of reactive oxygen species (ROS))	Amphotericin B	
	Ergosterol (inhibition of lanosterol 14-α-demethylase)	Nystatin	
Echinocandines	Cell wall (block of β-1,3 glucan synthesis)	Caspofungin, Micafungin, Anidulafungin	
Other antifungals	Chelation of polyvalent metal cations	Ciclopirox	
	Microtubules (prevention of the formation of the mitotic spindle)	Griseofulvin	
	Ergosterol (inhibition of D14 reductase and D7-D8 isomerase)	Amorolfine	

Note: Adapted from Refs. [8, 9].

Table 8.2 New drug delivery systems classified according to composition or type

<table>
<tr><th colspan="5">According to Composition</th><th colspan="2"></th><th></th></tr>
<tr><td>Phospholipid-based vesicles</td><td>Non-phospholipid vesicles</td><td>Polymeric nanoparticles</td><td>Lipid nanoparticles</td><td>Others</td><td colspan="2"></td><td></td></tr>
<tr><td rowspan="3"></td><td rowspan="3"></td><td rowspan="3"></td><td rowspan="3"></td><td>Nano-emulsions</td><td colspan="2" rowspan="3">Colloidal carriers</td><td rowspan="11">According to Type</td></tr>
<tr><td>Micro-emulsions</td></tr>
<tr><td>Micelles</td></tr>
<tr><td>Liposomes</td><td>Niosomes</td><td rowspan="4"></td><td rowspan="4"></td><td rowspan="4"></td><td rowspan="4">Vesicular carriers</td><td>Rigid</td></tr>
<tr><td>Transfersomes</td><td rowspan="3">Spanlastics</td><td rowspan="3">Ultra-deformable</td></tr>
<tr><td>Ethosomes</td></tr>
<tr><td>Transethosomes</td></tr>
<tr><td rowspan="2"></td><td rowspan="2"></td><td>Nanospheres and nanocapsules
Polymeric micelles</td><td>Solid–lipid nanoparticles (SLNs)</td><td>Silica nanoparticles</td><td colspan="2" rowspan="2">Nano-particulated carriers</td></tr>
<tr><td>Dendrimers
Polymersomes</td><td>Nanostructured lipid carriers (NLCs)</td><td>Metallic nanoparticles</td></tr>
</table>

Note: Adapted from Refs. [20–22].

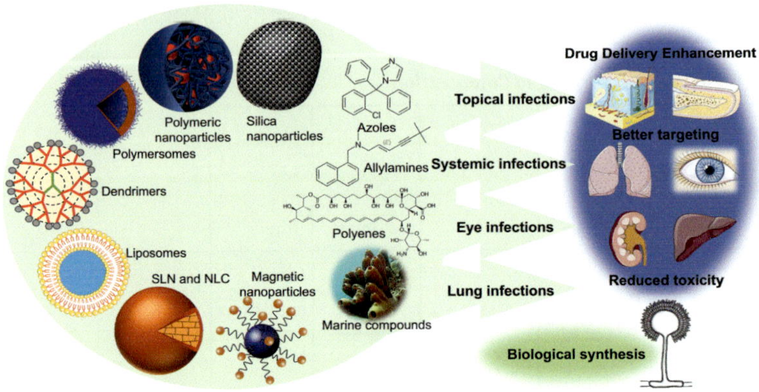

Figure 8.1 The new drug delivery systems based on nanotechnology that are currently being employed to enhance drug delivery, promote a better targeting, and reduce the toxicity of conventional antifungal drugs. It is also vital to point out the importance of the production of nanoparticles by fungi (biological synthesis) and the undeniable potential of the sea as a source of new molecules with antifungal activity. SLN: solid–lipid nanoparticles. NLC: nanostructured lipid carriers.

Nanoparticles have been employed in pharmaceutical formulations due to their ability to alter and improve the pharmacokinetic and pharmacodynamic properties of drugs. This is due to their ability to increase the solubility and stability of drugs, to allow a controlled release, and to exhibit biocompatibility with tissues and cells, which is reflected in an overall improvement in therapeutic efficacy [13, 14]. In addition, their subcellular size is compatible with intravascular injection, and their high surface area is amenable to modification with targeting moieties so that the drug is released in a specific cellular target, thus reducing systemic adverse effects and consequently increasing patient compliance, by decreasing the usual dose and frequency of administration [15, 16]. This target-specific action is possible since, at the nanomolecular level, it is possible to incorporate targeting ligands (*e.g.*, peptides, antibodies) that allow a preferential binding to specific receptors on the cell surface [17–19]. Hence, the development of new biopharmaceutical systems, especially nanoparticulate carriers, is a good strategy to improve the therapeutic efficacy, safety, and compliance of conventional antifungal drugs.

Table 8.3 gives an overview of the antifungal drug chemical classes and the new nanoparticulate drug delivery systems, as well as their route of administration and dosage form.

Table 8.3 Some of the novel nano-based drug delivery systems already developed for each antifungal drug

Antifungal Drug	Novel Drug Delivery System	Route of Administration	Dosage Form	Ref.
Miconazole	Niosomes	Transdermal	Gel	[23]
	SLN	Oral	N.A.	[24]
		Topical	Gel	[25]
	Microemulsion	Topical	N.A.	[26]
	Liposomes	Topical	Gel	[27]
	Nanoemulsion	Topical	N.A.	[28]
	Nanosponges	Vaginal	Gel	[29]
	Transfersomes	Topical	Gel	[30]
Econazole	Microemulsion	Percutaneous	N.A.	[31]
		Topical	Gel	[32]
	SLN	Topical	Gel	[33]
	NLC	Topical	Gel	[34]
	Liposomes	Topical	Gel	[35]
	Ethosomes	Topical	Gel	[36]
	Transethosomes	Transdermal	Gel	[37]
	Nanosponges	Topical	Hydrogel	[38]
	Niosomes	Transdermal	Gel	[39]
	Polymeric micelles	Topical	N.A.	[40]
	Nanoemulsion	Topical	N.A.	[41]
Ketoconazole	SLN/NLC	Topical	Gel	[42]
	Niosomes	Topical	Gel	[43]
	Microemulsion	Oral	N.A.	[44]
	Spanlastics	Ocular	N.A.	[45]
	Dendrimers	Topical	Hydrogel	[46]
	Liposomes	Topical	N.A.	[47]

(Continued)

Table 8.3 (*Continued*)

Antifungal Drug	Novel Drug Delivery System	Route of Administration	Dosage Form	Ref.
Clotrimazole	Liposomes	Topical	Gel	[48]
	Nanosponges	Topical	Hydrogel	[49]
	Ethosomes	Topical	Gel	[50]
	Niosomes	Topical	Gel	[51]
	Polymeric emulgel	Topical	Gel	[52]
	Polymeric micelles	Topical	N.A.	[40]
	SLN/NLC	Topical	N.A.	[53]
	Microemulsion	Buccal	Gel	[54]
		Vaginal	Gel	[55]
	Transfersomes	Transdermal/ Topical	N.A.	[56]
Itraconazole	Transfersomes	Transdermal	N.A.	[57]
	SLN	Ocular	N.A.	[58]
	NLC	Inhalation	N.A.	[59]
	Niosomes	Topical	N.A.	[60]
	Microemulsion	Transdermal	N.A.	[61]
	Liposomes	Topical		[62]
	Polymeric nanoparticles	Oral	N.A.	[63]
	Polymersome	Intravenous	N.A.	[59]
	Spanlastics	Ocular	N.A.	[64]
	Silica nanoparticles	Oral	N.A.	[65]
Fluconazole	Microemulsion	Vaginal	Gel	[66]
	Niosomes	Ocular	Gel	[67]
	Liposomes	Intravitral	N.A.	[68]
	SLN	Topical	Gel	[69]
	NLC	Oral	N.A.	[70]
	Microsponges	Topical	Gel	[71]
	Ethosomes	Topical	Gel	[72]
	Spanlastics	Ocular	N.A.	[73]
	Polymeric amphiphilogel	Topical	Gel	[74]
	Polymeric micelles	Topical	N.A.	[40]

Antifungal Drug	Novel Drug Delivery System	Route of Administration	Dosage Form	Ref.
Voriconazole	Microemulsion	Ocular	N.A.	[75]
	Polymeric nanoparticles	Ocular	N.A.	[76]
		Pulmonar	N.A.	[77]
	SLN	Topical	Gel	[78]
	Transethosome	Topical	N.A.	[79]
	Ethosome	Topical	N.A.	[80]
Terbinafine	Liposomes	Topical	Gel	[81]
	SLN	Topical	N.A.	[82]
	Transfersomes	Topical	N.A.	[83]
	Spanlastics	Transungual	N.A.	[84]
	Polymeric chitosan nanoparticles	Topical	Hydrogel	[85]
Naftifine	Microemulsion	Topical	N.A.	[86]
	Niosomes	Topical	Gel	[87]
Butenafine	Microemulsion	Topical	Hydrogel	[88]
Amphotericin B	Liposomes	Intravenous	N.A.	[89]
	SLN/NLC	Oral	N.A.	[90]
		Topical	N.A.	[91]
	Magnetic nanoparticles	Nasal instilation	N.A.	[92]
	Nanoemulsion	Topical	N.A.	[93]
	Polymeric nanoparticles	Intravenous	N.A.	[94]
		Oral	N.A.	[95]
	Polymersomes	Oral	N.A.	[96, 97]
	Transfersomes	Topical	N.A.	[98]
	Micelles	Intravenous	N.A.	[99]
	Silica nanoparticles	Intravenous	N.A.	[100]
Nystatin	SLN	Topical	N.A.	[101]
	Nanoemulsion	Topical	N.A.	[102]
	Liposomes	Intravenous	N.A.	[103]
	Niosomes	Parenteral	N.A.	[104]

(*Continued*)

Table 8.3 (*Continued*)

Antifungal Drug	Novel Drug Delivery System	Route of Administration	Dosage Form	Ref.
Griseofulvin	Niosomes	Oral	N.A.	[105]
Ciclopirox	Niosomes	Topical	Gel	[106]
Caspofungin, Micafungin, Anidulafungin, Amorolfine	No nanotech studies yet released			

Note: N.A. means that the dosage form is not mentioned in the reference cited; SLN: solid–lipid nanoparticles; NLC: nanostructured lipid carriers.

However, the efficacy and human safety of these new therapies remain uncertain in most of the articles found in literature. They generally lack controlled clinical trials, and sometimes the suggested routes of administration are less practical, or the production cost may hinder the replacement of the conventional treatment. Nevertheless, some nanoformulations have shown the potential to become a viable first-line treatment; for example, an amphotericin B lipid-based nanoformulation (Abelcet®) was approved in 1995 due to its reduced side effects and toxicity compared with conventional amphotericin formulations. The same happened with AmBisome®, a liposomal nanoformulation of amphotericin B, which showed reduced nephrotoxicity, while preventing and treating invasive fungal infections [107, 108]. Moreover, given the widespread use of antifungal agents and the limited therapeutic offer, fungi have developed resistance mechanisms, such as decrease in effective drug concentration inside the host cell (for example, through overexpression of efflux pump proteins, overexpression of drug targets, and drug sequestration in extra- or intracellular compartments, which may lead to biofilm formation), mutations in the target receptor that alter affinity to the drug, and metabolic bypasses [8, 109]. It is important to add that resistance is a cross-cutting issue to all the currently available classes of antifungal agents; therefore, overcoming antifungal resistance can be considered the mainstay for improving therapeutic strategies to treat antifungal infections [4, 110].

Despite the uprising of these issues in antifungal therapy, there are several mechanisms by which nanoparticles overcome the development of resistance mechanisms:

- Nanoparticles, such as nitric oxide, chitosan, and metallic nanoparticles, often act through multiple simultaneous antimicrobial mechanisms, which reduce the likelihood of resistance development (for example, through oxidative stress or osmotic damage) [111, 112].
- The resistance mechanisms can be prevented by packaging multiple antimicrobial drugs within the same nanoparticle because the likelihood of multiple simultaneous gene mutations in the same cell is low. The most striking examples are the encapsulation of antifungal drugs in chitosan or silver nanoparticles, combining the antifungal properties of both and decreasing the possibility of drug resistance [111, 113].
- Some nanoparticles, such as liposomes and dendrimers, can overcome the resistance mechanisms of decreased uptake and increased efflux of drug from the microbial cell. Liposomes can quickly fuse with the plasma membrane of the microbial cell and release a high concentration of drug into its plasma membrane or cytoplasm, thereby circumventing the decreased uptake mechanism of resistance. This means a faster delivery and avoidance of the transmembrane pumps that catalyze increased efflux of drugs. Dendrimers, on the other hand, are extensively branched molecules, whose surface can be filled with positively charged quaternary ammonium compounds, which bind to negatively charged microbial cell envelopes and increase membrane permeability. This allows the entry of more dendrimers to the microbial cell, the flow of its cytoplasmic contents to the exterior, and the ultimate destruction of the microbial cell membrane. This goes to show that dendrimers are also able to surpass the resistance mechanism of decreased uptake of drugs [114]. Other nanoparticles, such as nitric oxide–releasing silica nanoparticles and zinc oxide nanoparticles, are able to overcome biofilm formation by killing the microbes present in already formed biofilms or by inhibiting biofilm formation

through the generation of reactive oxygen species (ROS), respectively [115, 116].

- Nanoparticles have been used to target antifungal drugs to the specific site of infection, allowing the local release of high concentrations of drug, while keeping the total dose of drug administered low. This high local dose can destroy the infecting fungi before they can develop resistance, thereby overcoming this worrisome issue and translating into fewer side effects upon the patient [111].

That being said, it is important that future research expands beyond the mere formulation of these systems to also focusing on how to overcome the major challenges in their placing on the market, given the complexity of the formulations: the physical instability of nanoparticles, their small capacity for drug loading, the cytotoxicity/immunogenicity, and the high cost of production and standardization. Besides that, there is almost a complete lack of *in vivo* studies, as reaching the therapeutic range needed to perform these studies has proven to be an arduous job. That lies on the fact that, in many cases, there is an anticipated release of the drug, aggregation and precipitation of the nanoparticles, and the accumulation in non-target tissues.

8.2 Nanotechnology and Mycology

"Mycology" and "nanotechnology" have created a bidirectional relationship throughout the years. This dynamic interface led to the creation of the term "myconanotechnology" (Fig. 8.2) [117]. Nanotechnology has proven to be an interesting strategy to increase the potency and efficiency of conventional antifungals, to enable a decrease in toxicity and cost, to avoid an anticipated degradation, to ameliorate the drug distribution by increasing the circulation time and improving pharmacokinetics, and also to improve drug targeting, with promising *in vitro* and *in vivo* results [7]. Furthermore, many metallic nanoparticles have been used against human and plant pathogenic fungi in the light of their intrinsic antifungal activity, and a wide spectrum of fungi are able to biosynthesize nanoelemental particles [117].

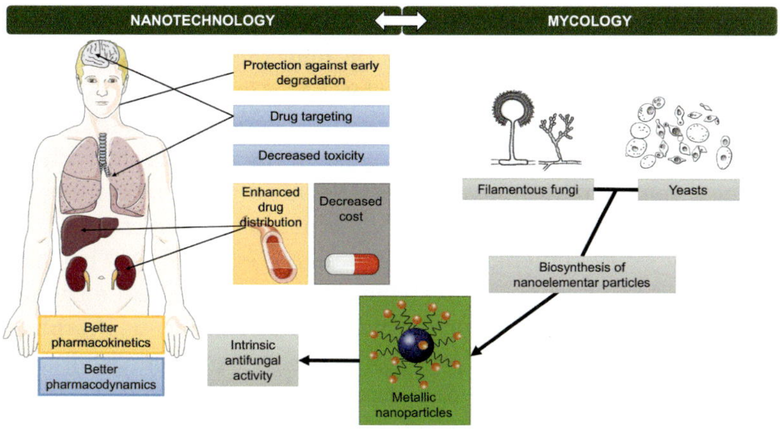

Figure 8.2 Bidirectional relationship of "nanoparticles" and "mycology": Nanotechnology has proven to be useful in improving the pharmacokinetics and pharmacodynamics of antifungals, and many fungi have been used to biologically synthetize nanoparticles.

8.2.1 Intrinsic Antifungal Potential of Nanoparticles

Metallic nanoparticles have been used to eliminate fungi that are pathogenic to humans and plants, because of their intrinsic antimicrobial activity [117]. The exact mechanisms through which this activity occurs are only hypothesized and can be explained by three main pathways: (i) direct uptake of nanoparticles; (ii) indirect activity of nanoparticles by the production of ROS; and (iii) impairment of cell wall/membrane through nanoparticle accumulation [118]. It is highly probable that a combination of these multiple pathways, rather than a single one, is responsible for the antimicrobial activity [119].

Metallic nanoparticles undergo dissolution processes thanks to their electrochemical potential [120]. This leads to their separation into ions within the microbial fluid or in the culture medium. These ions also accumulate in the interior or exterior of the microbial cells, causing an inhibitory answer against microtubules. The accumulation of nanoparticles outside the microtubules causes the formation of layers that block cellular respiratory chain and destroy the microtubules [118].

The electrical charge of the nanoparticle is vital for the interaction that occurs between it and the carried drug. The electrostatic mechanism justifies why the antimicrobial activity was first described in silver nanoparticles. It is widely accepted that the positive charge of the silver ion is crucial for the antimicrobial activity of these nanoparticles through electrostatic attraction between the negatively charged cellular membrane of microorganisms and the positively charged membrane of nanoparticles [118]. Ag^+ has high affinity for thiol groups in cysteine of respiratory chain enzymes; therefore, it uncouples the synthesis of adenosine triphosphate (ATP). Ag^+ also binds to transport proteins from the respiratory chain, causing the leakage of protons and thus the collapse of the proton motive force. Furthermore, Ag^+ obstructs the uptake of phosphate and so promotes the efflux of intracellular phosphate [120].

Silver nanoparticles exhibit potent antifungal activity against clinical isolates and laboratory strains of *Trichophyton mentagrophytes* at concentrations of 1–7 μg/mL, and a minimum inhibitory concentration (MIC) of 25 μg/mL against *Candida albicans* [121]. Silver nanoparticles also reveal good antifungal activity against *Aspergillus niger*, by inhibiting spore germination and preventing biofilm formation. When combined with simvastatin, there is an additive and synergistic antifungal effect, perhaps because simvastatin, as an ergosterol synthesis inhibitor, disrupts the fungal cell membrane, which allows the entry of silver nanoparticles [122].

Metals present in nanoparticles can act as catalysts, reacting with biomolecules thanks to their high specific surface area, inducing the direct production of free radicals when exposed to the acidic environment of lysosomes or interacting with oxidative organelles [119, 120].

ROS, such as superoxide anions (O_2^-), hydroxyl radicals (HO·), and hydrogen peroxide (H_2O_2), are oxygen-derived byproducts formed when a material is exposed to an oxygenated environment. Although the antioxidant cell defense prevents the effects of ROS to some extent, excessive ROS production may cause oxidative stress and lipid peroxidation, leading to membrane impairment, mitochondrial dysfunction, and DNA damage. Therefore, to avoid harmful effects in humans, cytotoxicity tests must be performed on nanoparticles that owe their antimicrobial effects to ROS production [118].

Chitosan and its chemical derivates have been used as building blocks for drug delivery nanoformulations in light of their biocompatibility and biodegradability, presenting some advantages, such as *in situ* gelling performance, mucoadhesive properties, and the ability to prolong the release of a wide range of compounds from low molecular weight compounds to macromolecular drugs [123]. Chitosan nanoparticles have proven to exhibit great antimicrobial activity against Candida infections [3]. According to the literature, this antimicrobial activity is attributed to positively charged amino groups that react with negatively charged groups of lipopolysaccharides and proteins on the surface of the microbial cells, resulting in the disintegration of the cell membrane. Chitosan nanoparticles are retained in cells for a long time and can become entrapped inside the nucleus due to nuclear envelope reassembly at the end of mitosis. By doing this, the nanoparticles can target the nucleus and disrupt the nuclear membrane, binding to DNA molecules and inhibiting mRNA and protein synthesis. In the specific case of fungi, chitosan acts by inhibiting the sporulation and germination of spores, by interfering with the activity of the growth-promoting enzymes [3, 124, 125].

Zinc oxide nanoparticles (ZnONPs) have also proved antifungal activity against dermatophyte infections and other pathogenic fungi, such as Candida and Aspergillus [3]. Meanwhile, it has been reported that the inhibitory efficiency of common antifungal drugs can be increased by combining them with ZnONPs. This synergistic antifungal effect could reduce the overuse of these drugs, decrease their toxicity, and increase their antifungal activity [126]. In addition, these nanoparticles might be an interesting and promising alternative to conventional preservatives in cosmetics in the future [127].

Along with the nanoparticles mentioned earlier, dendrimers also exhibit antifungal activity and provide the opportunity for complex therapy in which dendrimers can act both as the drug carrier and the adjunctive component of the dosage form [46].

8.2.2 Synthesis of Nanoparticles by Fungi

Thanks to their tolerance and ability to bioaccumulate metals, fungi now occupy a central role in the biological production of metallic nanoparticles (Table 8.4) [12]. Methods of biosynthesis are simpler and allow size and shape control of nanoparticles, in addition to being ecological by generating less solvents and chemical substances as waste. Besides fungi, other organisms are also used to synthesize nanoparticles, for instance bacteria, plants, or plant extracts [117]. Compared to bacteria, fungi produce a higher quantity of enzymes, which is translated into a higher yield in nanoparticle production. Furthermore, their growth, both in laboratory and at an industrial scale, is easier to control [117].

Biological synthesis can occur either intracellularly or extracellularly. In intracellular synthesis, the metal ions are transported to the inner part of microbial cells, where nucleation occurs, and nanoparticles are formed in the presence of an enzyme. These nanoparticles are smaller when compared to extracellularly produced ones. Extracellular synthesis is a much simpler process used by a majority of fungi as a result of their secretory components that participate in nanoparticles reduction and capping [12].

Both yeast and filamentous fungi can be used to synthetize nanoparticles. The process of synthetizing filamentous fungi-mediated nanoparticles is easy and cost-effective, since the mycelia in the biomass have high surface area and high intracellular metal absorption. Moreover, the fungi cell wall has many functional groups that facilitate the absorption of the metals [117].

Table 8.4 Some examples of metallic nanoparticles produced by fungi and their methods of synthesis

Fungal Species	Nanoparticles Type	Method of Synthesis
Phoma sp.	Silver	Extracellular
Fusarium oxysporum	Gold, magnetite	Extracellular
Verticillium sp.	Silver	Intracellular
Aspergillus fumigatus	Silver	Extracellular
Aspergillus niger	Silver	Extracellular
Fusarium semitectum	Silver	Extracellular
Trichoderma asperellum	Silver	Extracellular
Phaenerochaete chrysosporium	Silver	Extracellular

Note: Adapted from Refs. [12, 128].

Silver nanoparticles are the most fundamental among metallic nanoparticles that are involved in biomedical applications, especially cancer diagnosis and antimicrobial therapy [121]. One of the most remarkable examples is the extracellular synthesis of silver nanoparticles from filamentous fungi, like *Fusarium solani*, a pathogenic fungus isolated from infected onions. The silver nanoparticles obtained presented high stability due to capping by fungus-secreted proteins. Such biosynthetic procedures can potentially be applied for the production of other metal nanomaterials, such as gold and platinum, with countless applications in the medical field [129].

Many studies that followed reported the optimization of the biological production of silver nanoparticles using other fungi, such as *Fusarium oxysporum* [130], *Cochliobolus lunatus, Beauveria bassiana* [131], *Bipolaris maydis* [132], and many others. It was also reported that some of these biologically synthesized silver nanoparticles showed enhanced antifungal activity with fluconazole against *Phoma glomerata, Phoma herbarum, Fusarium semitectum, Trichoderma* sp., *Phoma glomerata, Phoma herbarum*, and *C. albicans* [121]. Surprisingly, it was additionally observed that the potential of silver nanoparticles is much wider than the inhibition of human and plant pathogenic fungi, as they also inhibit indoor fungal species such as *Penicillium brevicompactum, Aspergillus fumigatus, Cladosporium cladosporoides, Chaetomium globosum*, and *Stachybotrys chartarum* [133].

Likewise, yeasts can be useful to nanoparticle synthesis, as they produce enzymes responsible for the reduction of metallic salts and their conversion into elementary nanoparticles. Some examples are the biosynthesis of cadmium nanoparticles by *Candida glabrata* and *Schizosaccharomyces pombe* [117] and of selenium and silver nanocompounds by *Saccharomyces cerevisiae* in aerobic conditions [134, 135].

8.3 Routes of Administration of Antifungal Drugs

The administration of antifungal drugs is not restricted to oral or parenteral routes. Other routes such as transungual, pulmonary, and ocular have also acquired great importance in the treatment of certain fungal infections. The development of a drug delivery system for attaining therapeutic concentration at these target organs is a challenge that requires a comprehensive understanding of the dynamics and specific features of the nails, lungs, and eyes, respectively.

8.3.1 Transungual Route

This special topical route acquires singular importance in antifungal therapy, mainly when onychomycosis (*Tinea ungium*) caused by dermatophytes (such as *Trichophyton rubrum*) or yeasts is concerned [136]. These infections affect the nail plate and/or the nail bed and are very frequent [137].

Drug delivery through nails has its own challenges, and the use of nail penetration enhancers is compulsory in formulations. The nail plate is composed of cross-linked keratin linkages, an extensive bonding network responsible for its rigidity. Although there has been considerable research regarding new approaches for transungual drug delivery, topical permeability was limited by its barrier properties. Therefore, the lookout for novel approaches is important to enhance treatment efficacy and reduce treatment time and relapse rate [136]. There are still no nanotech solutions for this purpose, but medicated nail lacquers loaded with ciclopirox or amorolfine are the most feasible delivery vehicles [138]. The main hurdle in the development of nail lacquers for nail disorders is at delivering the therapeutically effective concentrations to the site of infection, which is often under the nail [137].

8.3.2 Pulmonary Delivery

Drug delivery to the lungs is challenging because the availability of therapeutic quantities of antifungal drugs at the site of infection is often inadequate on account of high blood flow turnover in the tissue. This can be problematic since a local and prolonged drug release is desirable when targeting this organ.

For pulmonary infections, formulating nanoparticles to be delivered by inhalation may help retaining the antifungals in the lungs over a prolonged period and may reduce the toxicity compared to the parenteral route [77, 139].

A study in which voriconazole was encapsulated within poly(lactic-co-glycolic acid) (PLGA) nanoparticles revealed that the nanoparticles were retained in the lungs for an extended period of time and presented higher lung biodistribution when compared to voriconazole alone, concluding that nanoparticles present a good alternative to deliver drugs in high concentration and in a sustained fashion to the deep lung tissue, providing a greater antifungal effect [77].

8.3.3 Ocular Route

The eye is a particularly puzzling organ for drug delivery systems. Physical barriers that hinder drug access into the eye limit the drug bioavailability. In addition, physiological processes such as blinking and tear drainage reduce the residence time of ocular drug delivery systems and drastically reduce the amount of transcorneal drug absorption [139, 140].

Fungal ocular infections are much less prevalent when compared to bacterial or viral infections. However, fungal keratitis is considered the second most common cause of blindness in developing countries and is an important cause of morbidity [22, 141]. People with prolonged use of corticosteroids or antibiotics, with diabetes mellitus, with corneal trauma or who underwent surgery, and even people wearing contact lenses are at a higher risk of having fungal keratitis [22].

New formulations such as polymer-based nanoparticles, liposomes, dendrimers, SLNs, spanlastics, and niosomes were developed to enhance drug bioavailability to the eye and to minimize antifungal adverse effects [142]. They also have the ability to overcome the disadvantages of conventional eye drops, such as short residence time, by prolonging the contact time at the corneal surface and achieving a sustained release of the drug. In addition, the encapsulation into these systems protects antifungals from degradation promoted by the metabolic enzymes on conjunctival and corneal surfaces and in tear fluids [22].

A study assessed the efficacy of a fluconazole liposomal formulation for candidal keratitis that aimed at prolonging the antifungal action by increasing the contact time. The authors reported 86.4% healing observed in rabbits treated with fluconazole-encapsulated liposomes, presumably because of higher viscosity and lipid solubility of fluconazole-loaded liposomes [143].

Furthermore, a fluconazole-loaded niosomal gel can also be successfully used as a topical drug delivery system for corneal fungal infections. Ethosomes are known to improve both transcorneal permeability and ocular bioavailability of poorly water-soluble drugs, which is a tremendous improvement in topical ocular drug delivery [144].

Spanlastics also enhance permeability and bioavailability of antifungal drugs, and their nano-size makes them able to reach the posterior segment of the eye, when eye drops are used [145]. For instance, itraconazole-loaded spanlastics proved to be safe and non-irritant to the eyes; moreover, the elasticity of these vesicles makes them suitable for drug delivery in both anterior and posterior eye diseases [64].

8.4 An Overview of Nanoparticle Types and Their Applicability on Antifungal Therapy

8.4.1 Lipid Nanoparticles

Although the safety and tolerability of systemic antifungal therapy have improved considerably, a rising number of immunocompromised patients are receiving systemic antifungal agents for progressively longer treatment courses, which increases the probability of longer-term risks, nonspecific drug interactions, and other toxicity-related reactions in different organs (Fig. 8.3) [146].

AmBisome® was the first successful example of an antifungal nanomedicine, and it was manufactured by NeXstar Pharmaceuticals (later acquired by Gilead Sciences) and approved by the FDA and the EMA in the 1990s [89, 147]. In this formulation, amphotericin B was incorporated in a unilamellar liposomal bilayer of approximately 45–80 nm. When compared to the conventional formulation, AmBisome® showed less toxicity and prolonged circulation time, which suggests a higher distribution rate [89].

Despite this, its daily usage was still limited by its cost [7]. Thus, other lipid-based nanoparticle formulations of amphotericin B were latter developed, as for example, lipid nanoparticles composed of phospholipids and cholesterol, which showed lower renal cell toxicity, decreased hematotoxicity to red blood cells, and higher antifungal activity *in vitro* and *in vivo* when compared to commercialized formulations such as Fungizone® or AmBisome® [148].

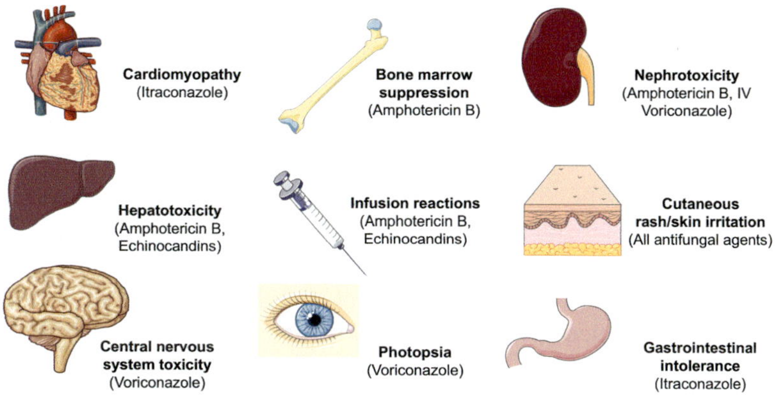

Figure 8.3 Common toxicities of antifungal agents. The main organs affected are the heart, the kidney, the liver, and the skin.

Besides amphotericin B, some other drugs have also been incorporated into liposomes with promising results. A good example is nystatin, a drug used mainly topically because of its poor oral bioavailability, thus excluded for the treatment of systemic fungal infections [22]. A daily injection of liposomal nystatin, however, showed great efficacy in the treatment of invasive aspergillosis, although with some mild renal toxicity and some infusion-related events [103]. Also, a multilamellar liposome gel system containing clotrimazole revealed to be useful in the treatment of vaginal candidiasis, providing a sustaining release in the vaginal fluid and a reducing dosing interval [48].

Econazole is a topical antifungal that may be irritant to the skin when conventional formulations are employed. Applying a liposomal formulation has revealed to be a good strategy to minimize the irritating potential and to increase compliance. Bioavailability studies of the liposomal gel revealed a seven-fold increase in the drug concentration in the epidermis, when compared to the control. Thus, the authors showed that it is possible to reduce the amount of drug applied and keep the therapeutic efficacy, minimizing the cutaneous irritation [149].

Ultra-deformable liposomes or transfersomes: These are generally regarded as an upgrade to liposomes because they overcome their poor penetration through the stratum corneum by modifications in the bilayer composition [22]. They are elastic nanovesicles composed of phospholipids and edge activators, most often surfactants or hydrophilic detergents with high mobility. These activators respond to mechanical stress by relocating within the vesicles to zones with smaller curvature, reducing the elastic energy of the membrane to a minimum. In this way, transfersomes can change their shape and squeeze through channels with one-tenth of the vesicles' diameter and cross the stratum corneum driven by an osmotic transdermal gradient [150]. These particles are essentially studied as carriers for dermal and transdermal delivery but have also shown to be effective carriers of genetic material and vaccines [21]. Transfersomes were tested for the topical delivery of amphotericin B and were found to enhance the drug penetration toward deep skin layers in a scale 40 times higher than AmBisome®, making this system clinically significant [98].

Despite all the advantages previously mentioned, they still present an important drawback, that is, the difficulty in loading hydrophobic drugs into the vesicles without impairing their deformability and elastic characteristics [150].

Ethosomes: These are essentially a type of transfersome that employs ethanol instead of the edge activator molecules; the main difference lies in the fact that ethanol evaporates once applied on the skin, whereas the edge activator molecules remain on the skin surface after water evaporation from the formulations [150].

Ethanol also plays an essential role in enhancing the delivery of both hydrophilic and lipophilic drugs without impairing their deformability and elastic characteristics, which was a significant disadvantage of transfersomes [21]. Hence, ethosomes show great potential as topical delivery systems for antifungal drugs [22]. For example, the formulation of an ethosomal gel containing econazole nitrate allowed controlled drug release, increased antifungal activity, and good stability after storage [36].

Fluconazole is used both for local and systemic fungal infections. Despite its worldwide usage, it shows low penetration rate because of its high solubility and low permeability. The preparation of an ethosomal gel for topical delivery of fluconazole has proved to be an efficient way to overcome the issues associated with bioavailability, degradation, stability, and side effects of this therapeutic agent [72].

Despite the great potential of ethosomes, it is worth mentioning that when the concentration of ethanol used is above 30–40% (w/w), the vesicular membrane tends to become more permeable and leaky, leaking out the entrapped drug, specially hydrophilic/ionized drugs [50, 151]. Furthermore, this high ethanolic content may affect the skin and cause irritation or contact dermatitis. To solve that, Cavamax W7, a permeation enhancer, was developed to improve topical delivery and reduce the amount of ethanol used, therefore reducing the risk of adverse effects [50]. The Cavamax W7 ethosomes were able to reach the last layer of epidermis (stratum basale), which turned them into a very valuable tool in the treatment of deep-seated fungal infections [22]. A clotrimazole-encapsulated Cavamax W7 composite ethosome gel presented a more stable and efficient vesicular system than the conventional ethosomal formulation, along with a better antifungal activity against *C. albicans* and *Aspergillus niger* [50].

Transethosomes: These are lipid vesicles with irregular shapes that represent a combination of the concepts applied to transfersomes and ethosomes, having a high content of ethanol (up to 30%) along with an edge activator molecule. This combination causes a rearrangement in the lipid bilayer of transethosomes, allowing them to have both higher deformability and skin permeation/penetration than the other deformable molecules. The formulation of these vesicles also presents the advantage of being quite easy to scale up, which constitutes an enormous advantage for industrial purposes [150]. Contrary to other deformable vesicles, transethosomes improve skin delivery of drugs both under occlusive and non-occlusive conditions [152]. For these reasons, they are considered the most promising lipid nanoparticles in antifungal therapy.

An econazole nitrate–loaded transethosome gel for transdermal delivery has been developed, which presented lower *ex vivo* skin penetration, higher *ex vivo* skin retention, and higher *in vitro* antifungal activity, when compared to the marketed cream of econazole nitrate [37]. Transethosomes containing terbinafine, amphotericin B, and ketoconazole also disclosed enhanced permeation [152]. In addition, voriconazole transethosomes significantly enhanced skin permeation and deposition of the drug when compared to conventional liposomes, transfersomes, and ethosomes [79].

Although transethosomes represent the state of the art in ultra-deformable vesicles, they still have some drawbacks that are important to mention, such as the solubility that the drugs need to fulfil in both lipophilic and aqueous environments so that they can reach the dermal microcirculation and access the systemic circulation. In addition, the molecular size of the drug needs to be reasonably high to make the percutaneous absorption possible (40–200 nm) [152].

SLNs and NLCs: These have become very attractive options for the topical delivery of drugs and as an alternative to overcome the drawbacks of liposomes. They provide an occlusive effect by virtue of their ability to form a surface film that reduces the transepidermal water loss, which can reduce the atopic eczema symptoms and improve the skin appearance [153]. Besides that, they exhibit an excellent tolerability profile, and their reduced size promotes a closer contact with the stratum corneum, intensifying the skin penetration of drugs. These particles are also able to increase the chemical stability of compounds sensitive to light, oxidation, and hydrolysis [154]. In Fig. 8.4, the advantages of NLCs over SLNs are outlined, namely, a disordered crystalline structure that avoids losses during drug loading, the possibility of loading both hydrophilic and hydrophobic drugs, biodegradability, and low *in vivo* toxicity [155]. On the other hand, SLNs present many limitations, namely the fact that a limited number of drugs are soluble in lipids, and these lipids may crystalize in more stable structures, causing premature drug release [21]. A schematic representation of these structures is outlined in Fig. 8.4.

An application of SLNs and NLCs is the incorporation of azoles to treat cutaneous fungal infections. The most common azoles (clotrimazole, miconazole, econazole) are extremely water insoluble, being very difficult to administrate and release those drugs at infection sites. However, compounds with lipophilic character can be efficiently encapsulated in SLNs [114].

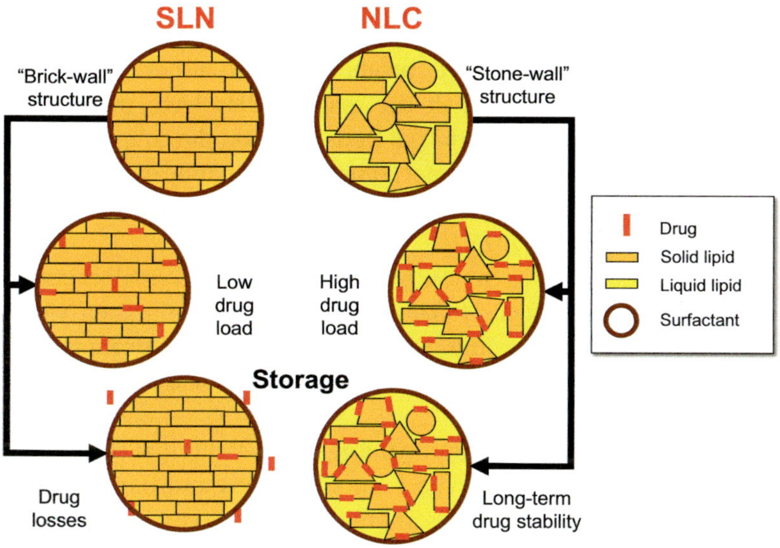

Figure 8.4 SLNs are formed by a lipid matrix that becomes solid at room and body temperatures, while NLCs contain both solid and liquid lipids. The formation of an almost perfect crystalline structure on SLNs is translated into a limited capacity of drug loading, while a disordered solid matrix on NLCs translates into enhanced drug-loading ability.

In 2010, a formulation was developed in which the SLNs were dispersed in a hydrogel to carry miconazole nitrate (MN), an azole frequently used to treat cutaneous fungal infections [156]. The studies indicated that this MN-loaded SLN-bearing hydrogel formulation was less prone to cause cutaneous irritation in comparison to the MN hydrogel, and the MN suspension formulation, showing an even greater efficacy in animal models (male albino rabbits). Furthermore, the SLN-bearing hydrogel provided sustained release of miconazole (slow initial release with a lag time of 1 h and a sustained drug release over a 24 h period) with greater drug deposition on the skin. This happens because gels help evenly disperse the matrix carriers, increasing the contact time and the deposition of carriers on the skin, resulting in a higher cutaneous penetration of the drug [18, 156].

8.4.2 Polymeric Nanoparticles

Polymeric nanoparticles are produced from either natural or synthetic polymers, in which the drug is dissolved, trapped, or bound to the nanoparticle matrix. Depending on the composition of the organic phase and the preparation method, nanocapsules (core–shell morphology) or nanospheres (matrix-like structure) can be obtained (Fig. 8.5) [117, 157].

These nanoparticles can carry proteins, DNA, and small chemical drugs, such as antifungals, to cells and specific target organs, which increases the safety profile. Their nanometric size promotes an effective permeation through the cellular membranes and increases their stability in blood flow.

It is also important to mention the gelling systems that can be used to formulate polymeric carriers: microsponges and nanosponges, amphiphilic gels, and emulgels or gellified emulsions, which play an important role in cutaneous drug delivery [21].

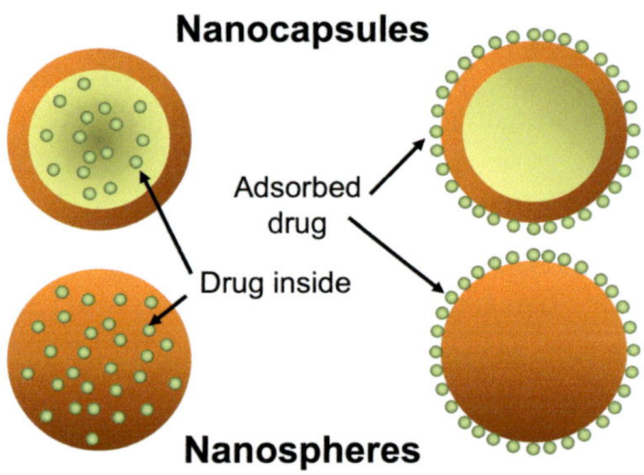

Figure 8.5 Schematic representation of the differences between nanocapsules and nanospheres.

Microsponge and nanosponge systems: These are polymer-based spheres that can encapsulate or adsorb many substances and then be incorporated into a dosage form (hydrogel) or be used for oral delivery [49, 71, 158].

A polymeric microsponge–based gel system was successfully produced for the topical delivery of fluconazole, with great entrapment efficiency, production yield, and extended drug release, which allows a reduction in application frequency—a feature with great importance in recidivist and prolonged fungal infections [71].

A nanosponge-based gel formulation loaded with clotrimazole showed controlled release with reduced side effects, bespeaking the safe and effective profile of these colloidal carriers for topical use [49]. An econazole nanosponge hydrogel was developed to overcome the poor solubility of the drug, showing the ability of forming a local depot for sustained drug release [38].

Amphiphilic gels: These are semisolid systems of nonionic nature that can be used as topical/transdermal carriers without promoting irritation of the skin. They aim at delivering the antifungal in a level within the therapeutic window for as long as possible and to avoid fluctuations in plasma drug level [74].

An amphiphilogel of fluconazole was successfully formulated for topical application, presenting overall stability and cumulative drug release within the range expected [74].

Emulgels: They have dual release control system (emulsion/microemulsion and gel), which increases the overall stability of the antifungal formulations [21]. In comparison to conventional creams or ointments, they exhibit better application properties as well as faster and more complete drug-release profile and they lack on greasiness and residues upon application [52].

Emulgels are suitable solutions for the incorporation of hydrophobic drugs in water-soluble gel bases; for instance, the incorporation of clotrimazole in an emulgel formulation prepared using either Carbopol 934 or HMPC 2910 showed good physical properties, stability, and antifungal activity [52].

A formulation with amphotericin B encapsulated in PLGA in association with ε-caprolactone was developed, which proved the potential of polymeric nanoparticles in preventing cytotoxicity yet preserving the fungicidal efficacy. With this polymeric combination, the authors described an amphotericin B encapsulation efficacy of 84% and a fungicidal effect for *C. albicans* equal to the free drug, with the advantage of presenting less toxicity and less associated mortality [96].

Desoxycholate amphotericin B was also formulated in a PLGA-DMSA (dimercaptosuccinic acid) polymeric blend. DMSA exhibits tropism to lungs, enabling the nanoparticles to be preferentially released at this target site to treat conditions such as paracoccidioidomycosis (PCM). Intraperitoneally administered nanoparticles presented equivalent therapeutic efficacy and decreased genotoxicity when compared to the free drug. The greatest advantage of this nano-based formulation was that it only required to be taken every 3 days, given the slow release of amphotericin B from the nanoparticles, contrary to the daily administration required by the currently commercialized formulation [159].

As mentioned previously, chitosan is a very versatile polymer owing to the ease of chemical modification and deacetylation [160]. Some studies have shown that this polymer also possesses antioxidant, antimicrobial, and anti-inflammatory properties, becoming very attractive from a biopharmaceutical viewpoint [161]. Chitosan nanoparticles on their own have been suggested to have a huge potential to become safe and effective antifungal agents [162].

Chitosan nanoparticles containing amphotericin B were initially developed with the aim of avoiding the drug gastrointestinal degradation, increasing the stability, and the bioavailability in target organs, such as the lungs, the liver, and the spleen, while decreasing renal exposition. When administered by the oral route, these nanoparticles were effective in treating visceral leishmaniosis, aspergillosis, and candidiasis, showing an efficacy comparable to the parenteral AmBisome® formulation. Moreover, this natural polysaccharide was also employed for the cutaneous release of amphotericin B in wounds caused by burns infected with *C. albicans*, promoting a better tissue healing and increased antifungal activity, when compared to the conventional formulation [163].

Polymeric micelles: These are amphiphilic nanosized colloidal carriers containing a hydrophobic core and hydrophilic shell that self-assemble in water above their critical association concentration. They present a unique architecture where the hydrophobic core can incorporate hydrophobic drugs, such as antifungals, which leads to a very significant improvement in their aqueous solubility [22]. These systems have attracted attention in drug delivery because their critical micelle concentration (CMC) is 1000-fold lower (<10 mg/L) than that of phospholipid micelles, since ionic surfactants have more easiness in self-assembling in water to form micelles [164].

Amphotericin B has been incorporated in polymeric micelles to treat brain fungal infections. The system was intended to improve drug solubility and permeability through biological membranes, as well as obviate issues like high toxicity and low efficacy against *Candida meningoencephalitis* [99].

Polymersomes: These are spherical structures comprising an aqueous core surrounded by a polymeric bilayer membrane, being viewed as synthetic analogues of liposomes. They are very versatile structures, which spontaneously self-assemble from amphiphilic diblock copolymers. This capacity increases drug efficacy and enables them to encapsulate both hydrophilic and hydrophobic drugs, including antifungals (Fig. 8.6) [165, 166]. Furthermore, polymersomes can present stimuli-responsive drug release, which means that their physical and chemical properties can change in response to certain triggers (changes in pH, temperature, redox conditions, light, magnetic field, ionic strength, or even concentration). This ability is promising in drug-controlled release, an important feature in antifungal drug therapy [166, 167]. In Fig. 8.6, the process by which a polymersome is formed is represented, as well as its ability to carry different biological molecules.

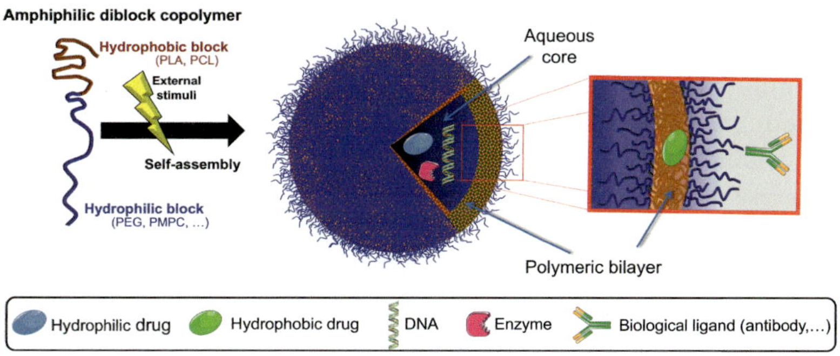

Figure 8.6 Schematic representation of the formation of a polymersome and its versatile properties. Polymersomes are generally self-assembled from block copolymers, presenting a unique structure that can encapsulate different biological and chemical molecules. Stimuli-sensitive polymersomes, whose drug release can be modulated by stimulus, have emerged as novel programmable delivery systems.

Amphotericin B–loaded $(PEG)_3$-PLA polymersomes were prepared by the solvent injection method, and this formulation was compared with two marketed formulations (Fungizone® and AmBisome®) in terms of release, molecular organization of amphotericin B, and hemolysis. The results were similar to those of marketed formulations, which indicated the potential for further *in vivo* development [97].

Dendrimers: These are homogenous polymeric tridimensional nanoarchitectures characterized by a highly branched and symmetrical structure consisting of a central core (a single atom or a group of atoms), building blocks of repeating monomeric units growing from the core (generations), and a high density of water-soluble functional groups on the surface (terminal groups) (Fig. 8.7) [168–170]. Hydrophobic drugs can be entrapped into the void spaces of the dendrimer, whereas hydrophilic drugs or biological entities such as DNA can be complexed with the terminal groups [171]. Their nano and uniform size structure (2–10 nm in diameter), high degree of branching, water solubility, multivalence, well-defined molecular weight, and available internal cavities make them extremely valuable as carrier systems for antifungal agents [168, 172].

Albeit different types of dendrimers have been described, the most frequently used for antifungal therapy are polyamidoamine (PAMAM) and polypropylene imine (PPI) dendrimers [170]. PAMAM dendrimers containing ketoconazole were shown to improve the solubility and the *in vitro* release of the drug, besides enhancing its antifungal activity [46].

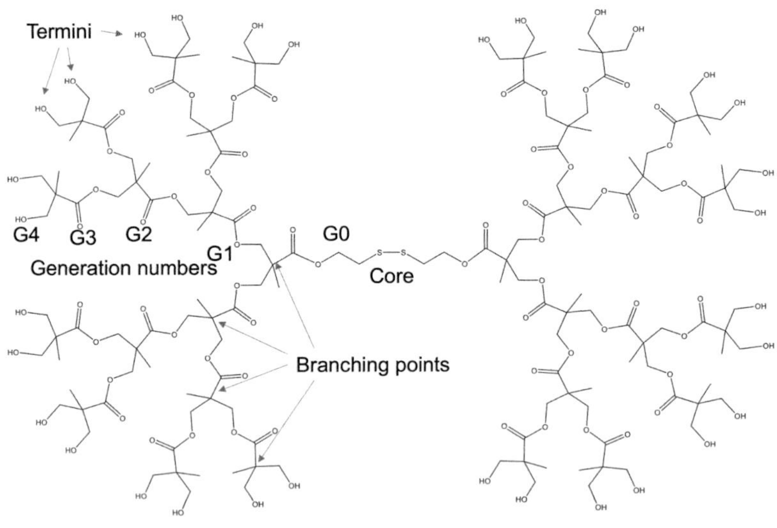

Figure 8.7 Schematic representation of a dendrimer, outlining its central core, branches, and terminal groups.

8.4.3 Metallic Nanoparticles

Mainly three groups of metallic nanoparticles—gold, silver, and iron oxide—have been used for the vectorization of antifungal drugs.

There are two main strategies for nanomaterial synthesis: the top-down approach, wherein a larger structure is broken down into smaller pieces, and the bottom-up approach, wherein the material is synthesized from the atomic level to make a large nanostructure [173].

Metallic nanoparticles can be synthetized in three different methods: chemically, physically, and biologically. Evaporation–condensation and laser ablation are the most relevant physical approaches for the preparation of metallic nanoparticles, presenting clear advantages such as having neither solvent contamination nor chemical reagents and further allowing uniformity of nanoparticles distribution [174]. Chemical synthesis by means of chemical or photochemical reduction, co-precipitation, or electrochemistry has been associated with many side effects related to the absorption of toxic chemical particles on the surface of nanoparticles. Furthermore, they face development issues related to stability and aggregation, control of crystal growth, morphology, and size distribution [12, 173], thereupon biological synthesis has acquired some expression as an attractive alternative, due to its reliability, cleanliness, and environment-friendly features [175].

Gold nanoparticles: These are used in immunochemical studies to identify protein interactions and in DNA fingerprinting to detect fungal DNA in a sample. They are also used to detect aminoglycosides such as streptomycin, gentamicin, and neomycin [12]. Gold nanodiscs have been successfully synthesized by the sonochemical reduction of gold salt solution and have revealed excellent antifungal activity against Candida isolates [176].

Silver nanoparticles: These are the most efficient on account of their antimicrobial efficacy against bacteria, viruses, and some eukaryotic microorganisms. In fact, they are the most widely used nanomaterials in consumer products, being applied as antimicrobial agents in the textile industry, for water treatment, in sunscreens to protect from UVB radiation, etc. [12].

Research has confirmed that silver and gold nanocubes present great antifungal properties against Candida isolates, especially *C. albicans*, without any resistant isolate. The same cannot not be said for silver and gold nanospheres or nanowires, which displayed not-so-large inhibition zones and a relevant percentage of resistant isolates. The highest percentages of resistant isolates belonged to *Candida tropicalis* against gold and silver nanospheres [177].

A combination treatment of ketoconazole with silver nanoparticles disclosed promising results both in terms of fungicidal activity and in preventing the relapse of Malassezia's skin fungal infection. Taking advantage of these results, a Carbopol hydrogel containing silver nanoparticles and ketoconazole was prepared, and the results revealed a small increase in the inhibition zones, without a decrease in the antifungal activity [178].

Albeit nanostructures of iron, cobalt, and nickel exhibit superparamagnetic properties and high magnetic susceptibility, superparamagnetic iron oxide nanoparticles (SPIONs) such as magnetite (Fe_3O_4), hematite (α-Fe_2O_3), and maghemite (γ-Fe_2O_3) are the most studied types of magnetic nanoparticles. This type of nanoparticles has received special attention by virtue of their capacity to be responsive to external magnetic fields, therefore being easily directed and released at a specific site of the organism [12, 179]. In addition, these particles present low toxicity and biocompatibility, having been extensively applied in the fields of diagnostics, imaging, hyperthermia, magnetic separation, cell proliferation, tissue repair, and drug delivery [12, 179].

A study has been conducted with itraconazole-loaded SPIONs prepared by the co-precipitation method, and it proved that the core-shells inhibited the fungal activity of the *C. albicans* fungi. These synthesized core-shells were encapsulated by a natural biodegradable polymer (polyhydroxybutyrate), which confined the tagged drug onto them. These nanoparticles displayed steady and slow release kinetics profile. Furthermore, chicken eggs were treated with the chemically synthesized naked SPIONs to check their suitability as contrasting agents, which was indeed confirmed, since they improved X-ray imaging visualization [180].

8.4.4 Other Nanoparticle Drug Delivery Systems

Niosomes: These are vesicles made of nonionic surfactants that serve as drug depots in the body, as they release the drug in a sustained fashion through their bilayers. They are also able to improve the oral bioavailability of poorly soluble drugs, enhance the skin permeability of topical drugs, protect the enclosed active substance from deleterious factors, and increase the stability of the entrapped drug [181].

Nistatin was successfully encapsulated in niosomes, providing a safe and effective formula for parenteral administration. This formulation provided reduced nephrotoxicity and hepatoxicity *in vivo* and showed pronounced efficacy against *C. albicans* with a higher level of drug in vital organs when compared to the free drug [104]. The encapsulation of ciclopirox in niosomes increased the half-life, promoted a prolonged drug release, and minimized the side effects [106]. Given the poor water solubility of griseofulvin and its slow absorption from oral route, a niosome was produced to enhance the bioavailability of this active substance to fasten its absorption and to obtain a sustained release, acting as a depot inside the body [105].

Ketoconazole niosomal gel showed more prolonged action than conventional ketoconazole formulations; hence, it can be developed to improve the antifungal activity [43].

Spanlastics: These are modified niosomes that present better permeability because they have edge activators in their composition, such as transfersomes and transethosomes [22, 151]. These systems show great flexibility, allowing them to squash through fenestrations smaller than their own radius to enter the cell [145]. Spanlastics have been used for the ocular delivery of ketoconazole [45] and loaded with terbinafine hydrochloride for the treatment of onychomycosis [84].

Oil-in-water nanoemulsions: These are thermodynamically unstable colloidal dispersions consisting of two immiscible liquids, with one of the liquids being dispersed as small spherical droplets (r < 100 nm) in the other liquid [182]. The capacity of nanoemulsions to dissolve large quantities of drugs with low solubility, their compatibility, their ability to protect drugs from enzymatic degradation and hydrolysis, as well as their capacity to penetrate the deeper skin layers, turn them into ideal drug delivery vectors [21, 183]. A nystatin nanoemulsion developed for topical application was found to have a higher antifungal effect than nystatin itself, representing a therapeutic improvement [102].

Oil-in-water microemulsions: These are a thermodynamically stable colloidal dispersions consisting of small spheroid particles (composed of oil, surfactant, and possibly co-surfactant) dispersed within an aqueous medium [182]. Because of this, they can incorporate drugs of different lipophilicity [184]. They are promising colloidal carriers due to their transparency, ease of preparation, and long-term stability [185].

When it comes to drug delivery through stratum corneum, microemulsions are very resourceful considering the ability of the oils and surfactants included in their composition to act as skin-penetration enhancers [184, 186]. Furthermore, they are also good candidates for oral delivery of poorly water-soluble drugs as they can improve their solubilization. Terbinafine hydrochloride is an example of a slightly water-soluble drug, which presented higher solubility, an increase in the dissolution rate, and better efficacy when incorporated within a microemulsion [184, 185]. Likewise, a microemulsion system of voriconazole also showed a significant increase in the antifungal activity against *C. albicans*, along with better drug skin penetration [187].

Silicon dioxide nanoparticles: These have been studied as drug carriers to enhance the antimicrobial activity of some drugs, because of their biodegradability, low toxicity, capacity to stimulate macrophages, and the ease of synthesis and functionalization. The most notable advantage of these nanoparticles is that they can be loaded with large amounts of drug [100, 188].

Silica nanoparticles can be modified or loaded with different materials or molecules that provide them antimicrobial properties, along with silica precursors. Despite this, their specific antifungal potential is yet to be studied in some cases. Examples of silica nanoparticles with antimicrobial activity include the following:

- Nitric oxide–silica nanoparticles with proven anti-biofilm activity [115]
- Metal-modified silica nanoparticles, which can include silver or copper, metals that have a well-documented antimicrobial effect by causing cell membrane and DNA damage, interacting with thiol groups of enzymes, and generating free radicals [189]
- Surface-modified and molecule-loaded silica nanoparticles, including silica nanoparticles functionalized with quaternary ammonium compounds and loaded with antifungal agents [190–192]
- Bioglasses and bioceramics with antimicrobial bioactive glass activity due to the release of silica and their other ionic compounds over time, which increases the pH environment, something not well tolerated by bacteria. In addition, different antibacterial drugs such as gentamicin and rifampicin/isoniazide have been successfully incorporated into mesoporous bioactive glasses and bioceramic/silica nanoparticles, respectively [192].

In particular, mesoporous silica nanoparticles (Fig. 8.8) have become the most prominent candidates for many biomedical applications, specifically for antifungal therapy, because of their uniform mesoporous tunnels and narrow pore size distribution. Besides that, they present outstanding biocompatibility and chemical stability and can be degraded and metabolized over a relatively short term. These aspects are translated into high drug-loading capacities and reduced probability of particle-induced toxicity [100, 193].

In a study, one group of mice with candidiasis induced by *C. albicans* was treated with amphotericin B–loaded silica nanoparticles and another group was treated with standard amphotericin B. The first group showed a significant increase in survival rates from 55% (when treated with the officinal drug) to 90% (when treated with the nanoformulation) [100].

Other studies showed that silica nanoparticles coated with quaternary ammonium surfactants have higher fungicidal and fungistatic effect against *C. albicans* than colloidal silver [190, 191].

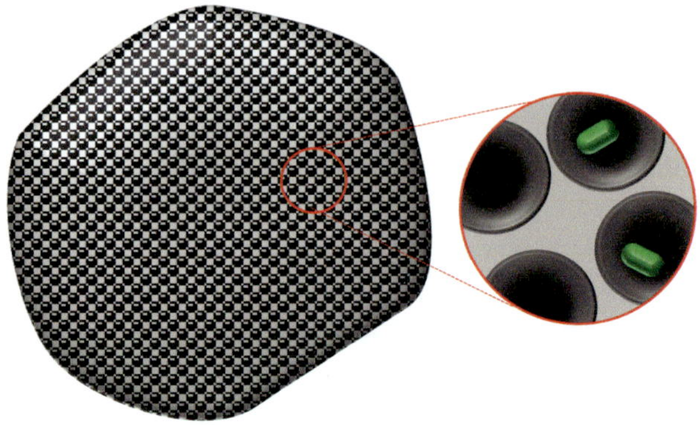

Figure 8.8 Schematic representation of a mesoporous silica nanoparticle, outlining its pore size distribution and high drug-loading capacity. These nanoparticles have a solid framework with pores in its structure (represented by the black spheres) and a large and active surface area, which lets the linkage of therapeutic molecules (represented by the green cylinders).

Mesoporous silica nanoparticles have also been used to increase the oral bioavailability of itraconazole, a poorly water-soluble antifungal drug [65]. The application of mesoporous silica nanoparticles as antifungal agents goes beyond the biomedical field, and in fact, these particles have been shown to be an effective and safe alternative to fungicides for plant disease management [193].

8.5 Preclinical Models to Study Fungal Infections and Antifungal Drugs

The main goal while developing *in vivo* and *in vitro* models in clinical mycology is to mimic, as closely as possible, the progression and clinical manifestations of the infection in human patients. A meticulous study in this phase gives precious insights into infection mechanisms, host resistance, fungi–host interactions, antifungal drug assessment, investigation of fungus virulence factors, pathogenicity of different fungus species, biofilm formation, assessment of potential vaccines, and, ultimately, finding a cure for the fungal infection [2, 194, 195].

8.5.1 *In Vivo* Models

In vivo models render the advantage of studying different features of the disease and the drugs being tested in a whole organism, delivering the most complex interactions that can be attained experimentally [2]. Animal models allow the study of both pharmacokinetics and pharmacodynamics of novel fungal compounds [196].

A diversity of animals, such as rats, rabbits, guinea pigs, hamsters, dogs, and birds, along with toads, cats, or bats, have been used in mycology research. Despite this variety, the mouse *Mus musculus* remains the elected species in mycology [197].

There are numerous advantages in the usage of murine models in mycology: They mimic fungal infections; share great similarity with the human organ system, biochemistry, pathology, and also genome; require less specialized animal care and personnel; predict clinical results; gather *in vivo* data on drug activity, pharmacology, safety, and drug interactions; and allow multiple variables to be controlled by the investigator [197]. These variables include the animal species, the fungal strain, the inoculum size, the immunosuppressive regimen, the host, the infection progression, the administration route, and the therapy duration [198].

Rats: They are essentially employed to study antifungal efficacy, particularly when the infection under study is vaginal candidiasis, since it offers more accurate data compared to the mouse [199]. A study used rats for *in vivo* testing of mucoadhesive liposomes coated with pectin as an alternative to conventional drug gel formulations. They observed, through histopathological examination, that while the Candida-infected untreated group presented inflammatory cells infiltrations with damaged mucosa, the vagina of the rats group receiving sertaconazole liposomal gel showed intact mucosa. In addition, the *ex vivo* study showed increased drug tissue retention and reduced drug tissue permeability of mucoadhesive liposomal gel compared to control gel [199].

Rabbits: They are also commonly used in mycology as they permit more in-depth observation of clinical signs and symptoms. They also allow evaluation of data difficult to obtain with small mammals, which requires, for instance, serial bleeding. Moreover, it is relatively easy to obtain larger volumes of cerebrospinal fluid from rabbits intended for the study of central nervous system (CNS) infections [197]. Rabbits have already been used as models of CNS (*e.g.*, cryptococcal meningitis), systemic, and eye infections caused by different fungi. They have also been used as *in vivo* models to study drug efficacy, for instance, with isavuconazole [200]. Rabbits are, indeed, quite valuable in this field, despite being more expensive, requiring special facilities and more vigilant monitoring and care [197].

Guinea pigs: They are the next most frequently used species in medical mycology, being the standard model for cutaneous (skin, hair, and nail) infections, as they are more prone to dermal fungal infections, not needing any pretreatment or preconditioning [196]. They have been tested as models for cutaneous infections caused by *Candida* [201] and seborrheic dermatitis caused by *Malassezia* sp. [202], along with systemic aspergillosis [203] and cryptococcosis [204]. Guinea pigs are also the preferred models for testing some triazoles, such as voriconazole and itraconazole, owing to their pharmacokinetics. In mice, voriconazole is rapidly metabolized and itraconazole reveals limited absorption, contrary to what happens in guinea pigs [198]. A disadvantage of guinea pigs is their complex social structure, which originates anxiety under unfamiliar environments [197].

Birds: Aspergillosis is one of the fungal infections that cause respiratory disease in birds. Therefore, infected birds, such as quails, turkeys, guinea fowl, and chicken, have been used to study aspergillosis pathogenesis, therapy, and vaccine development [197].

When an antifungal drug proceeds to preclinical studies, several parameters should be followed so that the models chosen are able to replicate all (or at least most) clinical outcomes of the disease and the information obtained is of value [197]. The primary parameters of fungal infection to be followed are fungal burden in tissues (colony-forming unit (CFU) count), survival (expressed in percentage), and mean survival time (MST), usually plotted as Kaplan–Meyer curves. In some experiments, evaluation may also include histopathological examination of tissues [194].

The standard animal models used to screen antifungal drugs present several weaknesses, namely, the substantial variability in drug penetration or metabolism from one experiment to the next, and the sometimes disproportionately acute progression of the disease

[197]. These models imply a prior animal's immunosuppression with a neutropenic agent (cyclophosphamide) to permit a quick dissemination. When the immunosuppression peak is achieved, the mice are inoculated intravenously with the fungal pathogen (this inoculum is highly dependent on the fungus species and strain) [194]. The drugs under study might be injected anywhere from 2 to 24 h after infection agents, being the infection course monitored for a period of up to 30 days [205]. The efficacy of the antifungal drug can be quantified through two different methods, both requiring a positive control (*e.g.*, fluconazole) for the comparison of results: (i) The animals are euthanized at different time points after infection (24–72 h). Afterward, the total fungal burden in target organs (*e.g.*, kidney) is quantified in CFU through the homogenization of the organ followed by serial dilution and plating on a fungal growth medium; or (ii) the animal mortality is evaluated over a period of time, in what is called a survival study [197, 205].

However, these methods require a large number of animals so that statistical significance is achieved, and they raise critical ethical issues related to animal suffering and criteria for euthanasia. In addition, they often imply diet changes, genetic manipulations, surgical alterations, estrogenization, pretreatment to trigger diseases, and, lastly, administration of immunosuppressive agents alone or in combination with antibiotics to prevent secondary bacterial infections [194, 197]. Several researchers have undertaken measures to employ fewer animals, improve the efficacy of screening, and refine the existing procedures to minimize the stress inflicted to the animal. These new alternative biomarkers of fungal infection, which serve as diagnostic tools and are alternatives to survival and infectious burden, include radiographic imaging, polymerase chain reaction (PCR), and assays for antigenemia (galactomannan and glucan) [197, 205].

In vivo imaging approaches, such as fluorescence and bioluminescence, make it possible to follow the location and multiplication of the invading organism, as well as perform longitudinal experiments in single animals [196]. These techniques have been applied to visualize and track *C. albicans* into the larvae of the *Galleria* moth [206] and *Aspergillus* in rabbits [207]. Despite their functionality, the drugs are mainly tested in a qualitative essay, where the output is survival [205].

Firefly luciferase has been optimized for bioluminescence imaging to monitor several types of infections in mice [208]. Nonetheless, a drawback of the bioluminescence approach is that oxygen is a prerequisite for light generation by luciferase and so this approach will not correctly estimate the fungal burden in fungal lesions (with hypoxia) [196].

Ryan et al. have developed a fluorescence-based approach in which they use *in vivo* imaging of a fluorescent strain of *C. albicans* (that expresses a red fluorescent protein) in a mouse strain deficient in the host defense peptide murine β-defensin 1 (mBD-1) and in honeycomb moth. When the fluorescent strain was injected into mBD-1 deficient mice, a nonlethal disseminated infection was observed, which allowed for real-time quantification. The results demonstrated the possibility of obtaining quantitative data, without requiring the second injection of luciferase substrate or immunosuppressive agents [205]. This technique has also been applied to assess pulmonary *Aspergillus* and *Cryptococcus* infections in live animals [196]. Other techniques, such as computed tomography and magnetic resonance, have also been used to evaluate invasive pulmonary aspergillosis [209, 210].

Deciding the infection route used in an animal model fungal infection can be challenging because animal hosts might present different susceptibilities when exposed to the same fungus by different routes. On the other hand, animals often require higher doses of antifungal drug than humans to achieve similar drug exposure [196]. It is also imperative to remember that each fungus shows tropism to a particular animal organ, which may differ from the primary organ infected in humans. Therefore, before advancing to *in vivo* studies, one should address and answer an extensive list of questions: How is the infection behavior and which model would be more reproducible? How is the host immune system? Is the infection chronic or acute? What is the therapeutic goal of the treatment, including optimal doses, duration, frequency, and administration route? Which infection parameters (survival, CFU, surrogate markers) should be followed? Which software is going to be used for statistical analysis [197]?

Mammalian models remain the benchmark for *in vivo* biological studies and will continue to be a key tool in mycology, essentially in solving riddles around fungal infections, such as their onset, progress, control, transmission, prevention, and elimination. Although animal models mirror human infections remarkably well, scientists today are faced with strong opposition to the use of laboratory animals in experimental studies [197].

Over the past decades, classical mammalian models have been increasingly replaced by alternative animals, the "mini hosts" or "non-conventional animal models," such as nematodes, insects, fish, and chicken embryos (Table 8.5). This replacement is due to several reasons: (i) Their less developed neural systems diminish their ability to feel pain; (ii) the possibility of employing a greater number of animals in each experiment; (iii) the maintenance is not as costly as for traditional animals; (iv) the correlation between these animals and mammals is good; and(v) there is an economy of human resources [8, 194].

Table 8.5 Alternative preclinical models for the study of fungal infections

In Vivo Model	Infection Procedure	Fungal Infection Studied and/or Antifungal Tested	Study Goal
Chicken embryos	Intraperitoneal injection via the choriollantoic membrane [211, 221–224]	*Candida albicans* [222], *Rhizopus* [224], *Aspergillus fumigatus* [221]	**Infection:** Fungal virulence and pathogenesis [211, 221–224]
Danio rerio	Anesthesy by immersion in water with tricaine + intraperitoneal microinjection [212–214]	*C. albicans* [213], *Cryptococcus neoformans* [214]	**Infection:** Toxicity assay, teratology testing, fungal virulence and pathogenesis [212–214]
	Anesthesy by immersion in water with tricaine + microinjection into the hindbrain ventricle [215, 225]	Voriconazole [225], Fluconazole, Nystatin, and Chitosan silver nanocomposite [215]	**Treatment:** Efficacy of antifungal drugs [215, 225]

(*Continued*)

Table 8.5 (*Continued*)

In Vivo Model	Infection Procedure	Fungal Infection Studied and/or Antifungal Tested	Study Goal
Drosophila melanogaster	Injection, ingestion, or rolling contact [216, 218, 226]	*Candida* sp. [218], *C. neoformans* [226], *Aspergillus flavus* [216]	**Infection:** Fungal virulence and pathogenesis [216, 218, 226]
	Ingestion, mixed with food [218, 219, 227]	Fluconazole [218], Posaconazole [219], Voriconazole [227]	**Treatment:** Efficacy of antifungal drugs [218, 219, 227]
Galleria mellonella	Injection in the ventral face of the last proleg [217, 223, 224, 228–232]	*Candida* sp. [230], *Cryptococcus* spp. [231], *Paracoccidioides lutzii* [229], *Histoplasma capsulatum* [229], *Fusarium* sp. [228], *Rhizopus* [224]	**Infection:** Fungal virulence and pathogenesis [217, 224, 228–231]
		Fluconazole [232], Liposomal amphotericin B, Posaconazole, Isavuconazole, and Nystatin [223]	**Treatment:** Efficacy of antifungal drugs [217, 223, 232]
Caenorhabditis elegans	Propagation on nematode growth medium seeded with *Escherichia coli* (food source) + population synchronization + inoculum spread into tissue culture plate allowing ingestion [220, 233–235]	*Candida* sp. [233], *Cryptococcus* spp. [235], *H. capsulatum* [234]	**Infection:** Fungal virulence and pathogenesis [220, 233–235]
		Caspofungin, Fluconazole, and Micafungin [233]	**Treatment:** Efficacy of antifungal drugs [220, 233]

Chicken embryos: Since 1971, chicken embryos have been employed to study fungal virulence and as a screening tool to define the most interesting mutants for subsequent testing in mice. In addition, this model allows histological evaluation of invasiveness as well as quantification of fungal burden and host immune response [211]. Although the ethical concerns around this type of experimentation are less problematical, it is never too much to underline that the euthanasia after experiments should be administered with anesthesia [8].

Zebrafish embryos, larvae, and adults (*Danio rerio*): These are also widely employed to perform toxicity assays, test for teratogenicity, and analyze virulence and pathogenesis. Their tiny size, low maintenance cost, high reproductive rate, fecundity, along with their optical transparency, which permits live imaging of host–pathogen interactions either through conventional microscopy or time-lapse fluorescence microscopy, make them an attractive alternative [212–214]. This model has been used to screen the efficacy of antifungal agents, including novel chitosan silver nanocomposites with proven *in vitro* antifungal activity [215].

Insects: Insects, such as the fruit fly *Drosophila melanogaster* and the larvae moth *Galleria mellonella,* show promise for identifying novel fungal virulence genes, testing the efficacy of antifungal drugs, and deciphering conserved antifungal innate immunity mechanisms [197, 216]. They are small and cheap and are easy to manipulate even at large numbers. In addition, since their immune systems share a great similarity with that of mammals, they have been used as models to assess virulence and pathogenicity mechanisms of different fungal species [194]. The larvae moth has been a valuable model for the virulence evaluation of *Cryptococcus neoformans* due to its thermotolerance and high susceptibility to infection [217]. These insects can also be useful to check the *in vivo* efficacy of new antifungal candidates and to assess their pharmacokinetics [8, 194]. Different antifungals have been tested for the treatment of different diseases using *D. melanogaster*, displaying both good correlation with mammalian models and increased survival rates on infected flies [218, 219]. Research groups have also studied the efficacy of different antifungal drugs, mainly fluconazole and amphotericin B, against many *Candida* species with the larvae moth [217]. Despite its usefulness as an alternative model, some argue that the fruit fly could be a potential vector of opportunistic pathogens and that the larvae moth could transfer immunity to bacteria [194].

Nematode: *Caenorhabditis elegans* is a eukaryotic multicellular nematode, which presents a rapid life cycle, produces many descendants, is easily observed, and has its genome fully sequenced [194]. Since it is susceptible to infection by human pathogenic yeasts and filamentous fungi, it can be employed for high-throughput screening of antifungal drug candidates and to study antifungal immune defenses [8, 194]. One of the main weaknesses of this model is that it cannot grow at 37°C [220].

A major challenge of preclinical testing is the establishment of clinically relevant models. Despite their unquestionable usefulness, non-mammalian models are still uncapable of fully replacing mammalian ones. Indeed, there is no such thing as "a perfect model," and the best model is, instead, the one that can provide the most meaningful answer to a specific research question. Often, animal models are used in combination, for example, by using non-mammals as a screening tool to study host–parasite interactions prior to additional studies on mammals.

8.5.2 *In Vitro* Models

Although *in vitro* models might lack the complexity of *in vivo* models, they still allow broad control over external growth conditions of cells concerning O_2 and CO_2 saturation, temperature, pH, and nutrients. Furthermore, both their manipulation and quantitative/qualitative examination are relatively easy, enabling the testing of parameters that cannot be studied in *in vivo* models previously mentioned, such as rats or rabbits [2].

In vitro models can be used to simulate fungal infections of diverse organs and tissues and the corresponding immune responses at conditions close to the physiological ones [2]. In addition, *in vitro* testing for drug efficacy is the first step of preclinical trials and allows the determination of the effective concentration of a drug against an organism [197].

There are numerous *in vitro* models, which can be enumerated from low to high complexity: monolayers in well plates, transwell systems, tridimensional tissue structures, and organ-on-chip [2], which are used to a higher or lesser degree to simulate infections in various organs, such as the skin, liver, lung, and gut [236].

Cell cultures: These cultures reproduce an infection in a well-controlled environment, where microorganism growth is evaluated as a response to different drug concentrations [197]. They are extensively employed to study the interaction between the microorganism and a target tissue or a specific cell type, due to their highly reproducible results. These cells can be either primary cells isolated from blood/tissues/biopsies and cultured for a limited time, or immortalized cell lines [2]. Ideally, to obtain a good correlation to *in vivo* events, the same culture ought to include several cell types. Then, the individual impact of each cell type can be assessed by introducing and/or omitting that cell type in the culture [2].

Transwell system: This system consists of upper and lower compartments separated by inserts, in which cells are cultured under static conditions on a porous membrane allowing apical–basal polarization and a natural tropism through the cell layers [2, 237]. A highly reproducible transwell model of alveolar surface has been developed to study host immunity in invasive aspergillosis [237].

Currently, cell culturing systems have advanced to tridimensional cultures, which consist of a rotating-wall vessel whose turbulence allows cell association and the formation of 3D structures [197].

Organ-on-chip (OOC): It is a tridimensional miniature organ generated from stem cells on a microfluidic biochip with a porous membrane and two individually accessible channels.

Organoid models: They pose the same challenges as OOC models in the creation of an adequate physiological microenvironment, particularly the incorporation of additional cell types, such as immune cells, endothelial cells, and extracellular matrix components. Contrary to OOC, stem cell–derived organoids currently lack perfusion and thus deprive epithelial cells of shear stress and metabolites removal. An emerging idea has successfully combined self-assembling organoids with microfluidic OOC techniques, resulting in what has been coined "organoids-on-a-chip" [238].

Multi-organ-on-chip: A new generation of complex *in vitro* systems combining individual OOCs into multi-organ-on-chip (MOC) models has been gaining momentum for mimicking and tracking certain steps of fungal dissemination in key target organs. In these models, issues related to organ scaling, relative physiological size, and adherence of fungi to tubing and subsequent clogging are yet to be resolved [197].

Figure 8.9 gives a summary of the *in vitro* models previously discussed from low to high complexity, along with different models to study fungi–host interactions in many targeted organs.

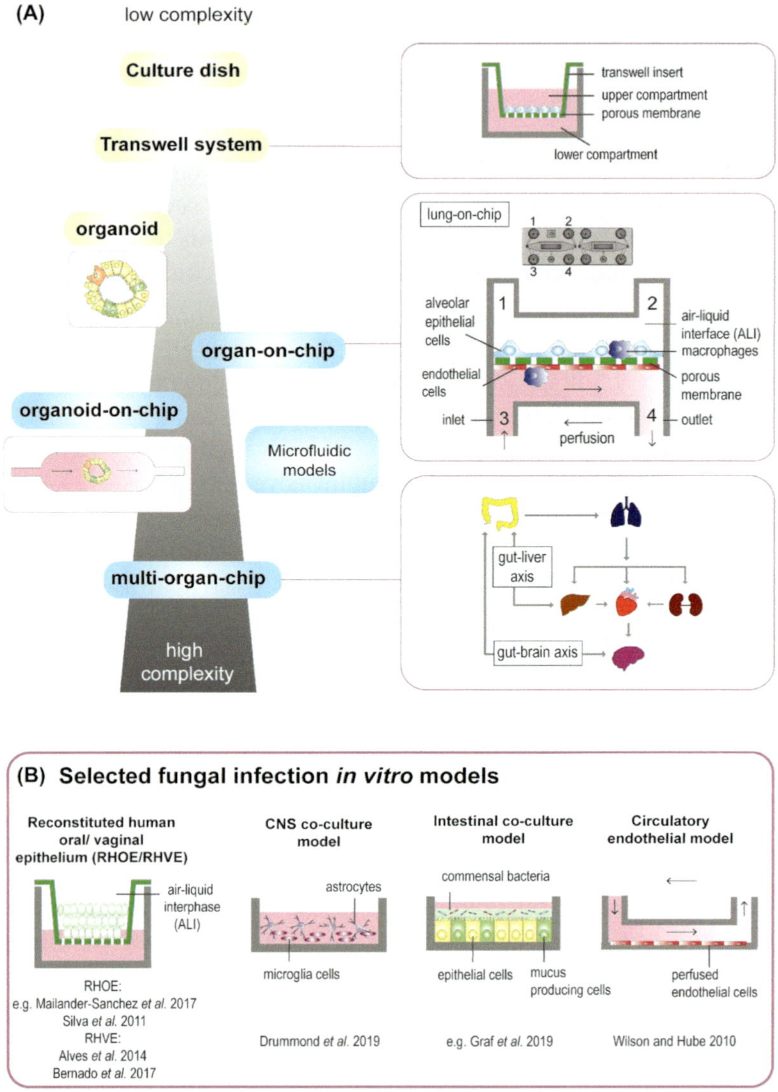

Figure 8.9 **(A)** Evolution of *in vitro* models from low to high complexity. Culture dish: one cell type cultured in media. Transwell system: transwell inserts separate the culture area into upper and lower compartments; cells are cultured under static conditions on a porous membrane allowing apical–basal polarization. Organoid: 3D miniature organ generated out of intestinal stem cells. Organ-on-chip (example): 3D lung-on-chip model on a microfluidic biochip holding a porous membrane and two individually accessible channels with one inlet and outlet each; pulmonary epithelial cells are cultured in the upper compartment in an air–liquid interface; and endothelial cells in the lower compartment are perfused with cell culture medium enabling the removal

of metabolites. Organoid-on-chip: maturation of organoids within a dynamic culture environment. Principle of a multi-organ-on-chip: interconnected organ-on-chip models of gut and liver, or gut and brain or other combinations of lung, intestine, liver, brain, and/or kidneys. Such combinations can, for example, mimic certain steps of fungal dissemination throughout the body. The intestine and lung serve as primary infection sites. **(B)** Selected *in vitro* models to study host–fungi interactions. 3D reconstituted human oral (RHOE) or vaginal (RHVE) epithelium grown at an air–liquid interface. Central nervous system (CNS) co-culture model including microglia cells and astrocytes. Intestinal co-culture model including epithelial cells, goblet cells, and bacteria. Circulatory model with perfused endothelial cells. Reprinted with permission from Ref. [2].

Antifungal activity: The antifungal activity of a drug candidate, relative to a standard antifungal, is typically quantified in terms of IC_{50} or IC_{90} (MIC of the drug or standard antifungal that inhibits fungi growth by 50% or 90%, respectively). Instead, MIC_{50} and MIC_{90} (antifungal concentration at which 50% or 90% of the organisms tested are inhibited, respectively) should be used for standardized antifungal susceptibility testing (AFST) and clinical breakpoints (efficacy of a given antifungal) [196]. The MIC values for these assays are well-defined in EUCAST (European Committee on Antimicrobial Susceptibility Testing) quality control tables, with both ranges and target listed [239]. These AFST methodologies present downsides, for instance, being time consuming, having a long turnaround time, the subjectivity of MIC endpoints, and the interlaboratory variability of MIC values [196].

Alternatives to these classical phenotypic AFST methodologies are on the horizon, including phenotype-centered approaches, which combine a culture step with molecular analysis by real-time PCR or MALDI-TOF MS (matrix-assisted laser desorption ionization time-of-flight mass spectrometry) [203, 240].

Fungicidal activity: The minimum fungicidal/lethal concentration (MFC/MLC), defined as the minimum concentration resulting in a 99.9% reduction in fungal cells, is considered the most rigorous *in vitro* criterion for determining fungicidal activity. Likewise, time-kill assays may also provide useful information [196]. According to the reference AFST methods, both MIC and MEC endpoints need to be clearly determined for studies about antifungal therapy efficacy, resistance mechanisms, and epidemiology [239].

Despite all the benefits of these models, the simplicity of an *in vitro* environment does not accurately replicate the complexity of an *in vivo* environment. Additionally, the *in vitro* activity of a majority of antifungal drugs is yet to be correlated with clinical outcome.

The main parameters of both models (*in vivo* and *in vitro*) are expressed in Table 8.6.

Table 8.6 Most important parameters for experimental testing *in vitro* and *in vivo*

	***In Vivo* Models**				
	Inoculation site in animal models	**Primary organ tropism in animal models**	**Evaluation methods**	***In vitro* Models and Cell Lines**	**Some Examples of Infections**
Systemic Infections	Mice: lateral tail veins [241]; intraperitoneal [194] Rabbits and guinea pigs: lateral ear vein; penile vein [197]	Mice: kidney* [241]	Survival curves; animal weight and/or tissue burden; histopathology; PCR [242, 243]	Cell culture [243] Human umbilical vein endothelial cells (HUVECS) combined or not with neutrophils [187]	Candidiasis [241]; aspergillosis; cryptococcosis; coccidiosis; histoplasmosis; fusariosis; trichosporonosis [196]
Central Nervous System Infections	Mice/guinea pigs: brain parenchyma (intracranial); lumbar puncture (intrathecal) [197, 204] Rabbits: sedation + injection into the cisterna magna [244]	Mice/rabbits: brain [188]	Dissemination studies; histopathology [244]	Monolayers on transwell inserts or cell culture plates; microfluidic blood brain barrier-on-chip with neurovascular unit cells (NVU) in co-culture Brain vascular endothelial cells HBMEC and HCMEC/D3, in co-culture or not with THP-1 cells or primary monocytes	Cryptococcosis (meningitis) [204]; aspergillosis; coccidioidomycosis; scedosporiosis; histoplasmosis [196]
Pulmonary Infections	Mice/guinea pigs: intranasal instilation; inhalation of dry fungi (aerosol chambers); intratracheal administration [197, 203]	Mice: lung** [194]	Determination of organ fungal burden through qPCR; analysis of epithelial damage by lactate dehydrogenase quantification; fungal growth by galactomannan quantification [237]	Transwell system [237]; lung-on-chip; alveolus-on-chip Pulmonary epithelial cell line A549, combined or not with dendritic cells or human alveolar/bronchial endothelial/epithelial cells [2]	Blastomycosis; aspergillosis [194], cryptococcosis; histoplasmosis; paracoccidioidomycosis; zygomycosis [196]

	***In Vivo* Models**				
	Inoculation site in animal models	**Primary organ tropism in animal models**	**Evaluation methods**	***In vitro* Models and Cell Lines**	**Some Examples of Infections**
Gastrointestinal Infections	Mice: antibiotic administration to clear endogenous microbiota + oral gavage (drinking/ feeding) [197]	Mice: stomach [242]	Mice: stool sample analysis to determine colonization, followed by harvesting and analysis for CFUs of gastrointestinal tract segments [196]	Transwell system; 2D intestinal models; intestine-on-chip; human intestinal organoids [245] Caco-2 combined or not with Raji B cells; HT-29; Caco-2 subclone C2BBe1 mixed or not with HT-29-MTX cells [196, 246]	Candidiasis [194]
Cutaneous Infections	Guinea pigs: skin trauma prior to the inoculum [196]	Guinea pigs: skin [196]	Guinea pigs: microscopic or culture examination of skin scrapings; skin biopsies with histopathology; scales and/or hair samples; scoring system [196]	—	Candidiasis; aspergillosis; mucormycosis; onychomycoses [196]
Vaginal Infections	Mice/rats: Estrogenization + lavage into the vagina (intravaginal route) [194, 197]	Mice/rats: vaginal wash [199]	Mice/rats: microbial recovery (CFUs) from vaginal swabs; histopathology; and immunological analysis [196]	Reconstituted vaginal epithelium (RHVE); organ-on-chip model of the vaginal mucosa [196] VK2/E6E7; A431 [196, 246]	Vaginal candidiasis [199]

(*Continued*)

Table 8.6 (*Continued*)

	***In Vivo* Models**				
	Inoculation site in animal models	**Primary organ tropism in animal models**	**Evaluation methods**	***In vitro* Models and Cell Lines**	**Some Examples of Infections**
Oral Infections	Mice: local anesthesia + placement of a swab impregnated with the inoculum under the tongue [2, 196]	Mice: level of microbial recovery and histopathology of infected tissue [196]	CFU count from oral cavity swab; histology [247]	Monolayer models; 3D reconstituted human oral epithelium (RHOE) combined or not with human fibroblasts and keratinocytes [196]	Oropharyngeal candidiasis [247]
				Oral epithelial cells TR146; oral mucosal cells OKF&/TERT-2; tongue cells SCC15 [196]	
			Palate tissue/ denture analysis by confocal microscopy; histology [248]	Mono- or multilayered mucosal models (mucosa-on-chip), which can include dendritic cells [196]	*Candida*-mediated stomatitis [248]
				Primary human palate epithelial cells; oral epithelial cells TR146 combined with primary fibroblasts [196]	

Note: *Different from tropism in humans; **Equal to tropism in humans, but in mice there is no dissemination to other tissues

8.6 Ongoing Clinical Trials on Myconanotechnology

Despite the considerable progress that we have witnessed in the development of nano-enabled technologies for drug development and delivery, only a few dozens of nanomedicines have been clinically approved. Clinical trials are time consuming and expensive, and the translation into the market has been limited due not only to inconspicuous clinical improvement but also to reproducibility issues that entangle compliance with GMP guidelines. In the mycology field, amphotericin B is the active substance that dominates the already marketed nanomedicines: Amphotec® (Three Rivers Pharmaceuticals, approved in 1992); Abelcet® (Sigma-Tau, approved in 1995); and AmBisome® (Gilead Sciences, approved in 1997), all lipid-based formulations [108, 249]. These nanotechnology-driven formulations arose to solve issues associated with the poor water solubility, low tolerance, and side effects that the drug presents when administered on its own. An econazole liposomal formulation (Pevaryllipogel®) is also available on the market [250]. Table 8.7 presents some examples of recent clinical trials.

8.7 Conclusion

The present rhythm of antifungal drug development is highly unlikely to keep up with the clinical needs, especially with the uprising of resistance to current agents. Therefore, additional therapeutic solutions to fungal illnesses are urgently needed.

The subject of sustainability and environmental impact has gained considerable relevance in today's society and scientific research, motivating a heated debate on whether research efforts should be focused on actively searching for new natural compounds with biological activity or, instead, on using high-throughput screening methodologies to discover new compounds (natural or synthetic). In either case, the focus nowadays should be on developing green technologies for their synthesis/extraction and processing. In fact, two of the three major classes of antifungal drugs (polyenes and echinocandins) were screened from natural products. The utilization of natural compounds, such as chitosan, is irrefutably

Table 8.7 Some examples of clinical trials on myconanotechnology [251]

Trade name/ Sponsor	ClinicalTrials. gov identifier	Antifungal	Nanoformulation	Clinical Phase	Disease
Sara Botros, Minia University	NCT04110834	Itraconazole	Nanoemulsion gel	II	Tinea versicolor
Sara Botros, Minia University	NCT04110860	Voriconazole	Nanoemulsion gel	II	Tinea versicolor
Matinas BioPharma	NCT02971007	Amphotericin B	Cochleate lipid-crystal nanoparticle	II	Vulvovaginal candidiasis
Matinas BioPharma	NCT02629419	Amphotericin B	Cochleate lipid-crystal nanoparticle	II	Mucocuta-neous candidiasis
Ahmed Abdellatif, Al-Azhar University	NCT03752424	—	Silver nanoparticle gel	I	Mycosis
Celtic Pharma Development Services	NCT01145807	Terbinafine (TDT067)	Transfersome	III	Onycho-mycosis
Matinas BioPharma	NCT04031833	Amphotericin B	Cochleate lipid-crystal nanoparticle	I/II	Cryptococcal meningitis

a more sustainable option and brings both environmental and pharmacokinetic advantages to nanotechnological approaches. However, despite exhibiting an undeniable potential, the activity and toxicity mechanisms of compounds from marine origin are yet to be clarified.

Nanoparticles have been presented as promising solutions, mainly due to their ability to target-specific sites where fungi are harbored and their capacity to enhance the pharmacological effect of drugs. These features enable lower dosing, more comfortable regimens (in terms of drug administration), increased bioavailability, and less serious adverse effects.

Certain classes of nanoparticles have been given special attention due to the outstanding features they exhibit: (i) magnetic nanoparticles and their capacity to directly restrict fungal growth; (ii) ultra-deformable vesicles (transethosomes) and their ease of scaling-up; (iii) mesoporous silica nanoparticles and their high drug-loading capacity; (iv) polymersomes and their ability to carry both hydrophobic and hydrophilic substances and to respond to external stimuli; (v) PAMAM dendrimers and their versatile and biocompatible structure. We speculate that future antifungal therapies will mostly rely on these five types of nanoparticles and will take advantage of the existing knowledge regarding their use for other therapeutic purposes (for instance, as anticancer agent carriers).

These new nanotechnological systems can have a real impact on fighting antifungal resistance by presenting a broad spectrum of activity, increased potency, and decreased host toxicity.

Interdisciplinary cooperation, as well as proper exploitation of natural medicinal products, may be the key to the success of myconanotechnology. Physicists, chemists, biologists, and healthcare/clinical researchers have complementary knowledge that should be put together (*e.g.*, for developing innovative screening strategies or novel chemical libraries) so that more effective, practical, and safe nanoparticles are designed. That would likely spur the interest of pharmaceutical industry and, ultimately, encourage a rethinking of the current antifungal arsenal, while providing a better quality of life for patients.

Funding

This work was supported by the Foundation for Science and Technology, I.P. (FCT), Portugal, through grant reference 2020.05884.BD, and by the Applied Molecular Biosciences Unit – UCIBIO, which receives financial support from Portuguese National Funds (FCT/MCTES, Fundação para a Ciência e Tecnologia and Ministério da Ciência, Tecnologia e Ensino Superior) through grant UIDB/04378/2020.

Conflicts of Interest

The authors declare no conflict of interest.

References

1. F. Sousa, D. Ferreira, S. Reis, and P. Costa, Current insights on antifungal therapy: Novel nanotechnology approaches for drug delivery systems and new drugs from natural sources, *Pharmaceuticals* **13**, 248 (2020).
2. A. Last, M. Maurer, A. S. Mosig, M. S. Gresnigt, and B. Hube, *In vitro* infection models to study fungal-host interactions, *FEMS Microbiol. Rev.* **45**, 1 (2021).
3. M. Rai, A. P. Ingle, R. Pandit, P. Paralikar, I. Gupta, N. Anasane, and M. Dolenc-Voljč, Nanotechnology for the treatment of fungal infections on human skin, in *The Microbiology of Skin, Soft Tissue, Bone and Joint Infections*, edited by K. Kon and M. Rai (Academic Press, 2017), pp. 169.
4. K. M. Pianalto and J. A. Alspaugh, New horizons in antifungal therapy, *J. Fungi (Basel)* **2** (4): 26 (2016).
5. Y. L. Chang, S. J. Yu, J. Heitman, M. Wellington, and Y. L. Chen, New facets of antifungal therapy, *Virulence* **8**, 222 (2017).
6. S. Nami, A. Aghebati-Maleki, H. Morovati, and L. Aghebati-Maleki, Current antifungal drugs and immunotherapeutic approaches as promising strategies to treatment of fungal diseases, *Biomed. Pharmacother.* **110**, 857 (2019).
7. A. C. Souza and A. C. Amaral, Antifungal therapy for systemic mycosis and the nanobiotechnology era: Improving efficacy, biodistribution and toxicity, *Front. Microbiol.* **8**, 336 (2017).

8. L. Scorzoni, A. C. A. de Paula e Silva, C. M. Marcos, P. A. Assato, W. C. M. A. de Melo, H. C. de Oliveira, C. B. Costa-Orlandi, M. J. S. Mendes-Giannini, and A. M. Fusco-Almeida, Antifungal therapy: New advances in the understanding and treatment of mycosis, *Front. Microbiol.* **8**, 36 (2017).

9. S. N. Sirish Sadhna Khatry and M. Sadanandan, Novel drug delivery systems for antifungal therapy, *Int. J. Pharm. Pharm. Sci.* **2**, 6 (2010).

10. J. R. Perfect, Is there an emerging need for new antifungals? *Expert Opin. Emerg. Drugs* **21**, 129 (2016).

11. R. J. Archna Panday, A review of kinetics of nanoparticulated delayed release formulations, *J. Nanomed. Nanotechnol.* **6** (4), 1000304 (2015).

12. S. Hasan, A review on nanoparticles: Their synthesis and types, *Res. J. Recent Sci.* **4**, 1 (2015).

13. B. V. N. Nagavarma, A. H. K. S. Y. Ayaz, L. S. Vasudha, and H. G. Shivahumar, Different techniques for preparation of polymeric nanoparticles: A review, *Asian J. Pharm. Clin. Res.* **5**, 16 (2012).

14. P. Bhatt, R. Lalani, I. Vhora, S. Patil, J. Amrutiya, A. Misra, and R. Mashru, Liposomes encapsulating native and cyclodextrin enclosed Paclitaxel: Enhanced loading efficiency and its pharmacokinetic evaluation, *Int. J. Pharm.* **536**, 95 (2018).

15. J. H. Lee and Y. Yeo, Controlled drug release from pharmaceutical nanocarriers, *Chem. Eng. Sci.* **125**, 75 (2015).

16. W. T. Wang, J. G. Liu, W. H. Sun, W. Y. Wang, S. L. Wang, and N. H. Zhu, Widely tunable single bandpass microwave photonic filter based on Brillouin-assisted optical carrier recovery, *Opt. Express* **22**, 29304 (2014).

17. R. Goyal, L. K. Macri, H. M. Kaplan, and J. Kohn, Nanoparticles and nanofibers for topical drug delivery, *J. Controlled Release* **240**, 77 (2016).

18. A. T. Rangari, Polymeric nanoparticles based topical drug delivery: An overview, *Asian J. Biomed. Pharm. Sci.* **5**, 5 (2015).

19. R. A. Siegel and M. J. Rathbone, Overview of controlled release mechanisms, in *Fundamentals and Applications of Controlled Release Drug Delivery, Advances in Delivery Science and Technology*, edited by C. R. Society (Springer, 2012), pp. 19.

20. Z. P. Aguilar, *Nanomaterials for Medical Applications* (Elsevier, 2013).

21. K. P. Dhawade Nakusha and V. Pande, Emerging trends in topical antifungal therapy: A review, *Inventi J.* **2015**, 1 (2015).

22. G. M. Soliman, Nanoparticles as safe and effective delivery systems of antifungal agents: Achievements and challenges, *Int. J. Pharm.* **523**, 15 (2017).
23. P. U. M. Firthouse, S. M. Halith, S.U.Wahab, M.Sirajudeen, and S. K. Mohideen, Formulation and evaluation of miconazole niosomes, *Int. J. Pharmtech Res.* **3**, 1019 (2011).
24. B. M. Aljaeid and K. M. Hosny, Miconazole-loaded solid lipid nanoparticles: Formulation and evaluation of a novel formula with high bioavailability and antifungal activity, *Int. J. Nanomed.* **11**, 441 (2016).
25. M. R. Bhalekar, V. Pokharkar, A. Madgulkar, N. Patil, and N. Patil, Preparation and evaluation of miconazole nitrate-loaded solid lipid nanoparticles for topical delivery, *AAPS Pharmscitech.* **10**, 289 (2009).
26. I. Shahzadi, M. I. Masood, F. Chowdhary, A. A. Anjum, M. A. Nawaz, I. Maqsood, and M. Q. Zaman, Microemulsion formulation for topical delivery of miconazole nitrate, *Int. J. Pharm. Sci. Rev. Res.* **24**, 30 (2014).
27. R. M. Elmoslemany, O. Y. Abdallah, L. K. El-Khordagui, and N. M. Khalafallah, Propylene glycol liposomes as a topical delivery system for miconazole nitrate: Comparison with conventional liposomes, *AAPS Pharmscitech.* **13**, 723 (2012).
28. H. L. Maha, K. R. Sinaga, and M. Masfria, Formulation and evaluation of miconazole nitrate nanoemulsion and cream, *Asian J. Pharm. Clin. Res.* **11** (2018).
29. P. S. Kumar, N. Hematheerthani, J. V. Ratna, and V.Saikishore, Design and characterization of miconazole nitrate loaded nanosponges containing vaginal gels, *Int. J. Pharm. Anal. Res.* **5**, 410 (2016).
30. M. Qushawy, A. Nasr, M. Abd-Alhaseeb, and S. Swidan, Design, optimization and characterization of a transfersomal gel using miconazole nitrate for the treatment of candida skin infections, *Pharmaceutics* **10** (2018).
31. S. Ge, Y. Lin, H. Lu, Q. Li, J. He, B. Chen, C. Wu, and Y. Xu, Percutaneous delivery of econazole using microemulsion as vehicle: Formulation, evaluation and vesicle-skin interaction, *Int. J. Pharm.* **465**, 120 (2014).
32. D. Evelyn, C. C. Wooi, J. R. Kumar, S. Muralidharan, and S. A. Dhanaraj, Development and evaluation of microemulsion based gel (MBGs) containing econazole nitrate for nail fungal infection, *J. Pharm. Res.* **5**, 2385 (2012).
33. V. Sanna, E. Gavini, M. Cossu, G. Rassu, and P. Giunchedi, Solid lipid nanoparticles (SLN) as carriers for the topical delivery of econazole

nitrate: *In vitro* characterization, *ex vivo* and *in vivo* studies, *J. Pharm. Pharm.* **59**, 1057 (2007).

34. L. Keshri and K. Pathak, Development of thermodynamically stable nanostructured lipid carrier system using central composite design for zero order permeation of econazole nitrate through epidermis, *Pharm. Dev. Technol.* **18**, 634 (2013).

35. Q. Xianrong, M. H. Liu, H. Y. Liu, Y. Maitani, and T. Nagai, Topical econazole delivery using liposomal gel, *S.T.P. Pharma Sci.* **13**, 241 (2003).

36. P. Verma and K. Pathak, Nanosized ethanolic vesicles loaded with econazole nitrate for the treatment of deep fungal infections through topical gel formulation, *Nanomedicine* **8**, 489 (2012).

37. S. Verma and P. Utreja, Transethosomes of econazole nitrate for transdermal delivery: Development, *in vitro* characterization, and *ex vivo* assessment, *Pharm. Nanotechnol.* **6**, 171 (2018).

38. R. Sharma, R. B. Walker, and K. Pathak, Evaluation of the kinetics and mechanism of drug release from econazole nitrate nanosponge loaded carbapol hydrogel, *Indian J. Pharm. Educ. Res.* **45**, 25 (2011).

39. Y. P. Kumar, K. V. Kumar, R. R. Shekar, M. Ravi, and V. S. Kishore, Formulation and evaluation of econazole niosomes, *Sch. Acad. J. Pharm.* **2**, 315 (2013).

40. Y. G. Bachhav, K. Mondon, Y. N. Kalia, R. Gurny, and M. Moller, Novel micelle formulations to increase cutaneous bioavailability of azole antifungals, *J. Controlled Release* **153**, 126 (2011).

41. M. P. Piemi, D. Korner, S. Benita, and MartyJp, Positively and negatively charged submicron emulsions for enhanced topical delivery of antifungal drugs, *J. Controlled Release* **58**, 177 (1999).

42. E. B. Souto and R. H. Muller, SLN and NLC for topical delivery of ketoconazole, *J. Microencapsul.* **22**, 501 (2005).

43. S. Shirsand, M. Para, D. Nagendrakumar, K. Kanani, and D. Keerthy, Formulation and evaluation of Ketoconazole niosomal gel drug delivery system, *Int. J. Pharm. Investig.* **2**, 201 (2012).

44. N. Tiwari, A. Sivakumar, A. Mukherjee, and N. Chandrasekaran, Enhanced antifungal activity of Ketoconazole using rose oil based novel microemulsion formulation, *J. Drug Deliv. Sci. Technol.* **47**, 434 (2018).

45. S. Kakkar and I. P. Kaur, Spanlastics: A novel nanovesicular carrier system for ocular delivery, *Int. J. Pharm.* **413**, 202 (2011).

46. K. Winnicka, M. Wroblewska, P. Wieczorek, P. T. Sacha, and E. Tryniszewska, Hydrogel of ketoconazole and PAMAM dendrimers: Formulation and antifungal activity, *Molecules* **17**, 4612 (2012).

47. S. Ashe, D. Nayak, G. Tiwari, P. R. Rauta, and B. Nayak, Development of liposome-encapsulated ketoconazole: Formulation, characterisation and evaluation of pharmacological therapeutic efficacy, *Micro. Nano Lett.* **10**, 126 (2015).

48. M. Ning, Y. Guo, H. Pan, X. Chen, and Z. Gu, Preparation, *in vitro* and *in vivo* evaluation of liposomal/niosomal gel delivery systems for clotrimazole, *Drug Dev. Ind. Pharm.* **31**, 375 (2005).

49. A. S. Kumar, S. P. S., and M. A. Kuriachan, Formulation and evaluation of antifungal nanosponge loaded hydrogel for topical delivery, *Int. J. Pharm. Pharm. Res.* **13**, 362 (2018).

50. N. Akhtar and K. Pathak, Cavamax W7 composite ethosomal gel of clotrimazole for improved topical delivery: Development and comparison with ethosomal gel, *Am. Assoc. Pharm. Sci.* **13**, 344 (2012).

51. S. Shirsand, R. Kumar, G. Keshavshetti, S. S. Bushetti, and V. S. Padivala, Formulation and evaluation of clotrimazole niosomal gel for topical application, *Rajiv Gandhi Univ. Health Sci. J. Pharm. Sci.* **5**, 32(2015).

52. G. E. Yassin, Formulation and evaluation of optimized clotrimazole emulgel formulations, *Br. J. Pharm. Res.* **4**, 1014 (2014).

53. E. B. Souto, S. A. Wissing, C. M. Barbosa, and R. H. Muller, Development of a controlled release formulation based on SLN and NLC for topical clotrimazole delivery, *Int. J. Pharm.* **278**, 71 (2004).

54. J. Kaewbanjong, P. Wan Sia Heng, and P. Boonme, Clotrimazole microemulsion and microemulsion-based gel: Evaluation of buccal drug delivery and irritancy using chick chorioallantoic membrane as the model, *J. Pharm. Pharm.* **69**, 1716 (2017).

55. Y. G. Bachhav and V. B. Patravale, Microemulsion-based vaginal gel of clotrimazole: Formulation, *in vitro* evaluation, and stability studies, *AAPS Pharmscitech.* **10**, 476 (2009).

56. R. G. Maheshwari, R. K. Tekade, P. A. Sharma, G. Darwhekar, A. Tyagi, R. P. Patel, and D. K. Jain, Ethosomes and ultradeformable liposomes for transdermal delivery of clotrimazole: A comparative assessment. *Saudi Pharm. J. Spj. Off. Publ. Saudi Pharm. Soc.* **20**, 161 (2012).

57. W. S. Zheng, X. Q. Fang, L. L. Wang, and Y. J. Zhang, Preparation and quality assessment of itraconazole transfersomes, *Int. J. Pharm.* **436**, 291 (2012).

58. B. Mohanty, D. K. Majumdar, S. K. Mishra, A. K. Panda, and S. Patnaik, Development and characterization of itraconazole-loaded solid lipid nanoparticles for ocular delivery, *Pharm. Dev. Technol.* **20**, 458 (2015).
59. J. Pardeike, S. Weber, T. Haber, J. Wagner, H. P. Zarfl, H. Plank, and A. Zimmer, Development of an itraconazole-loaded nanostructured lipid carrier (NLC) formulation for pulmonary application, *Int. J. Pharm.* **419**, 329 (2011).
60. V. D. Wagh and O. J. Deshmukh, Itraconazole niosomes drug delivery system and its antimycotic activity against *Candida albicans*, *Isrn. Pharm.* **2012**, 653465 (2012).
61. A. Chudasama, V. Patel, M. Nivsarkar, K. Vasu, and C. Shishoo, Investigation of microemulsion system for transdermal delivery of itraconazole, *J. Adv. Pharm. Technol. Res.* **2**, 30 (2011).
62. A. F. Leal, M. C. Leite, C. S. Medeiros, I. M. Cavalcanti, A. G. Wanderley, N. S. Magalhaes, and R. P. Neves, Antifungal activity of a liposomal itraconazole formulation in experimental *Aspergillus flavus* keratitis with endophthalmitis, *Mycopathologia* **179**, 225 (2015).
63. X. Ling, Z. Huang, J. Wang, J. Xie, M. Feng, Y. Chen, F. Abbas, J. Tu, J. Wu, and C. Sun, Development of an itraconazole encapsulated polymeric nanoparticle platform for effective antifungal therapy, *J. Mater. Chem. B* **4**, 1787 (2016).
64. A. N. ElMeshad and A. M. Mohsen, Enhanced corneal permeation and antimycotic activity of itraconazole against *Candida albicans* via a novel nanosystem vesicle, *Drug Deliv.* **23**, 2115 (2016).
65. R. Mellaerts, R. Mols, J. A. Jammaer, C. A. Aerts, P. Annaert, J. Van Humbeeck, G. Van den Mooter, P. Augustijns, and J. A. Martens, Increasing the oral bioavailability of the poorly water soluble drug itraconazole with ordered mesoporous silica, *Eur. J. Pharm. Biopharm.* **69**, 223 (2008).
66. Y. G. Bachhav and V. B. Patravale, Microemulsion based vaginal gel of fluconazole: Formulation, *in vitro* and *in vivo* evaluation, *Int. J. Pharm.* **365**, 175 (2009).
67. O. A. E. Soliman, E. A. Mohamed, and N. A. A. Khatera, Enhanced ocular bioavailability of fluconazole from niosomal gels and microemulsions: Formulation, optimization, and *in vitro-in vivo* evaluation, *Pharm. Dev. Technol.* **24**, 48 (2019).
68. S. K. Gupta, N. Dhingra, T. Velpandian, and J. Jaiswal, Efficacy of fluconazole and liposome entrapped fluconazole for C. albicans induced experimental mycotic endophthalmitis in rabbit eyes, *Acta Ophthalmol. Scand.* **78**, 448 (2000).

69. S. El-Housiny, M. A. Shams Eldeen, Y. A. El-Attar, H. A. Salem, D. Attia, E. R. Bendas, and M. A. El-Nabarawi, Fluconazole-loaded solid lipid nanoparticles topical gel for treatment of pityriasis versicolor: Formulation and clinical study, *Drug Deliv.* **25**, 78 (2018).

70. H. R. Kelidari, M. Moazeni, R. Babaei, M. Saeedi, J. Akbari, P. I. Parkoohi, M. Nabili, A. A. Gohar, K. Morteza-Semnani, and A. Nokhodchi, Improved yeast delivery of fluconazole with a nanostructured lipid carrier system, *Biomed. Pharmacother.* **89**, 83 (2017).

71. A. Moin, T. K. Deb, R. A. Osmani, R. R. Bhosale, and U. Hani, Fabrication, characterization, and evaluation of microsponge delivery system for facilitated fungal therapy, *J. Basic Clin. Pharm.* **7**, 39 (2016).

72. N. Indora and D. Kaushik, Design, development and evaluation of ethosomal gel of fluconazole for topical fungal infection, *Int. J. Eng. Sci. Invent. Res. Dev.* **1**, 280 (2015).

73. I. P. Kaur, C. Rana, M. Singh, S. Bhushan, H. Singh, and S. Kakkar, Development and evaluation of novel surfactant-based elastic vesicular system for ocular delivery of fluconazole, *J. Ocul. Pharm.* **28**, 484 (2012).

74. S. K. Lalit, A. S. Panwar, G. Darwhekar, and D. K. Jain, Formulation and evaluation of fluconazole amphiphilogel, *Der. Pharm. Lett.* **3**, 125 (2011).

75. R. Kumar and V. R. Sinha, Preparation and optimization of voriconazole microemulsion for ocular delivery, *Colloids Surf. B Biointerfaces* **117**, 82 (2014).

76. E. Basaran, H. K. Gencer, E. Yenilmez, and U. M. Guven, Voriconazole incorporated polymeric nanoparticles for ocular application, *Lat. Am. J. Pharm.* **36**, 1983 (2017).

77. P. J. Das, P. Paul, B. Mukherjee, B. Mazumder, L. Mondal, R. Baishya, M. C. Debnath, and K. S. Dey, Pulmonary delivery of voriconazole loaded nanoparticles providing a prolonged drug level in lungs: A promise for treating fungal infection, *Mol. Pharm.* **12**, 2651 (2015).

78. D. K. Pandurangan, P. Bodagala, V. K. Palanirajan, and S. Govindaraj, Formulation and evaluation of voriconazole ophthalmic solid lipid nanoparticles *in situ* gel, *Int. J. Pharm. Investig.* **6**, 56 (2016).

79. C. K. Song, P. Balakrishnan, C. K. Shim, S. J. Chung, S. Chong, and D. D. Kim, A novel vesicular carrier, transethosome, for enhanced skin delivery of voriconazole: Characterization and *in vitro/in vivo* evaluation, *Colloids Surf. B Biointerfaces.* **92**, 299 (2012).

80. W. Faisal, G. M. Soliman, and A. M. Hamdan, Enhanced skin deposition and delivery of voriconazole using ethosomal preparations, *J. Liposome. Res.* **28**, 14 (2018).

81. S. T. Tanriverdi, S. Hilmioglu Polat, D. Yesim Metin, G. Kandiloglu, and O. Ozer, Terbinafine hydrochloride loaded liposome film formulation for treatment of onychomycosis: *In vitro* and *in vivo* evaluation, *J. Liposome Res.* **26**, 163 (2016).

82. Y. C. Chen, D. Z. Liu, J. J. Liu, T. W. Chang, H. O. Ho, and M. T. Sheu, Development of terbinafine solid lipid nanoparticles as a topical delivery system, *Int. J. Nanomed.* **7**, 4409 (2012).

83. M. Ghannoum, N. Isham, W. Henry, H. A. Kroon, and S. Yurdakul, Evaluation of the morphological effects of TDT 067 (terbinafine in transfersome) and conventional terbinafine on dermatophyte hyphae *in vitro* and *in vivo*, *Antimicrob. Agents Chemother.* **56**, 2530 (2012).

84. N. I. Elsherif, R. N. Shamma, and G. Abdelbary, Terbinafine hydrochloride trans-ungual delivery via nanovesicular systems: *In vitro* characterization and *ex vivo* evaluation, *AAPS Pharmscitech.* **18**, 551 (2017).

85. I. Ozcan, O. Abaci, A. H. Uztan, B. Aksu, H. Boyacioglu, T. Guneri, and O. Ozer, Enhanced topical delivery of terbinafine hydrochloride with chitosan hydrogels, *AAPS Pharmscitech.* **10**, 1024 (2009).

86. M. S. Erdal, G. Ozhan, M. C. Mat, Y. Ozsoy, and S. Gungor, Colloidal nanocarriers for the enhanced cutaneous delivery of naftifine: Characterization studies and *in vitro* and *in vivo* evaluations, *Int. J. Nanomed.* **11**, 1027 (2016).

87. H. S. Barakat, I. A. Darwish, L. K. El-Khordagui, and N. M. Khalafallah, Development of naftifine hydrochloride alcohol-free niosome gel, *Drug Dev. Ind. Pharm.* **35**, 631 (2009).

88. A. B. Pillai, J. V. Nair, N. K. Gupta, and S. Gupta, Microemulsion-loaded hydrogel formulation of butenafine hydrochloride for improved topical delivery, *Arch Derm. Res.* **307**, 625 (2015).

89. R. J. Hay, Liposomal amphotericin B, *AmBisome*, *J. Infect.* **28 Suppl 1**, 35 (1994).

90. P. Jansook, W. Pichayakorn, and G. C. Ritthidej, Amphotericin B-loaded solid lipid nanoparticles (SLNs) and nanostructured lipid carrier (NLCs): Effect of drug loading and biopharmaceutical characterizations, *Drug Dev. Ind. Pharm.* **44**, 1693 (2018).

91. D. Butani, C. Yewale, and A. Misra, Topical amphotericin B solid lipid nanoparticles: Design and development, *Colloids Surf. B Biointerfaces* **139**, 17 (2016).

92. C. A. Saldanha, M. P. Garcia, D. C. Iocca, L. G. Rebelo, A. C. Souza, A. L. Bocca, F. Almeida Santos Mde, P. C. Morais, and R. B. Azevedo, Antifungal activity of amphotericin B conjugated to nanosized magnetite in the treatment of paracoccidioidomycosis, *PLoS Negl. Trop Dis.* **10**, e0004754 (2016).

93. L. Sosa, B. Clares, H. L. Alvarado, N. Bozal, O. Domenech, and A. C. Calpena, Amphotericin B releasing topical nanoemulsion for the treatment of candidiasis and aspergillosis, *Nanomedicine* **13**, 2303 (2017).

94. A. C. Souza, A. L. Nascimento, N. M. de Vasconcelos, M. S. Jeronimo, I. M. Siqueira, L. R-Santos, D. O. Cintra, L. L. Fuscaldi, O. R. Pires Junior, R. Titze-de-Almeida, M. F. Borin, S. N. Báo, O. P. Martins, V. N. Cardoso, S. O. Fernandes, M. R. Mortari, A. C. Tedesco, A. C. Amaral, M. S. S. Felipe, and A. L. Bocca, Activity and *in vivo* tracking of amphotericin B loaded PLGA nanoparticles, *Eur. J. Med. Chem.* **95**, 267 (2015).

95. J. L. Italia, M. N. Kumar, and K. C. Carter, Evaluating the potential of polyester nanoparticles for per oral delivery of amphotericin B in treating visceral leishmaniasis, *J. Biomed. Nanotechnol.* **8**, 695 (2012).

96. X. Tang, H. Zhu, L. Sun, W. Hou, S. Cai, R. Zhang, and F. Liu, Enhanced antifungal effects of amphotericin B-TPGS-b-(PCL-ran-PGA) nanoparticles *in vitro* and *in vivo*, *Int. J. Nanomed.* **9**, 5403 (2014).

97. N. K. Jay Prakash Jain, Development of amphotericin B loaded polymersomes based on (PEG)3-PLA co-polymers: Factors affecting size and *in vitro* evaluation, *Eur. J. Pharm. Sci.* **40**, 456 (2010).

98. A. P. Perez, M. J. Altube, P. Schilrreff, G. Apezteguia, F. S. Celes, S. Zacchino, C. I. de Oliveira, E. L. Romero, and M. J. Morilla, Topical amphotericin B in ultradeformable liposomes: Formulation, skin penetration study, antifungal and antileishmanial activity *in vitro*, *Colloids Surf. B Biointerfaces* **139**, 190 (2016).

99. A. C. Moreno-Rodriguez, S. Torrado-Duran, G. Molero, J. J. Garcia-Rodriguez, and S. Torrado-Santiago, Efficacy and toxicity evaluation of new amphotericin B micelle systems for brain fungal infections, *Int. J. Pharm.* **494**, 17 (2015).

100. A. Lykov, K. Gaidul, I. Goldina, V. Konenkov, V. Kozlov, N. Lyakhov, and A. Dushkin, Silica nanoparticles as a basis for efficacy of antimicrobial drugs, in *Nanostructures for Antimicrobial Therapy*, edited by A. Ficai and A. M. Grumezescu (Elsevier 2017), pp. 551.

101. R. M. Khalil, A. A. A. E. Rahman, M. A. Kassem, M. S. E. Ridi, M. M. A. Samra, G. E. A. Awad, and S. S. Mansy, Preparation and *in vivo* Assessment of nystatin-loaded solid lipid nanoparticles for topical delivery against cutaneous candidiasis, *Int. J. Pharmacol. Pharm. Sci.* **8** ,401 (2014).

102. F. Fernandez-Campos, B. Clares Naveros, O. Lopez Serrano, C. Alonso Merino, and A. C. Calpena Campmany, Evaluation of novel nystatin nanoemulsion for skin candidosis infections, *Mycoses* **56**, 70 (2013).

103. F. Offner, V. Krcmery, M. Boogaerts, C. Doyen, D. Engelhard, P. Ribaud, C. Cordonnier, B. de Pauw, S. Durrant, J. Marie, P. Moreau, H. Guiot, G. Samonis, R. Sylvester, and R. Herbrecht, Liposomal nystatin in patients with invasive aspergillosis refractory to or intolerant of amphotericin B, *Antimicrob. Agents Chemother.* **48**, 4808 (2004).

104. M. S. El-Ridy, A. Abdelbary, T. Essam, R. M. El-Salam, and A. A. Kassem, Niosomes as a potential drug delivery system for increasing the efficacy and safety of nystatin, *Drug Dev. Ind. Pharm.* **37**, 1491 (2011).

105. P. S. Jadon, V. Gajbhiye, R. S. Jadon, K. R. Gajbhiye, and N. Ganesh, Enhanced oral bioavailability of griseofulvin via niosomes, *AAPS Pharmscitech.* **10**, 1186 (2009).

106. S. B. Shirsand and G. G. Keshavshetti, Formulaiton and characterization of drug loaded niosomes for antifungal activity, *Sper. J. Adv. Nov. Drug Deliv.* **1**, 12 (2016).

107. S. M. Jaya Raja Kumar and S. Parasuraman, Antifugal agents: New approach for novel delivery systems, *J. Pharm. Sci. Res.* **6**, 229 (2014).

108. N. Gupta, D. B. Rai, A. K. Jangid, and H. Kulhari, Use of nanotechnology in antimicrobial therapy, in *Methods in Microbiology*, edited by V. Gurtler, A. S. Ball, and S. Soni (Elsevier, 2019), pp. 143.

109. D. Sanglard, Emerging threats in antifungal-resistant fungal pathogens, *Front. Med.* **3**, 11 (2016).

110. A. Srinivasan, J. L. Lopez-Ribot, and A. K. Ramasubramanian, Overcoming antifungal resistance, *Drug Discov. Today Technol.* **11**, 65 (2014).

111. R. Y. Pelgrift and A. J. Friedman, Nanotechnology as a therapeutic tool to combat microbial resistance, *Adv. Drug Deliv. Rev.* **65**, 1803 (2013).

112. D. O. Schairer, J. S. Chouake, J. D. Nosanchuk, and A. J. Friedman, The potential of nitric oxide releasing therapies as antimicrobial agents, *Virulence* **3**, 271 (2012).

113. K. Blecher, A. Nasir, and A. Friedman, The growing role of nanotechnology in combating infectious disease, *Virulence* **2**, 395 (2011).

114. L. Zhang, D. Pornpattananangku, C. M. Hu, and C. M. Huang, Development of nanoparticles for antimicrobial drug delivery, *Curr. Med. Chem.* **17**, 585 (2010).

115. E. M. Hetrick, J. H. Shin, H. S. Paul, and M. H. Schoenfisch, Anti-biofilm efficacy of nitric oxide-releasing silica nanoparticles, *Biomaterials* **30**, 2782 (2009).

116. M. J. Hajipour, K. M. Fromm, A. A. Ashkarran, D. Jimenez de Aberasturi, I. R. de Larramendi, T. Rojo, V. Serpooshan, W. J. Parak, and M. Mahmoudi, Antibacterial properties of nanoparticles, *Trends Biotechnol.* **30**, 499 (2012).

117. F. Asghari, Z. Jahanshiri, M. Imani, M. Shams-Ghahfarokhi, and M. Razzaghi-Abyaneh, Antifungal nanomaterials: Synthesis, properties, and applications, in *Nanobiomaterials in Antimicrobial Therapy*, edited by A. M. Grumezescu (Elsevier, 2016), pp. 343.

118. M. D. Mashitah, Y. S. Chan, and J. Jason, Antimicrobial properties of nanobiomaterials and the mechanism, in *Nanobiomaterials in Antimicrobial Therapy*, edited by A. M. Grumezescu (Elsevier, 2016), pp. 261.

119. Y. N. Slavin, J. Asnis, U. O. Hafeli, and H. Bach, Metal nanoparticles: Understanding the mechanisms behind antibacterial activity, *J. Nanobiotechnol.* **15**, 65 (2017).

120. A. Qidway, A. Pandey, R. Kumar, S. K. Shukla, and A. Dikshit, Advances in biogenic nanoparticles and the mechanisms of antimicrobial effects, *Indian J. Pharm. Sci.* **80**, 592 (2018).

121. X. F. Zhang, Z. G. Liu, W. Shen, and S. Gurunathan, Silver nanoparticles: Synthesis, characterization, properties, applications, and therapeutic approaches, *Int. J. Mol. Sci.* **17** (2016).

122. K. P. Bocate, G. F. Reis, P. C. de Souza, A. G. Oliveira Junior, N. Duran, G. Nakazato, M. C. Furlaneto, R. S. de Almeida, and L. A. Panagio, Antifungal activity of silver nanoparticles and simvastatin against toxigenic species of Aspergillus, *Int. J. Food Microbiol.* **291**, 79 (2019).

123. N. L. Calvo, S. Sreekumar, L. A. Svetaz, M. C. Lamas, B. M. Moerschbacher, and D. Leonardi, Design and characterization of chitosan nanoformulations for the delivery of antifungal agents, *Int. J. Mol. Sci.* **20** (2019).

124. M. Kucharska, M. Sikora, K. Brzoza-Malczewska, and M. Owczarek, Antimicrobial properties of chitin and chitosan, in *Chitin and Chitosan: Properties and Applications*, edited by L. A. M. van den Broek and C. G. Boeriu (John Wiley & Sons Ltd., 2019), pp. 169.

125. N. Aibani, R. Rai, P. Patel, G. Cuddihy, and E. K. Wasan, Chitosan nanoparticles at the biological interface: Implications for drug delivery, *Pharmaceutics* **13**, 1686 (2021).

126. Q. Sun, J. Li, and T. Le, Zinc oxide nanoparticle as a novel class of antifungal agents: Current advances and future perspectives, *J. Agric. Food Chem.* **66**, 11209 (2018).

127. P. N. Singh and A. Nanda, Antimicrobial and antifungal potential of zinc oxide nanoparticles in comparison to conventional zinc oxide particles, *J. Chem. Pharm. Res.* **5**, 457 (2013).

128. M. Rai, A. Yadav, P. Bridge, and A. Gade, Myconanotechnology: A new and emerging science, in *Applied Mycology*, edited by M. Rai and P. D. Bridge (CABI, 2009), pp. 258

129. A. Ingle, M. Rai, A. Gade, and M. Bawaskar, Fusarium solani: A novel biological agent for the extracellular synthesis of silver nanoparticles, *J. Nanoparticle Res.* **11**, 2079 (2009).

130. H. Korbekandi, Z. Ashari, S. Iravani, and S. Abbasi, Optimization of biological synthesis of silver nanoparticles using *Fusarium oxysporum*, *Iran J. Pharm. Res.* **12**, 289 (2013).

131. D. Kamil, T. Prameeladevi, S. Ganesh, N. Prabhakaran, R. Nareshkumar, and S. P. Thomas, Green synthesis of silver nanoparticles by entomopathogenic fungus *Beauveria bassiana* and their bioefficacy against mustard aphid (*Lipaphis erysimi* Kalt.), *Indian J. Exp. Biol.* **55**, 555 (2017).

132. W. Huang, X. Fang, H. Wang, F. Chen, H. Duan, Y. Bi, and H. Yu, Biosynthesis of AgNPs by *B. maydis* and its antifungal effect against *Exserohilum turcicum*, *IET Nanobiotechnol.* **12**, 585 (2018).

133. A. Ogar, G. Tylko, and K. Turnau, Antifungal properties of silver nanoparticles against indoor mould growth, *Sci. Total Environ.* **521–522**, 305 (2015).

134. H. Hariharan, N. A. Al-Harbi, P. Karuppiah, and S. K. Rajaram, Microbial synthesis of selenium nanocomposite using *Saccharomyces cerevisiae* and its antimicrobial activity against pathogens causing nosocomial infection, *Chalcogenide Lett.* **9**, 509 (2012).

135. T. A. Muthupandian Saravanan, L. Negash, A. Gebreyesus, A. Selvaraj, V. Rayar, and K. Dheekonda, Extracellular biosynthesis and biomedical application of silver nanoparticles synthesized from baker's yeast, *Int. J. Res. Pharm. Biomed. Sci.* **4**, 822 (2013).

136. P. P. Shanbhaga and U. Janib, Drug delivery through nails: Present and future, *New Horiz. Transl. Med.* **3**, 252 (2017).

137. P. Palekar Shanbhag and U. Jani, Drug delivery through nails: Present and future, *New Horiz. Transl. Med.* **3**, 252 (2017).

138. A. K. Tiwary and B. Sapra, High failure rate of transungal drug delivery: Need for new strategies, *Ther. Deliv.* **8**, 239 (2017).

139. A. Beloqui, M. A. Solinis, A. Rodriguez-Gascon, A. J. Almeida, and V. Preat, Nanostructured lipid carriers: Promising drug delivery systems for future clinics, *Nanomedicine* **12**, 143 (2016).

140. C. Puglia, P. Blasi, C. Ostacolo, E. Sommella, C. Bucolo, C. B. M. Platania, G. L. Romano, F. Geraci, F. Drago, D. Santonocito, B. Albertini, P. Campiglia, G. Puglisi, and R. Pignatello, Innovative nanoparticles enhance N-palmitoylethanolamide intraocular delivery, *Front. Pharmacol.* **9**, 285 (2018).

141. L. Kumar, S. Verma, A. Bhardwaj, S. Vaidya, and B. Vaidya, Eradication of superficial fungal infections by conventional and novel approaches: A comprehensive review, *Artif. Cells Nanomed. Biotechnol.* **42**, 32 (2014).

142. G. P. Mishra, M. Bagui, V. Tamboli, and A. K. Mitra, Recent applications of liposomes in ophthalmic drug delivery, *J. Drug Deliv.* **2011**, 863734 (2011).

143. F. S. Habib, E. A. Fouad, M. S. Abdel-Rhaman, and D. Fathalla, Liposomes as an ocular delivery system of fluconazole: *In vitro* studies, *Acta Ophthalmol.* **88**, 901 (2010).

144. G. Fetih, Fluconazole-loaded niosomal gels as a topical ocular drug delivery system for corneal fungal infections, *J. Drug Deliv. Sci. Technol.* **35**, 8 (2016).

145. P. Mohanta, N. Pandey, D. N. Kapoor, S. Singh, Y. Sarvi, and P. Sharma, Development of surfactant-based nanocarrier system for delivery of an antifungal drug, *J. Pharm. Res.* **11**, 1153 (2017).

146. A. R. A. S. Baishakhi Bhowmik, D. V. Gowda, A. Singh, A. Srivastava, and R. A. M. Osmani, Recent trends and advances in fungal drug delivery, *J. Chem. Pharm. Res.* **8**, 169 (2016).

147. G. M. Jensen, The care and feeding of a commercial liposomal product: Liposomal amphotericin B (AmBisome®), *J. Liposome Res.* **27**, 173 (2017).

148. S. H. Jung, D. H. Lim, S. H. Jung, J. E. Lee, K. S. Jeong, H. Seong, and B. C. Shin, Amphotericin B-entrapping lipid nanoparticles and their *in vitro* and *in vivo* characteristics, *Eur. J. Pharm Sci.* **37**, 313 (2009).

149. H. C. Korting, W. Klövekorn, and G. Klövekorn, Comparative efficacy and tolerability of econazole liposomal gel 1%, branded econazole conventional cream 1% and generic clotrimazole cream 1% in tinea pedis, *Clin. Drug Investig.* **14**, 286 (1997).

150. A. Ascenso, S. Raposo, C. Batista, P. Cardoso, T. Mendes, F. G. Praca, M. V. Bentley, and S. Simoes, Development, characterization, and skin delivery studies of related ultradeformable vesicles: Transfersomes, ethosomes, and transethosomes, *Int. J. Nanomed.* **10**, 5837 (2015).

151. S. Verma and P. Utreja, Vesicular nanocarrier based treatment of skin fungal infections: Potential and emerging trends in nanoscale pharmacotherapy, *Asian J. Pharm. Sci.* **14**, 117 (2019).

152. J. Shaji and R. Bajaj, Transethosomes: A new prospect for enhanced transdermal delivery, *Int. J. Pharm. Sci. Res.* **9**, 2681 (2018).

153. T. Ramasamy, U. S. Khandasami, H. Ruttala, and S. Shanmugam, Development of solid lipid nanoparticles enriched hydrogels for topical delivery of anti-fungal agent, *Macromol. Res.* **20**, 682 (2012).

154. J. Pardeike, A. Hommoss, and R. H. Muller, Lipid nanoparticles (SLN, NLC) in cosmetic and pharmaceutical dermal products, *Int. J. Pharm.* **366**, 170 (2009).

155. Q. Li, T. Cai, Y. Huang, X. Xia, S. P. C. Cole, and Y. Cai, A Review of the structure, preparation, and application of NLCs, PNPs, and PLNs, *Nanomaterials* **7**, 122 (2017).

156. S. Jain, S. Jain, P. Khare, A. Gulbake, D. Bansal, and S. K. Jain, Design and development of solid lipid nanoparticles for topical delivery of an anti-fungal agent, *Drug Deliv.* **17**, 443 (2010).

157. C. I. C. Crucho and M. T. Barros, Polymeric nanoparticles: A study on the preparation variables and characterization methods, *Mater Sci. Eng. C Mater Biol. Appl.* **80**, 771 (2017).

158. E. K. Patel and R. Oswal, Nanosponge and micro sponges: A novel drug delivery system, *Int. J. Res. Pharm. Chem.* **2**, 2281 (2012).

159. A. C. Amaral, A. L. Bocca, A. M. Ribeiro, J. Nunes, D. L. G. Peixoto, A. R. Simioni, F. L. Primo, Z. G. M. Lacava, R. Bentes, R. Titze-de-Almeida, A. C. Tedesco, P. C. Morais, and M. S. S. Felipe, Amphotericin B in poly(lactic-co-glycolic acid) (PLGA) and dimercaptosuccinic acid (DMSA) nanoparticles against paracoccidioidomycosis, *J. Antimicrob. Chemother.* **63**, 526 (2009).

160. N. Kamaly, B. Yameen, J. Wu, and O. C. Farokhzad, Degradable controlled-release polymers and polymeric nanoparticles: Mechanisms of controlling drug release, *Chem. Rev.* **116**, 2602 (2016).

161. T. Caon, L. C. Porto, A. Granada, M. P. Tagliari, M. A. Silva, C. M. Simoes, R. Borsali, and V. Soldi, Chitosan-decorated polystyrene-b-poly(acrylic acid) polymersomes as novel carriers for topical delivery of finasteride, *Eur. J. Pharm. Sci.* **52**, 165 (2014).

162. L. Y. Ing, N. M. Zin, A. Sarwar, and H. Katas, Antifungal activity of chitosan nanoparticles and correlation with their physical properties, *Int. J. Biomater.* **2012**, 632698 (2012).

163. D. R. Serrano, A. Lalatsa, M. A. Dea-Ayuela, P. E. Bilbao-Ramos, N. L. Garrett, J. Moger, J. Guarro, J. Capilla, M. P. Ballesteros, A. G. Schätzlein, F. Bolás, J. J. Torrado, and I. F. Uchegbu., Oral particle uptake and organ targeting drives the activity of amphotericin B nanoparticles, *Mol. Pharm.* **12**, 420 (2015).

164. B. S. Makhmalzade and F. Chavoshy, Polymeric micelles as cutaneous drug delivery system in normal skin and dermatological disorders, *J. Adv. Pharm. Technol. Res.* **9**, 2 (2018).

165. T. Thambi, V. G. Deepagan, H. Ko, D. S. Lee, and J. H. Park, Bioreducible polymersomes for intracellular dual-drug delivery, *J. Mater. Chem.* **22**, 22028 (2012).

166. J. S. Lee and J. Feijen, Polymersomes for drug delivery: Design, formation and characterization, *J. Controlled Release.* **161**, 473 (2012).

167. H. Sun, F. Meng, R. Cheng, C. Deng, and Z. Zhong, Reduction and pH dual-bioresponsive crosslinked polymersomes for efficient intracellular delivery of proteins and potent induction of cancer cell apoptosis, *Acta Biomater.* **10**, 2159 (2014).

168. A. P. Sherje, M. Jadhav, B. R. Dravyakar, and D. Kadam, Dendrimers: A versatile nanocarrier for drug delivery and targeting, *Int. J. Pharm.* **548**, 707 (2018).

169. A. M. Caminade, R. Laurent, and J. P. Majoral, Characterization of dendrimers, *Adv. Drug Deliv. Rev.* **57**, 2130 (2005).

170. U. Gupta and O. Perumal, Dendrimers and its biomedical applications, in *Natural and Synthetic Biomedical Polymers*, edited by S. G. Kumbar, C. T. Laurencin, and M. Deng (Elsevier, 2014), pp. 243.

171. S. Nimesh, Dendrimers, in *Gene Therapy: Potential Applications of Nanotechnology* (Woodhead Publishing Ltd., 2013), pp. 259.

172. B. K. Nanjwade, H. M. Bechra, G. K. Derkar, F. V. Manvi, and V. K. Nanjwade, Dendrimers: Emerging polymers for drug delivery systems, *Eur. J. Pharm. Sci.* **38**, 185 (2009).

173. D. Zhang, X. L. Ma, Y. Gu, H. Huang, and G. W. Zhang, Green synthesis of metallic nanoparticles and their potential applications to treat cancer, *Front. Chem.* **8**, 799 (2020).

174. S. Iravani, H. Korbekandi, S. V. Mirmohammadi, and B. Zolfaghari, Synthesis of silver nanoparticles: Chemical, physical and biological methods, *Res. Pharm. Sci.* **9**, 385 (2014).

175. E. Abbasi, M. Milani, S. F. Aval, M. Kouhi, A. Akbarzadeh, H. T. Nasrabadi, P. Nikasa, S. W. Joo, Y. Hanifehpour, K. Nejati-Koshki, M. Samiei, Silver nanoparticles: Synthesis methods, bio-applications and properties, *Crit. Rev. Microbiol.* **42**, 173 (2016).

176. I. A. Wani, T. Ahmad, and N. Manzoor, Size and shape dependant antifungal activity of gold nanoparticles: A case study of Candida, *Colloids Surf. B Biointerfaces* **101**, 162 (2013).

177. A. Jebali, F. H. Hajjar, F. Pourdanesh, S. Hekmatimoghaddam, B. Kazemi, A. Masoudi, K. Daliri, and N. Sedighi, Silver and gold nanostructures: Antifungal property of different shapes of these nanostructures on Candida species, *Med. Mycol.* **52**, 65 (2014).

178. J. E. Mussin, M. V. Roldan, F. Rojas, M. L. A. Sosa, N. Pellegri, and G. Giusiano, Antifungal activity of silver nanoparticles in combination with ketoconazole against *Malassezia furfur*, *AMB Express.* **9**, 131 (2019).

179. N. V. S. Vallabani and S. Singh, Recent advances and future prospects of iron oxide nanoparticles in biomedicine and diagnostics, *3 Biotech.* **8**, 279 (2018).

180. A. V. Samrot , C. S. Sahithya, D. Sruthi P, J. Selvarani A , Raji P, Prakash P, P. Ponnaiah, I. Petchi, S. Pattammadath, S. K. Purayil, P. T. M. Javad, Thirumurugan R, and K. Kasipandian, Itraconazole coated super paramagnetic iron oxide nanoparticles for antimicrobial studies, *Biointerface Res. Appl. Chem.* **10**, 6218 (2020).

181. A. Sankhyan and P. Pawar, Recent trends in niosome as vesicular drug delivery system, *J. Appl. Pharm. Sci.* **2**, 20 (2012).

182. D. J. McClements, Nanoemulsions versus microemulsions: Terminology, differences, and similarities, *Soft Matter* **8**, 1719 (2012).

183. M. L. de Souza, D. D. Oliveira, N. P. Pereira, and D. M. Soares, Nanoemulsions and dermatological diseases: Contributions and therapeutic advances, *Int. J. Derm.* **57**, 894 (2018).

184. P. Boonme, J. Kaewbanjong, T. Amnuaikit, T. Andreani, A. M. Silva, and E. B. Souto, Microemulsion and microemulsion-based gels for topical antifungal therapy with phytochemicals, *Curr. Pharm. Des.* **22**, 4257 (2016).

185. K. Mehta and D. Bhatt, Preparation, optimization and *in vitro* microbiological efficacy of antifungal microemulsion, *Int. J. Pharm. Sci. Res.* **2**, 2424 (2011).

186. S. Güngör, M. S. Erdal, and B. Aksu, New formulation strategies in topical antifungal therapy, *J. Cosmet. Dermatol. Sci. Appl.* **3**, 56 (2013).

187. G. N. El-Hadidy, H. K. Ibrahim, M. I. Mohamed, and M. F. El-Milligi, Microemulsions as vehicles for topical administration of voriconazole: Formulation and *in vitro* evaluation, *Drug Dev. Ind. Pharm.* **38**, 64 (2012).

188. H. Zazo, C. I. Colino, and J. M. Lanao, Current applications of nanoparticles in infectious diseases, *J. Controlled Release* **224**, 86 (2016).

189. J. Peszke, M. Dulski, A. Nowak, K. Balin, M. Zubko, S. Sułowicz, B. Nowak, Z. Piotrowska-Seget, E. Talik, M. Wojtyniak, A. Mrozek-Wilczkiewicz, K. Malarz, and J. Szade, Unique properties of silver and copper silica-based nanocomposites as antimicrobial agent, *RSC Adv.* **7**, 28092 (2017).

190. D. Botequim, J. Maia, M. M. Lino, L. M. Lopes, P. N. Simoes, L. M. Ilharco, and L. Ferreira, Nanoparticles and surfaces presenting antifungal, antibacterial and antiviral properties, *Langmuir* **28**, 7646 (2012).

191. C. S. Paulo, M. Vidal, and L. S. Ferreira, Antifungal nanoparticles and surfaces, *Biomacromolecules* **11**, 2810 (2010).

192. D. E. Camporotondi, M. L. Foglia, G. S. Alvarez, A. Mebert, L. E. Diaz, T. Coradin, and M. Desimone, Antimicrobial properties of silica modified nanoparticles, in *Microbial Pathogens and Strategies for Combating Them: Science, Technology and Education*, edited by A. Méndez-Vilas (Formatex Research Center, 2013), pp. 283.

193. A. Derbalah, M. Shenashen, A. Hamza, A. Mohamed, and S. El Safty, Antifungal activity of fabricated mesoporous silica nanoparticles against early blight of tomato, *Egypt. J. Basic Appl. Sci.* **5**, 145 (2019).

194. E. Segal and M. Frenkel, Experimental *in vivo* models of candidiasis, *J. Fungi* **4**, 21 (2018).

195. J. E. Nett and D. R. Andes, Fungal biofilms: *In vivo* models for discovery of anti-biofilm drugs, in *Microbial Biofilms,* edited by M. Ghannoum, M. Parsek, M. Whiteley, and P. K. Mukherjee (ASM Press, 2015), pp. 33.

196. P. Van Dijck, J. Sjollema, B. P. A. Cammue, K. Lagrou, J. Berman, C. d'Enfert, D. R. Andes, M. C. Arendrup, A. A. Brakhage, R. Calderone, E. Cantón, T. Coenye, P. Cos, L. E. Cowen, M. Edgerton, A. Espinel-Ingroff, S. G. Filler, M. Ghannoum, N. A. R. Gow, H. Haas, M. A. Jabra-Rizk, E. M. Johnson, S. R. Lockhart, J. L. Lopez-Ribot, J. Maertens, C. A. Munro, J. E. Nett, C. J. Nobile, M. A. Pfaller, G. Ramage, D. Sanglard, M. Sanguinetti, I. Spriet, P. E. Verweij, A. Warris, J. Wauters, M. R. Yeaman, S. A. J. Zaat, and K. Thevisse, Methodologies for *in vitro* and *in vivo* evaluation of efficacy of antifungal and antibiofilm agents and surface coatings against fungal biofilms, *Microb. Cell* 5, 300 (2018).

197. J. Capilla, K. V. Clemons, and D. A. Stevens, Animal models: An important tool in mycology, *Med. Mycol.* **45**, 657 (2007).

198. V. M. Stevens, S. W. Mueller, P. M. Reynolds, R. MacLaren, and T. H. Kiser, Extrapolating antifungal animal data to humans: Is it reliable? *Curr. Fungal Infect. Rep.* **14**, 50 (2020).

199. M. M. Abdellatif, I. A. Khalil, Y. E. Elakkad, H. A. Eliwa, T. M. Samir, and A. K. Al-Mokaddem, Formulation and characterization of sertaconazole nitrate mucoadhesive liposomes for vaginal candidiasis, *Int. J. Nanomedicine* **15**, 4079 (2020).

200. L. L. Kovanda, C. Giamberardino, L. McEntee, D. L. Toffaletti, K. S. Franke, A. Bartuska, G. Smilnak, G. C. de Castro, K. Ripple, W. W. Hope, and J. R. Perfect, Pharmacodynamics of isavuconazole in a rabbit model of cryptococcal meningoencephalitis, *Antimicrob. Agents Chemother.* **63**, e00546 (2019).

201. K. Maebashi, T. Itoyama, K. Uchida, N. Suegara, and H. Yamaguchi, A novel model of cutaneous candidiasis produced in prednisolone-treated guinea-pigs, *J. Med. Vet. Mycol.* **32**, 349 (1994).

202. H. Koga, Y. Munechika, H. Matsumoto, Y. Nanjoh, K. Harada, K. Makimura, and R. Tsuboi, Guinea pig seborrheic dermatitis model of *Malassezia restricta* and the utility of luliconazole, *Med. Mycol.* **58**, 820 (2020).

203. D. G. Hooper, V. E. Bolton, J. S. Sutton, F. T. Guilford, D. C. Straus, L. K. Najvar, N. P. Wiederhold, W. R. Kirkpatrick, and T. F. Patterson, Assessment of *Aspergillus fumigatus* in guinea pig bronchoalveolar lavages and pulmonary tissue by culture and realtime polymerase chain reaction studies, *Int. J. Mol. Sci.* **13**, 726 (2012).

204. W. R. Kirkpatrick, L. K. Najvar, R. Bocanegra, T. F. Patterson, and J. R. Graybill, New guinea pig model of cryptococcal meningitis, *Antimicrob. Agents Chemother.* **51**, 3011 (2007).

205. L. K. Ryan, A. G. Hise, C. M. Hossain, W. Ruddick, R. Parveen, K. B. Freeman, D. G. Weaver, H. P. Narra, R. W. Scott, and G. Diamond, A novel immunocompetent mouse model for testing antifungal drugs against invasive *Candida albicans* infection, *J. Fungi* **6**, 197 (2020).

206. E. Delarze, F. Ischer, D. Sanglard, and A. T. Coste, Adaptation of a *Gaussia princeps* luciferase reporter system in *Candida albicans* for *in vivo* detection in the *Galleria mellonella* infection model, *Virulence* **6**, 684 (2015).

207. M. Gunzer, C. R. Thornton, and N. Beziere, Advances in the *in vivo* molecular imaging of invasive aspergillosis, *J. Fungi* **6**, 338 (2020).

208. S. Dorsaz, A. T. Coste, and D. Sanglard, Red-shifted firefly luciferase optimized for *Candida albicans in vivo* bioluminescence imaging, *Front. Microbiol.* **8**, 1478 (2017).

209. D. Kabbani, L. Goldraich, H. Ross, C. Rotstein, and S. Husain, Outbreak of invasive aspergillosis in heart transplant recipients: The role of screening computed tomography scans in asymptomatic patients and universal antifungal prophylaxis, *Transpl. Infect. Dis.* **20**, e12808 (2018).

210. J. Poelmans, U. Himmelreich, L. Vanherp, L. Zhai, A. Hillen, B. Holvoet, S. Belderbos, M. Brock, J. Maertens, G. V. Velde, and K. Lagrou, A multimodal imaging approach enables *in vivo* assessment of antifungal treatment in a mouse model of invasive pulmonary aspergillosis, *Antimicrob. Agents Chemother.* **62**, e00240 (2018).

211. I. D. Jacobsen, K. Grosse, and B. Hube, Embryonated chicken eggs as alternative infection model for pathogenic fungi, *Methods Mol. Biol.* **845**, 487 (2012).

212. E. E. Rosowski, B. P. Knox, L. S. Archambault, A. Huttenlocher, N. P. Keller, R. T. Wheeler, and J. M. Davis, The zebrafish as a model host for invasive fungal infections, *J. Fungi* **4**, 136 (2018).

213. C. Chao, P. Hsu, C. Jen, I. Chen, C. Wang, H. Chan, P. Tsai, K. Tung, C. Wang, C. Lan, and Y. Chuang., Zebrafish as a model host for *Candida albicans* infection, *Infec. Immun.* **78**, 2512 (2010).

214. J. L. Tenor, S. H. Oehlers, J. L. Yang, D. M. Tobin, and J. R. Perfect, Live imaging of host–parasite interactions in a zebrafish infection model reveals cryptococcal determinants of virulence and central nervous system invasion, *mBio* **6**, e01425 (2015).

215. D. C. M. Kulatunga, S. H. S. Dananjaya, C. Nikapitiya, C.-H. Kim, J. Lee, and M. De Zoysa, *Candida albicans* infection model in zebrafish (*Danio rerio*) for screening anticandidal drugs, *Mycopathologia* **184**, 559 (2019).

216. M. S. Lionakis, Drosophila and galleria insect model hosts: New tools for the study of fungal virulence, pharmacology and immunology, *Virulence* **2**, 521 (2011).

217. S. Jemel, J. Guillot, K. Kallel, F. Botterel, and E. Dannaoui, *Galleria mellonella* for the evaluation of antifungal efficacy against medically important fungi, a narrative review, *Microorganisms* **8**, 390 (2020).

218. G. Chamilos, M. S. Lionakis, R. E. Lewis, J. L. Lopez-Ribot, S. P. Saville, N. D. Albert, G. Halder, and D. P. Kontoyiannis, *Drosophila melanogaster* as a facile model for large-scale studies of virulence mechanisms and

antifungal drug efficacy in Candida species, *J. Infect. Dis.* **193**, 1014 (2006).

219. S. Wurster, A. Bandi, N. D. Beyda, N. D. Albert, N. M. Raman, I. I. Raad, and D. P. Kontoyiannis, *Drosophila melanogaster* as a model to study virulence and azole treatment of the emerging pathogen *Candida auris, J. Antimicrob. Chemother.* **74**, 1904 (2019).

220. M. Muhammed, J. J. Coleman, and E. Mylonakis, *Caenorhabditis elegans*: A nematode infection model for pathogenic fungi, *Methods Mol. Biol.* **845**, 447 (2012).

221. I. D. Jacobsen, K. Große, S. Slesiona, B. Hube, A. Berndt, and M. Brock, Embryonated eggs as an alternative infection model to investigate *Aspergillus fumigatus* virulence, *Infect. Immun.* **78**, 2995 (2010).

222. I. D. Jacobsen, K. Große, A. Berndt, and B. Hube, Pathogenesis of *Candida albicans* infections in the alternative chorio-allantoic membrane chicken embryo model resembles systemic murine infections, *PLOS ONE* **6**, e19741 (2011).

223. E. Maurer, C. Hörtnagl, M. Lackner, D. Grässle, V. Naschberger, P. Moser, E. Segal, M. Semis, C. Lass-Flörl, and U. Binder, *Galleria mellonella* as a model system to study virulence potential of mucormycetes and evaluation of antifungal treatment, *Med. Mycol.* **57**, 351 (2018).

224. K. Kaerger, V. U. Schwartze, S. Dolatabadi, I. Nyilasi, S. A. Kovács, U. Binder, T. Papp, S. de Hoog, I. D. Jacobsen, and K. Voigt, Adaptation to thermotolerance in Rhizopus coincides with virulence as revealed by avian and invertebrate infection models, phylogeny, physiological and metabolic flexibility, *Virulence* **6**, 395 (2015).

225. E. E. Rosowski, J. He, J. Huisken, N. P. Keller, and A. Huttenlocher, Efficacy of voriconazole against *Aspergillus fumigatus* infection depends on host immune function, *Antimicrob. Agents Chemother.* **64**, e00917 (2020).

226. M. Ishii, Y. Matsumoto, and K. Sekimizu, Usefulness of silkworm as a host animal for understanding pathogenicity of *Cryptococcus neoformans, Drug Discov. Ther.* **10**, 9 (2016).

227. M. S. Lionakis and D. P. Kontoyiannis, *Drosophila melanogaster* as a model organism for invasive aspergillosis, *Methods Mol. Biol.* **845**, 455 (2012).

228. J. J. Coleman, M. Muhammed, P. V. Kasperkovitz, J. M. Vyas, and E. Mylonakis, Fusarium pathogenesis investigated using *Galleria mellonella* as a heterologous host, *Fungal Biol.* **115**, 1279 (2011).

229. L. Thomaz, R. García-Rodas, A. J. Guimarães, C. P. Taborda, O. Zaragoza, and J. D. Nosanchuk, *Galleria mellonella* as a model host to study *Paracoccidioides lutzii* and *Histoplasma capsulatum*, *Virulence* **4**, 139 (2013).

230. G. Sheehan and K. Kavanagh, Proteomic analysis of the responses of *Candida albicans* during infection of *Galleria mellonella* larvae, *J. Fungi* **5**, 7 (2019).

231. C. Firacative, A. Khan, S. Duan, K. Ferreira-Paim, D. Leemon, and W. Meyer, Rearing and maintenance of *Galleria mellonella* and its application to study fungal virulence, *J. Fungi* **6**, 130 (2020).

232. D.-D. Li, L. Deng, G.-H. Hu, L.-X. Zhao, D.-D. Hu, Y.-Y. Jiang, and Y. Wang, Using *Galleria mellonella–Candida albicans* infection model to evaluate antifungal agents, *Biol. Pharm. Bull.* **36**, 1482 (2013).

233. A. Hernando-Ortiz, E. Mateo, M. Ortega-Riveros, I. De-la-Pinta, G. Quindós, and E. Eraso, *Caenorhabditis elegans* as a model system to assess *Candida glabrata*, *Candida nivariensis*, and *Candida bracarensis* virulence and antifungal efficacy, *Antimicrob. Agents Chemother.* **64**, e00824 (2020).

234. C. H. Johnson, S. Ayyadevara, J. E. Mcewen, and R. J. Shmookler Reis, *Histoplasma capsulatum* and *Caenorhabditis elegans*: A simple nematode model for an innate immune response to fungal infection, *Med. Mycol.* **47**, 808 (2009).

235. E. Mylonakis, F. M. Ausubel, J. R. Perfect, J. Heitman, and S. B. Calderwood, Killing of *Caenorhabditis elegans* by *Cryptococcus neoformans* as a model of yeast pathogenesis, *Proc. Natl. Acad. Sci. U. S. A.* **99**, 15675 (2002).

236. A. Shanti, J. Teo, and C. Stefanini, *In vitro* immune organs-on-chip for drug development: A review, *Pharmaceutics* **10**, 278 (2018).

237. C. O. Morton, S. Wurster, M. Fliesser, F. Ebel, L. Page, K. Hünniger, O. Kurzai, A. L. Schmitt, D. Michel, J. Springer, H. Einsele, and J. Loeffler, Validation of a simplified *in vitro* Transwell® model of the alveolar surface to assess host immunity induced by different morphotypes of *Aspergillus fumigatus*, *Int. J. Med. Microbiol.* **308**, 1009 (2018).

238. Y. Wang, P. Wang, and J. Qin, Microfluidic organs-on-a-chip for modeling human infectious diseases, *Acc. Chem. Res.* **54**, 3550 (2021).

239. EUCAST, Antifungal susceptibility testing (AFST), https://www.eucast.org/ast_of_fungi (accessed on 14th July 2022)

240. R. Patel, A moldy application of MALDI: MALDI-ToF mass spectrometry for fungal identification, *J. Fungi* **5**, 4 (2019).

241. E. Klaile, M. M. Müller, C. Zubiría-Barrera, S. Brehme, T. E. Klassert, M. Stock, A. Durotin, T. D. Nguyen, S. Feer, B. B. Singer, P. F. Zipfel, S. Rudolphi, I. D. Jacobsen, and H. Slevogt, Unaltered fungal burden and lethality in human CEACAM1-transgenic mice during *Candida albicans* dissemination and systemic infection, *Front. Microbiol.* **10**, 2703 (2019).

242. H. Sandovsky-Losica, L. Barr-Nea, and E. Segal, Fatal systemic candidiasis of gastrointestinal origin: An experimental model in mice compromised by anti-cancer treatment, *J. Med. Vet. Mycol.* **30**, 219 (1992).

243. G. Desoubeaux and C. Cray, Rodent models of invasive aspergillosis due to *Aspergillus fumigatus*: Still a long path toward standardization, *Front. Microbiol.* **8**, 841 (2017).

244. J. R. Perfect, S. D. Lang, and D. T. Durack, Chronic cryptococcal meningitis: A new experimental model in rabbits, *Am. J. Pathol.* **101**, 177 (1980).

245. B. Baddal and P. Marrazzo, Refining host–pathogen interactions: Organ-on-chip side of the coin, *Pathogens* **10**, 203 (2021).

246. K. Sohn, I. Senyürek, J. Fertey, A. Königsdorfer, C. Joffroy, N. Hauser, G. Zelt, H. Brunner, and S. Rupp, An *in vitro* assay to study the transcriptional response during adherence of *Candida albicans* to different human epithelia, *FEMS Yeast Res.* **6**, 1085 (2006).

247. S. S. W. Wong, R. Y. T. Kao, K. Y. Yuen, Y. Wang, D. Yang, L. P. Samaranayake, and C. J. Seneviratne, *In vitro* and *in vivo* activity of a novel antifungal small molecule against candida infections, *PLOS ONE* **9**, e85836 (2014).

248. C. C. Johnson, A. Yu, H. Lee, P. L. Fidel, Jr., and M. C. Noverr, Development of a contemporary animal model of *Candida albicans*-associated denture stomatitis using a novel intraoral denture system, *Infect. Immun.* **80**, 1736 (2012).

249. D. Pooja, A. Kadari, H. Kulhari, and R. Sistla, Lipid-based nanomedicines: Current clinical status and future perspectives, in *Lipid Nanocarriers for Drug Targeting*, edited by A. M. Grumezescu (Elsevier, 2018), pp. 509.

250. P. Koppa Raghu, K. K. Bansal, P. Thakor, V. Bhavana, J. Madan, J. M. Rosenholm, and N. K. Mehra, Evolution of nanotechnology in delivering drugs to eyes, skin and wounds via topical route, *Pharmaceuticals (Basel)* **13**, 167 (2020).

251. ClinicalTrials (U. S. National Library of Medicine), https://clinicaltrials.gov/ct2/home (accessed on 14th July 2022).

Chapter 9

Biogenic and Biomimetic Nanocarriers

Chenwen Li,[a,b] Chunfan Zhang,[a] Gang Li,[a] and Jianxiang Zhang[a,b]
[a]*Department of Pharmaceutics, College of Pharmacy, Third Military Medical University (Army Medical University), Chongqing 400038, People's Republic of China*
[b]*State Key Laboratory of Trauma and Chemical Poisoning, Third Military Medical University (Army Medical University), Chongqing 400038, People's Republic of China*
jxzhang1980@gmail.com, jxzhang@tmmu.edu.cn

9.1 Introduction

The large-scale investigation into drug delivery systems for the treatment of infectious diseases has enabled the encapsulation of antiviral drugs and antibiotics and their transport to the infected organs and tissues with high efficiency, thereby achieving improved efficacy and decreased side effects as well as overcoming antibiotic resistance [1]. Recently, different biologically inspired strategies have been established to engineer biogenic and biomimetic nanoparticles

Nanotherapeutics for Infectious Diseases
Edited by Íris Luz Batalha

ISBN 978-981-5129-36-6 (Hardcover), 978-1-003-56678-6 (eBook)
www.jennystanford.com

with optimized biophysicochemical features for drug delivery and vaccine development. These novel systems may mimic the natural processes of the original pathogens or cells, thus prolonging blood circulation and enhancing the interactions with specific tissues or target cells. They may serve as drug nanocarriers for targeted delivery and/or be themselves employed as effective nanotherapies or nanovaccines against infectious diseases due to intrinsic immunomodulatory and antibacterial features. This chapter will introduce representative biogenic and biomimetic nanocarriers for the treatment of infectious diseases.

9.2 Biogenic Nanocarriers

9.2.1 Extracellular Vesicles or Exosomes

Extracellular vesicles (EVs), which are spherical biomembrane structures secreted by most eukaryotic and prokaryotic cells, have attracted increasing attention as promising therapeutic agents and novel drug delivery nanocarriers [2]. Based on their size and biogenesis, EVs are commonly classified into exosomes (50–200 nm), microvesicles (100 nm to 1 μm), and apoptotic bodies (1–5 μm) [3]. In nature, EVs serve as a mediator for intercellular communications through the transfer of lipids, nucleic acids, and proteins. In this regard, they have shown great importance in the regulation of several physiological processes as well as some pathological conditions, such as tumor progression [4] and infectious diseases [5, 6]. The potential of utilizing EVs for the treatment of infectious diseases has been exploited, including their innate therapeutic activity attributable to their bioactive cargos, use as vaccines due to their immunomodulatory effects, and delivery of therapeutic agents ranging from small molecules to nucleic acids.

9.2.1.1 Innate therapeutic activity

EVs derived from various types of cells, such as mesenchymal stem cells (MSCs) and immune cells, have recently been discovered as a mechanism for intercellular communication and present great anti-infectious potential. The function of EVs is largely dependent on their source cells. In particular, MSC-derived EVs (MSC-EVs) appear to have good antimicrobial effects, immunomodulatory characteristics, and capacity for injured tissue repair, making them promising entities for the treatment of infectious diseases [7].

The application of MSC-EVs in respiratory infection has been exploited in many studies. For example, intratracheally administered MSC-EVs significantly lessened the injury of hyperoxia-induced acute and chronic bronchopulmonary dysplasia in terms of improving alveolarization and pulmonary vascular remodeling [8, 9].

The effects of MSC-EVs on COVID-19 have also been investigated [10]. A prospective non-randomized open-label cohort study by Vikram et al. demonstrated that intravenous (IV) administration of allogeneic bone marrow MSC-derived exosomes (ExoFlo^{TM}) significantly improved the clinical status and oxygenation of severe COVID-19 patients by suppressing the cytokine storm and rebuilding immunity [11]. In addition, two clinical trials are undergoing to test the effects of MSC-exosomes on severe COVID-19 by IV injection (NCT04798716) or inhalation (NCT04276987).

MSC-EVs remarkably mitigated *Escherichia coli* pneumonia and lipopolysaccharide (LPS)-induced lung injury via transferring microRNA [12, 13]. The antiviral activity and immunomodulatory properties of MSC-EVs were also validated in a pig model of the influenza virus [14]. Furthermore, a number of studies have revealed the beneficial effects of MSC-EVs in animal models of sepsis. miR-146a and miR-21 were significantly upregulated in interleukin (IL)-1β-stimulated MSC-EVs. When these MSC-EVs were transferred to macrophages, they induced M2 polarization and, therefore, exerted enhanced therapeutic effects in cecal ligation and puncture septic models [15, 16]. miR 223–enriched MSC-exosomes contributed to MSC-mediated cardio-protective effects in an animal model of sepsis [17]. MSC-EVs also decreased the incidence and severity of necrotizing enterocolitis in terms of repairing the intestinal barrier [18–21].

Besides MSC-EVs, EVs derived from other cell types have also shown advantages in the treatment of infectious diseases. Bleier's group demonstrated that nasal mucosa–derived exosomes (NMDEs) played a vital role in the immunopathophysiology of inflammation in patients with chronic rhinosinusitis with nasal polyps via interepithelial transfer of P-glycoprotein [22]. In a follow-up study, they found that NMDEs could also eliminate airway pathogens and show passive epithelial immunoprotection in patients undergoing inferior turbinate removal and LPS exposure by transferring nitric oxide synthase and nitric oxide, which functioned through Toll-like receptor 4 and inhibited nuclear factor κB kinase activation [23]. Additionally, microvesicles derived from opsonized particle-stimulated neutrophils were able to suppress bacterial growth via β2 integrin–dependent binding onto the bacterial surface [24]. Exosomes isolated from reticulocytes infected with *Plasmodium yoelii* protected mice from lethal infection [25].

In another study, a wound dressing OxOBand consisting of adipose-derived stem cell (ADSC) exosomes and polyurethane-based oxygen-releasing antioxidant scaffolds (PUAO-CPO) significantly alleviated diabetic as well as *Pseudomonas aeruginosa*– and *Staphylococcus aureus*–induced infections [26]. Ladened exosomes endocytosed by cells increased the migration of human fibroblasts and keratinocytes, thereby improving the survival rate of human neuroblastoma cells exposed to hyperglycemia.

Most recently, Xie et al. engineered angiotensin-converting enzyme 2 (ACE2)-enriched EVs (PM-ACE2-EVs) by fusing the S-palmitoylation-dependent plasma membrane. It was found that PM-ACE2-EVs could bind with the receptor binding domain of SARS-CoV-2 S protein with high affinity and block its interaction with ACE2 on the cell surface, thus providing host protection against COVID-19 *in vivo* (Fig. 9.1) [27].

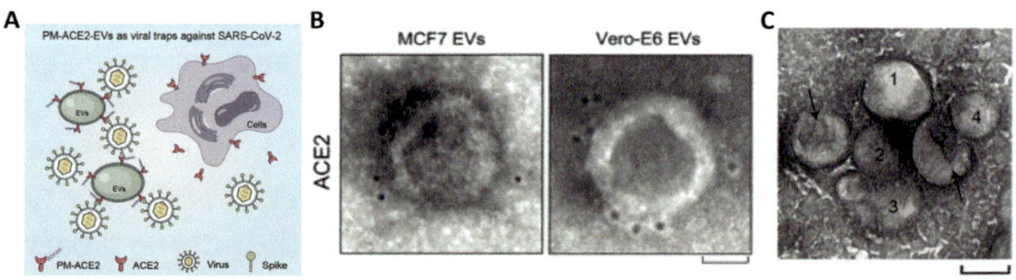

Figure 9.1 Engineering extracellular vesicles enriched with palmitoylated ACE2 as COVID-19 therapy. **(A)** The proposed model of a trap for SARS-CoV-2 by engineered PM-ACE2-enriched EVs. **(B)** Immunogold transmission electron microscopy (IG-TEM) images of EVs derived from MCF7 and Vero-E6 cell lines incubated with anti-ACE2 antibodies and immune-gold labeled by gold particles. Gold particles are depicted as black dots. Scale bar, 50 nm. **(C)** A representative TEM image of PM-ACE2-EVs incubated with SARS-CoV-2 pseudovirus. The black arrows indicate EVs, and the viruses are numbered. Reprinted with permission from Ref. [27].

9.2.1.2 Vaccination

EVs, which express adhesion molecules, could efficiently bind and be taken up by immune-competent cells. The immunomodulatory effects of EVs involve antigen transfer and presentation, making them promising candidates for vaccination [28]. *Leishmania major* antigen-loaded exosomes derived from dendritic cells (DCs) showed potential in mediating protective immunity against cutaneous leishmaniasis and were able to serve as an effective vaccine [29]. Diphtheria toxoid (DT)-pulsed bone marrow dendritic cell (BMDC)-derived exosomes stimulated a primary IgG response to DT *in vivo* [30]. Similarly, *Streptococcus pneumoniae* capsular polysaccharide 14 (Cps14) antigen-pulsed BMDC-derived exosomes could induce protective IgM and IgG responses against lethal *S. pneumoniae* infection [31]. Exosomes released by macrophages that were pretreated with *Mycobacterium tuberculosis* culture filtrate proteins induced antigen-specific interferon (IFN)-γ and IL-2 production in $CD4^+$ and $CD8^+$ T cells and exerted a protective immune response in low-dose *M. tuberculosis* mouse aerosol infection model [32]. Vaccination of mice, before pregnancy, with exosomes from DCs pulsed with *Toxoplasma gondii*–derived antigens could increase the neonatal survival rate of pups challenged with *T. gondii* [33]. In addition, EVs derived from macrophages showed long-lasting immunity against hepatitis B and C viruses [34, 35].

9.2.1.3 EV-mediated delivery of therapeutic molecules

EVs consist of a lipid bilayer membrane decorated with transmembrane proteins. Small EVs have shown great potential as targeted biogenic nanocarriers for drug delivery, owing to their high biocompatibility and efficacy as well as low immunogenicity. Dexamethasone was successfully encapsulated into human U937 monocyte-derived EVs by sonication and buoyant density gradient ultracentrifugation. These vesicles, designated as ghost ^{Dexa}NVs, demonstrated remarkable therapeutic effects in systemic inflammatory response syndrome induced by gram-negative bacteria [36]. In another study, exosomes derived from RAW264.7 mouse macrophages were loaded with the antibiotic linezolid by a co-incubation method. Free linezolid accumulated only in small amounts in eukaryotic cells, thus showing poor antibiotic activity against intracellular bacteria. By contrast, linezolid-loaded exosomes remarkably improved the accumulation of drug molecules in RAW264.7 macrophages, thereby affording better efficacy against intracellular methicillin-resistant *S. aureus* (MRSA) infections than free linezolid [37]. These studies revealed that EVs may be effective biogenic nanocarriers for antibiotics, generally exerting stronger antimicrobial activity than free drugs, especially for intracellular infections.

9.2.2 Outer Membrane Vesicles

Outer membrane vesicles (OMVs) contain cytoplasmic components, which are commonly secreted by gram-negative bacteria. They are spherical vesicles with diameters varying between 20 and 250 nm. Their biological functions involve the regulation of bacterial responses to stress conditions (*e.g.,* through the removal of toxic compounds such as misfolded proteins), as well as bacterial host lipid acquisition and exchange (*e.g.,* cholesterol in the outer membrane for optimal fitness). OMVs also mediate complex bacterial communities (*e.g.,* by promoting nutrient, iron, and zinc acquisition), thereby contributing to enhancing pathogenesis and antimicrobial resistance [38]. Recently, OMVs have sparked increasing interest in vaccination and targeted drug delivery as potent therapeutic tools.

9.2.2.1 Vaccination

OMVs can effectively induce host adaptive immune responses owing to the presentation of bacterial surface protective antigens combined with an adjuvant LPS [38, 39]. Antigen-presenting cells (APCs) recognize and uptake OMVs, promote antigen presentation, and then activate antigen-specific T cells. Subsequently, the activated $CD8^+$ T cells specifically kill bacteria, while $CD4^+$ T cells enhance antibody production by B cells. Several studies have explored OMVs as safe and efficient vaccines to defeat infectious diseases caused by various pathogens such as *Neisseria meningitidis* [40], *Salmonella* [41], *Vibrio cholerae* [42], *Shigella boydii* [43], and *Acinetobacter baumannii* [44]. Especially, OMV-based vaccines against meningococcal infection have been extensively explored for years. There are currently several OMV-based vaccines in clinical use, including MenBvac®, VA-Mengoc-BC®, MeNZB®, and Bexsero®. MenBvac® is the first OMV-based vaccine against P1.7,16 *N. meningitidis* serogroup B, which was developed in Norway in the 1990s [45]. This vaccine initially achieved a

57.2% protection rate, but large-scale studies suggested short-term persistence of response and less protection [40, 46]. VA-Mengoc-BC®, developed in Cuba, is the most widely used vaccine against serogroup B meningococcal disease in the world. The effectiveness of VA-Mengoc-BC® against homologous serogroup B has been above 80% in all age groups with high and long-lasting antibody titters, and it also demonstrated protection against a serogroup C strain in heterologous contexts [47–49]. Another OMV-based New Zealand vaccine MeNZB®, isolated from the strain NZ98/254, was developed to control epidemic disease caused by subtype P1.7b,4. It induced remarkable production of serum bactericidal antibodies in 96% of adults and 76% of children [50]. Due to the significantly reduced disease incidence and short duration of protection, the routine administration of MeNZB® ceased in 2008 [51]. To further broaden protection, a multicomponent meningococcal B vaccine, Bexsero®, was developed. Approved by the Food and Drug Administration (FDA) in 2015, it is composed of three surface-exposed recombinant proteins added to MeNZB® and provides 66–91% coverage against vaccine antigens worldwide [52]. A recent study demonstrated that Bexsero® also elicited antibodies to *Neisseria gonorrhoeae* [53].

Given the success of *N. meningitidis* serogroup B OMV vaccines, the application of the OMV platform has been expanded to other vaccines. OMVs derived from *Shigella flexneri* showed protection against shigellosis via conjunctival or nasal routes of administration, and the efficacy was enhanced when OMVs were encapsulated into poly(anhydride) nanoparticles [54]. A *Shigella sonnei* OMV vaccine examined in Phase II clinical studies has demonstrated a strong response of serum bactericidal antibodies in adults [55, 56]. Moreover, OMV-based vaccines against *Yersinia pestis*–induced plague [57], *Burkholderia pseudomallei*–induced pulmonary infection [58], and *M. tuberculosis*–induced tuberculosis [59] have been examined in preclinical studies.

9.2.2.2 OMVs as targeted drug delivery carriers

In recent years, the application of OMVs as nanocarriers for diagnostic and/or therapeutic delivery has been broadly explored, due to their multiple advantages such as nanoscale size, easy targeted delivery of cargos to specific sites, and efficient protection of loaded drugs from degradation. Nevertheless, most studies have focused on cancer therapy [60–62]. Although the use of OMVs as targeted delivery carriers for the treatment of infectious diseases is still in its infancy, some existing studies have demonstrated their promising prospects. Generally, two approaches are utilized to encapsulate drugs into OMVs: *in vitro* and *in vivo* loading. *In vitro* loading consists of loading drugs into isolated OMVs via ultrasonication, electroporation, extrusion, or incubation. *In vivo* loading is the strategy employed to encapsulate drugs into OMVs during their biogenesis. In this approach, bacterial cells are co-incubated with the desired drug and the resulting drug-containing OMVs are then harvested. Kadurugamuwa et al. reported that gentamicin-induced OMVs isolated from *S. flexneri* and containing 85 ± 2 ng of gentamicin per μg of OMV protein effectively reduced the intracellular pathogen load in *S. flexneri*–infected Henle cells. This suggested that OMVs can be used as antibiotic carriers [63]. In a recent study, Huang et al. identified a mechanism of OMV-mediated antibiotic efflux used by drug-resistant bacteria to generate antibiotic-loaded *Acinetobacter baumannii* OMVs (AbOMVs). AbOMVs were prepared under the stimulation of various antibiotics. It was found that levofloxacin evoked the most significant increase in the protein level and particle number of OMVs. The OMVs stimulated by levofloxacin at one-fourth minimum inhibitory concentration (MIC) loaded 167.67 μg levofloxacin per 10^{12} particles. Levofloxacin-loaded OMVs (LEF-OMVs) exhibited strong bactericidal ability against enterotoxigenic *E. coli* (ETEC), *Klebsiella pneumoniae*, and *P. aeruginosa in vitro*, by efficiently invading bacterial cells. *In vivo* imaging results demonstrated that LEF-OMVs were retained in the intestine for at least 36 h after oral administration and showed remarkable protection in a mouse intestinal ETEC infection model with minimal adverse effects on intestinal microbes [64]. In addition, various engineered OMVs have been developed to combat infectious diseases. In a pioneer work, Au nanoparticles (AuNPs) coated with OMVs derived from *E. coli* were prepared through repeated extrusion, and the obtained OMVs showed enhanced buffer stability. Bioengineered OMVs-AuNPs evoked instant activation and maturation of DCs via subcutaneous injection, generated durable and high antibody responses, and induced strong Th1 and Th17 biased cell responses [65]. Recently, to mimic the adhesion of *Helicobacter pylori* onto the gastric epithelium, the same group prepared *H. pylori* OMVs-coated poly(lactic-co-glycolic acid) (PLGA) nanoparticles (OM-NPs). OM-NPs could block the adhesion of *H. pylori* to gastric epithelial cells and decrease the attachment of *H. pylori* to stomach tissues, thereby demonstrating that OM-NPs can function as a promising antiadhesion agent against bacterial infections (Fig. 9.2) [66]. In a similar study, bovine serum albumin (BSA) nanoparticles reinforced OMVs (BN-OMVs) were constructed by hydrophobic interaction, which effectively prevented mice from fatal infection of carbapenem-resistant *K. pneumoniae* [67].

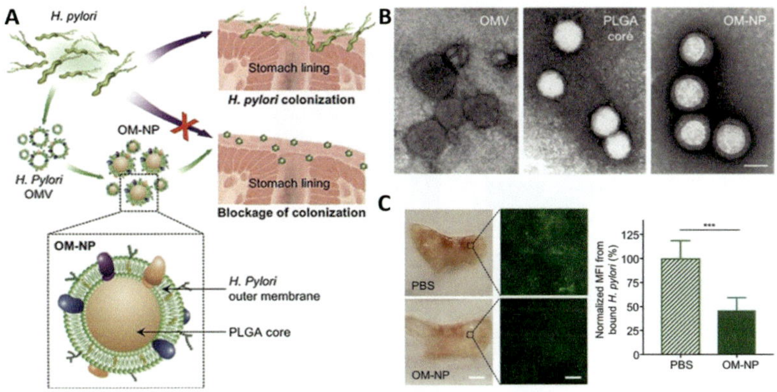

Figure 9.2 Inhibition of pathogen adhesion by *H. pylori* outer membrane–coated nanoparticles. **(A)** Schematic representation of using OM-NPs to inhibit *H. pylori* adhesion on the stomach lining. **(B)** TEM images of OMVs (left), PLGA cores (middle), and OM-NPs (right). Scale bar, 50 nm. **(C)** OM-NPs reduce *H. pylori* colonization on the mouse stomach tissue. Reprinted with permission from Ref. [66].

9.3 Biomimetic Nanocarriers

9.3.1 Bacteria-based Nanocarriers

The use of bacteria, bacteria-derived EVs, and bacteria components has great potential for the development of targeted drug delivery vehicles to treat infectious diseases owing to their unique properties, such as rapid proliferation, gene manipulation, and bacterial host cell recognition interactions. In this section, different types of bacteria-based nanocarriers applied in infectious diseases are discussed.

9.3.1.1 Bacterial microbots

The motility and intrinsic chemotaxis behavior of certain bacteria can be employed to deliver nanotherapeutics to sites of interest. Specifically, cargo-loaded nanoparticles can be attached to bacteria to form the so-called "bacterial microbots." Recently, magnetotactic bacteria were used to improve accessibility to the infected

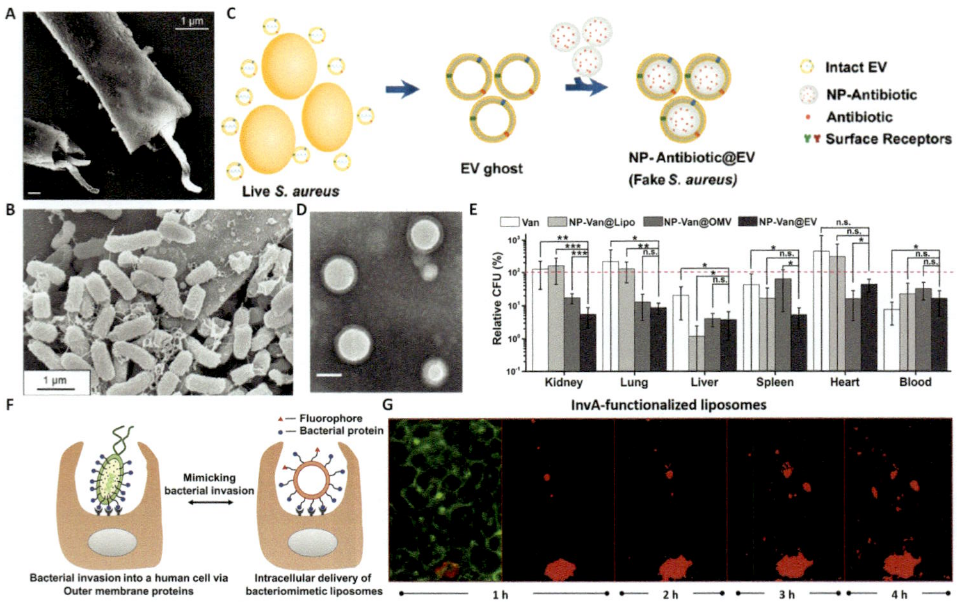

Figure 9.3 Engineering of bacteria-based nanocarriers. **(A, B)** Magnetotactic bacteria-powered biohybrids. **(A)** Scanning electron microscopy (SEM) of *Magnetosopirrillum gryphiswalense* MSR-1 cells captured within a drug-loaded mesoporous silica microtube. **(B)** SEM image of a biohybrid trapped in an *E. coli* biofilm. **(C–E)** Bacterial membrane–coated NPs for drug delivery. **(C)** Membrane-coated NPs were fabricated by coating *S. aureus*–derived extracellular vesicles around antibiotic-loaded NPs. **(D)** TEM image showing the spherical core-shell structure of the NP@EV. Scale bar, 50 nm. **(E)** NPs coated with *S. aureus* EVs and loaded with the antibiotic vancomycin (van) significantly reduced the bacterial burden in the major organs of mice infected with *S. aureus* compared with the formulations coated with a PEGylated lipid bilayer (Lipo) or *E. coli* OMVs. **(F)** Schematic illustration of bacteriomimetic invasin-functionalized nanocarriers for intracellular delivery. **(G)** Representative Z-projection images of the red channel of sequestered z-stacks of living HEp-2 cells incubated with InvA497-functionalized liposomes. Images in (A,B) are reprinted with permission from Ref. [68]. Images in (C–E) are reprinted with permission from Ref. [71]. Images in (F,G) are reprinted with permission from Ref. [75].

areas. Morgan et al. developed microswimmers to integrate the magnetotactic bacteria *Magnetospirillum gryphiswaldense* (MSR-1) with mesoporous silica microtubes as an antibiotic delivery system (Fig. 9.3A,B) [68]. This biohybrid system could be magnetically guided to *E. coli* biofilms where the loaded antibiotic ciprofloxacin was triggered to be released by the localized acidic microenvironment of the biofilm. MSR-1 bacteria enhanced the propulsion ability of the biohybrid to penetrate the biofilm, ensuring delivery of the antibiotic to the inner acidic environments. These bacteria-powered biohybrids demonstrated their potential for combating biofilm-related infections. Nevertheless, *in vivo* efficacy, biocompatibility, and dose-limiting toxicity need to be further validated.

9.3.1.2 Bacterial membrane–based biomimetic nanocarriers

Bacterial EVs secreted by gram-positive bacteria contain a myriad of immune stimulatory molecules with intrinsic adjuvant properties and pathogen-associated patterns that mediate innate immunity and stimulate adaptive immune responses. Hence, coating antibiotic-loaded nanoparticles with the membrane of bacterial EVs may be an attractive strategy to provide an antibacterial immune response and vaccination against bacterial infection [69, 70]. Taking advantage of *S. aureus*'s ability to invade infected host macrophages, Gao et al. constructed an active-targeting antibiotic nanocarrier via cloaking PLGA nanoparticles with the membrane of *S. aureus* EVs (NP@EV) (Fig. 9.3C–E) [71]. The NP@EV nanocarrier could be more efficiently internalized by *S. aureus*–infected murine Ana-1 macrophages in comparison to PLGA nanoparticles coated with liposomes (NP@Lipo) or coated with *E. coli* OMVs (NP@OMV), indicating the specificity of the bacterial host interactions via the membrane proteins of *S. aureus* EVs. Compared to non-infected mice, the nanocarrier NP@EV mainly accumulated in the infected major organs, including the lungs, heart, kidneys, and spleen after IV injection into *S. aureus* bacteremia-bearing mice, validating the active-targeting capacity of NP@EV *in vivo*. Additionally, NP@EV preloaded with vancomycin and rifampicin revealed striking therapeutic effects in alleviating the bacterial burden of *S. aureus*–infected organs. Taken together, these results suggested that bacterial EV-coated nanocarriers may be a promising drug delivery nanoplatform to manage infectious diseases. In a similar study, Chen et al. developed a multi-antigenic nanovaccine EV/ICG/MSN based on *S. aureus* EVs coating indocyanine green (ICG)-loaded magnetic mesoporous silica nanoparticles [72]. EV/ICG/MSN was efficiently taken up by DCs, improved DCs maturation, and promoted $CD8^+$ T cell responses by upregulating MHC-I expression upon laser irradiation. *Ex vivo* fluorescence images showed remarkably enhanced inguinal lymph node accumulation for EV/ICG/MSN after subcutaneous injection. Such engineered nanovaccine could prevent skin/soft tissue infections induced by drug-resistant *S. aureus* and reduce bacterial invasion with laser irradiation. These studies demonstrated that bacterial membrane–coated nanocarriers are effective platforms for the treatment of infectious diseases.

Apart from the use of EVs, one unique work developed bacterial protoplast-based nanovesicles (PDNVs) as adjuvant-free vaccine-delivery carriers [73]. Bacterial protoplasts were isolated from *E. coli* to deplete the noxious outer membrane components and then extruded through polycarbonate membrane filters to acquire PDNVs, which were less immunogenic and more suitable for drug delivery compared to OMVs. To examine the immunotherapeutic potential, *E. coli* antigen OmpA and *S. aureus*–specific coagulase were loaded into PDNVs, respectively. The antigen-loaded PDNVs evoked intense antigen-specific humoral and cellular immune responses and provided effective protection against gram-negative sepsis and gram-positive bacteria-induced pneumonia, respectively. Overall, bacterial protoplast–derived vesicles can serve as an effective and safe vaccine-delivery system.

9.3.1.3 Bacteriomimetic invasin-functionalized nanocarriers

The inherent capacity of bacteria to target and invade host cells has inspired the development of multiple engineered bacteriomimetic nanomedicines. One promising strategy is to decorate the surface of nanoparticles with the bacterial outer membrane protein invasin to simulate that of pathogenic bacteria [74]. In this regard, Lehr et al. developed bioinspired liposomes functionalized with InvA497, a C-terminal fragment of the invasin protein of *Yersinia pseudotuberculosis* (Fig. 9.3F,G) [75, 76]. In comparison to liposomes functionalized with BSA, InvA497-functionalized carriers showed enhanced adhesion and internalization in mammalian epithelial HEp-2 cells through the β1-integrin receptor-mediated mechanism. Furthermore, gentamicin-containing bioinspired liposomes remarkably alleviated the infection load of HEp-2 cells infected with *Y. pseudotuberculosis* and *Salmonella enterica*. In a follow-up study, they compared the intracellular bacteria-killing ability of spherical and aspherical InvA497-functionalized nanocarriers [77]. The results showed that bioinspired nanocarriers of both aspherical and spherical shapes improved cellular uptake of HEp-2 cells, compared to nonfunctionalized carriers. In addition, aspherical functionalized antibiotic-loaded nanocarriers demonstrated higher suppression of intracellular infection caused by *S. flexneri*. Most recently, the same group developed liposomes surface decorated with extracellular adherence protein (Eap), derived from *S. aureus*, to orally deliver the poorly permeable antibiotic colistin [78]. The engineered liposomes could protect colistin from degradation in the harsh gastrointestinal tract conditions and promote internalization into eukaryotic cells, thereby reducing intracellular infections by *S. enterica*. This series of studies revealed that bacterial protein-functionalized nanocarriers hold great promise as a new avenue for improving the therapeutic efficacy of antibiotics against intracellular infections.

9.3.2 Virus-based Nanocarriers

Viruses can attach to the surface of host cells via favorable interactions between virion proteins and lipids, carbohydrate moieties, or membrane proteins, subsequently entering host cells through fusion/penetration or phagocytosis/pinocytosis. In addition, viruses can evade the immune surveillance of the host and are characterized by high stability, diverse shape and size, and easy modification. Correspondingly, the inherent natural features of biomimetic virus-derived nanovesicles show great potential as vaccines and novel drug delivery nanocarriers.

9.3.2.1 Viral vectors

Virus-based vectors have been developed for applications in a variety of infectious diseases owing to their capacity to self-assemble into cords or liquid crystals, as well as their ability to target specific bacteria, and their amenability to surface modification. Adenovirus (Ad) vectors have been among the most widely used viral vectors for vaccine applications and gene therapy since they are highly infectious and easy to prepare. Two representative examples were studied utilizing Ad vector particles as a vaccine strategy for Ebola and human immunodeficiency virus (HIV) in nonhuman primate models [79, 80]. Vaccinated animals were protected from lethal doses of viruses in both studies. Of note, an Ad serotype 26 (Ad26)-based vector against the Ebola virus (Ad26.ZEBOV) has been approved in the European Union [81].

Vaccines against SARS-CoV-2 have been extensively studied during the global pandemic. Four types of Ad vector vaccines for COVID-19 have been approved, including Ad5-nCOV (China) [82], Sputnik V (Russia) [83], Ad26.COV2-S (Europe) [84], and ChAdOx1 nCov-19 (UK) [85]. These Ad vector vaccines effectively induced anti-spike protein antibodies and stimulated $CD8^+$ T cell immune responses through intramuscular administration. Additionally, Ad vector vaccines against the Zika virus, HIV, and influenza virus have been tested in clinical trials [86–88].

The high incidence of antibiotic resistance has captured much attention in the use of naturally obtained bacteriophages to defeat bacterial infections. Benhar et al. employed filamentous bacteriophages as targeted drug nanocarriers to deliver antibiotics for the treatment of infectious diseases [89]. To enhance the targeting efficiency to specific bacteria, the p8 coat protein of phages was genetically modified to display an *S. aureus*–binding peptide. In addition, an anti–*S. aureus* IgG antibody was attached to the minor coat protein of the phages through an Fc-binding ZZ domain. The antibiotic chloramphenicol was selected as a model drug to conjugate to the phages through a labile ester bond. After the engineered phages specifically bound the bacteria, the ester bond was cleaved by an esterase in the serum and the active form of chloramphenicol was released, thereby significantly inhibiting the growth of *S. aureus*. This tactic could specifically transport different drugs to target bacteria, leading to improved selectivity and lower toxicity of the loaded antibiotics. To increase the drug loading capacity, the same group used aminoglycoside antibiotics as solubility-enhancing branched linkers to connect the hydrophobic chloramphenicol with the phage. The established antibody-targeted phage nanomedicines loaded over 40,000 chloramphenicol molecules/phage and demonstrated dramatic antimicrobial activity against *E. coli*, *Streptococcus pyogenes*, and *S. aureus in vitro* and *in vivo* [90, 91].

Phages can also eliminate pathogens through gene delivery. As an interesting example, an M13 phagemid for encoding toxins Gef and ChpBK was designed, and this engineered phage remarkably decreased the growth of bacteria in the blood after injection into infected mice by expressing Gef/ChpBK and causing cell death during nutritional stress [92]. Similarly, using the ΦNM1 bacteriophage, Bikard et al. developed a phagemid-delivered CRISPR-Cas9 system that targeted the methicillin-resistant gene mecA to eradicate methicillin-resistant strains from a mixed bacterial population. This antimicrobial selectively killed antibiotic-resistant and virulent *S. aureus* strains in a mouse model of staphylococcal skin infection [93]. In another study, Krom et al. constructed bacterial phagemids to express antimicrobial peptides (AMPs) and protein toxins [94]. This system is more sophisticated and can easily regulate the expression of the type and number of AMPs and toxins. The engineered gene-carrying system successfully targeted *E. coli*–infected cells and showed effectiveness in a murine peritonitis infection model.

Moreover, plant viruses are also attractive nanocarriers owing to their advantages such as safety for humans, degradability, exact structure, availability for genetic and chemical modification, and excellent bioavailability [95]. The cowpea mosaic virus (CPMV) is an icosahedral plant virus, and it has been examined to deliver drugs to the injured central nervous system (CNS) for the treatment of neurodegenerative and infectious diseases. These CPMV nanoparticles could be internalized by bEND.3 brain microvascular endothelial cells (an *in vitro* model of the BBB) and be detected in the brain parenchyma in mice infected with neurotropic mouse hepatitis virus (MHV), suggesting that CPMV nanoparticles can serve as a potential vehicle to treat infectious diseases in the CNS [96]. Furthermore, CPMV exhibits tropism for dendritic cells and makes them an ideal platform to defeat viral infections. Wen et al. packaged an antiviral protease inhibitor PF-429242 in CMPV to inhibit the pivotal protein site-1 protease for viral infection [97]. The results demonstrated that PF-429242 efficiently protected the host from infection by a prototypic lymphocytic choriomeningitis virus.

9.3.2.2 Virus-like particles

Virus-like particles (VLPs) are self-assembled viral capsids or envelope proteins derived from viruses but in the absence of viral genome. Compared to viral vectors, VPLs are noninfective, and they can be easily prepared and scaled up at a low cost, presenting defined structures, high stability, and capacity for loading different cargo [98]. VLPs can be used as vaccines since they present pathogen-associated molecular patterns and can be recognized by the corresponding receptors on dendritic cells, resulting in the initiation of immune responses [99]. VLP vaccines against HPV (Cervarix® and Gardasil®), HBV (EngeriX-B®, Sci-B-VAC ®, and Recombivax HB®), and HEV (Hecolin®) have been approved for human use. VRC-CHKVLP059-00-VP against mosquito-borne Chikungunya virus has achieved a Phase II clinical trial (NCT02562482). A plant-derived quadrivalent VLP influenza vaccine for adults has also been developed (NCT03301051). Furthermore, many other VLP vaccines are also under investigation. For example, a modular murine polyomavirus (MuPyV) VLP was constructed to present a Group A streptococcus (GAS) antigenic peptide J8i. This engineered vaccine significantly induced systemic and mucosal responses via intranasal administration, thereby decreasing GAS colonization in the throat [100]. Likewise, an Ebola virus VLP vaccine (eVLP) has been engineered through the co-expression of the Ebola matrix protein VP40 and the virus glycoprotein in 293T cells. eVLP was highly immunogenic and could efficiently induce the activation of $CD19^+$ B cells, as well as $CD4^+$ and $CD8^+$ T cells. Vaccination with eVLPs protected mice from lethal Ebola virus challenges [101]. A combination of respiratory syncytial virus (RSV) fusion protein VLPs and glycoprotein VLPs demonstrated enhanced immune response and remarkably reduced lung viral replication and histopathology injury [102]. Kim et al. developed a VLP-based vaccine incorporating influenza virus heterologous M2 extracellular domains (M2e5x VLPs) without adjuvants, which was coated on microneedles for skin vaccination with minimal invasion. This vaccine was stable for 8 weeks and stimulated potent humoral and mucosal antibody responses after skin immunization, generating an expansive cross-protective effect against H1N1, H3N2, and H5N1 [103]. In addition to vaccines against viral infections, the plant Papaya mosaic virus (PapMV)-derived VLPs containing ssRNA have been successfully developed as vaccines against a lethal challenge with *S. pneumoniae* and *Listeria monocytogenes* [104, 105].

Beyond vaccines, VLPs are also highly suitable as nanocarriers for the delivery of different drugs, varying from small molecular drugs and peptides/proteins to nucleic acids, owing to their hollow structures and capability to escape endosomes before lysosomal degradation. Some VLPs can be used as targeted drug delivery systems due to their natural tropism toward a specific tissue. For instance, rotavirus-derived VLPs can be used for site-specific delivery of cargos to the intestine due to their affinity to the intestinal tissue, while HBV-derived VLPs can target liver cells owing to this inherent property of HBV [106]. In addition, site-specific drug delivery could be achieved by conjugating various targeting ligands on VLPs via chemical engineering [107].

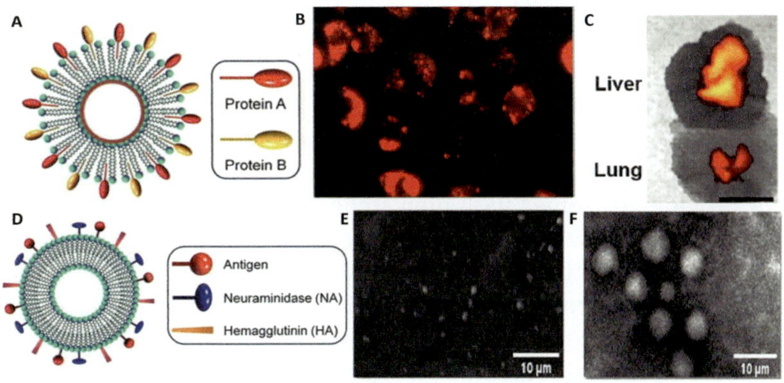

Figure 9.4 Engineering of virus-based nanocarriers. **(A)** Schematic illustration of virus-like particles (VLPs). **(B)** Representative microscopy images of RAW264.7 macrophages after 24 h of incubation with bacteriophage Qβ@mCherry-derived VLPs. **(C)** A representative image of isolated liver and lung tissues from mice administered Qβ-VLPs after 2 h. Scale bar, 2 cm. **(D)** Schematic illustration of virosomes. **(E,F)** SEM and TEM images of the prepared nanocurcumin-virosome. Images in (A and D) are reprinted with permission from Ref. [98]. Images in (B,C) are reprinted with permission from Ref. [108]. Images in (E,F) are reprinted with permission from Ref. [112].

Taking advantage of several macrolide antibiotics preferentially accumulated in the lung macrophages via engaging the mannose receptor, a series of bacteriophage Qβ VLP-macrolide conjugates were prepared by the copper(I)-accelerated azide–alkyne cycloaddition reaction. All conjugates showed remarkable cellular uptake in macrophages and polarized macrophages toward an M2 phenotype *in vitro*. Qβ VLP loaded with azithromycin notably accumulated in the lung tissue after IV injection in mice, while Qβ VLP alone did not traffic to the lung, indicating the guidance of macrolide on the VLP surface to lung-resident macrophages. This creative delivery system afforded a novel approach for intracellular drug delivery to treat pulmonary infections (Fig. 9.4A–C) [108].

9.3.2.3 Virosomes

Virosomes comprise virus-derived proteins incorporated in phospholipid bilayer membranes of unilamellar liposomes with a diameter of 20–200 nm [109]. A range of virosomes has been applied as effective vaccines due to their capability of delivering immunogens to host cells, as well as potent adjuvants to stimulate cellular and humoral immune responses without causing any side effects. Envelope proteins derived from the influenza virus, such as neuraminidase (NA) and hemagglutinin (HA), are most commonly used to construct virosomes for vaccination. HA and NA provide virosomes with high stability, homogeneity, and fusion activity and allow virosomes to enter the host cells by receptor-mediated endocytosis [110]. These two envelope proteins undergo conformational transition at the acidic pH of host cell endosomes, leading to the fusion of the virosomal membrane with the endosomal membrane. In addition, some other enveloped proteins derived from the RSV, the hemagglutinating virus of Japan (HVJ), the Chikungunya virus, and vesicular stomatitis virus (VSV) can be used to engineer virosomes. The vaccines based on virosomes against hepatitis A (Epaxal®) and influenza (Inflexal®V) have been licensed for dissemination.

Virosomes are also unique drug delivery systems, being able to deliver various payloads such as antibodies, nucleic acids, antigens, and other types of drugs to specific cells and tissues through receptor-mediated uptake. Virosomes possess multiple advantages, such as biocompatibility, nontoxicity, and reproducibility. They can mimic the receptor binding and membrane fusion properties of native viruses, which enable them to transport cargo for intracellular delivery in targeted cells. For example, reconstituted influenza virosomes loaded with a synthetic peptide comprising residues 446 to 490 of apical membrane antigen 1 of *Plasmodium falciparum* were constructed as an antigen delivery system to treat malaria [111]. The peptide incorporated in virosomes was protected from protease degradation after administration, and all immunized mice produced parasite growth–inhibitory antibodies. Kumar et al. compared the activities of virosomes and liposomes as drug delivery carriers. In this case, a herbal drug nanocurcumin (NC) was separately encapsulated into virosomes and liposomes. Both vehicles are spherical, and the diameter was 70 and 265 nm for NC-loaded virosomes and liposomes, respectively. An *in vitro* release study revealed that NC virosomes showed controlled and prolonged release of curcumin. Furthermore, the enhanced anti-inflammatory and antimigratory activities as well as cytotoxicity of NC in HepG2 and CaCo2 cells were achieved by loading into virosomes rather than liposomes. The collective results demonstrated that virosomes are a potential candidate as a promising drug delivery nanocarrier for different therapeutic applications (Fig. 9.4D–F) [112]. Amacker et al. constructed a chimeric virosome to deliver biomacromolecules (such as hepatitis C virus Core132 peptide) into the cytoplasm, which combined the membrane fusion feature of virosomes and the high peptide-loading ability of liposomes [113]. Such virosome showed potent immunogenicity and high efficiency in inducing specific cytotoxic T cells after vaccination.

9.3.3 Mammalian Cell–based Nanocarriers

Mammalian cell–based nanocarriers have shown terrific advantages in the treatment of infectious diseases, owing to their peculiar characteristics such as efficient targeting, prolonged circulation, flexible morphology, ability to cross impermeable biological barriers, and high biocompatibility. Various types of cells, including erythrocytes, leukocytes, stem cells, and platelets, as well as their membranes, have been employed as novel biomimetic carriers.

9.3.3.1 Erythrocyte-based nanocarriers

Erythrocytes or red blood cells (RBCs) are the most abundant circulatory cells in the body. With a half-life of 100 to 120 days and containing various bioactive ingredients on their membranes, RBCs can be used as ideal drug delivery systems. Two main preparation approaches have been adopted for this purpose. One is based on the encapsulation of drug molecules in RBCs, and another one relies on the coupling of therapeutic cargos to the RBC membrane through chemical cross-linking. The methods for the encapsulation of various drugs in RBCs include hypotonic dilution, electroporation, lipid fusion, and endocytosis. RBCs can be used to gradually release drugs in the bloodstream and maintain a constant drug concentration over time, limiting the side effects and prolonging the action time. Many different types of low molecular weight drugs have been encapsulated in RBCs. For example, frequent and high doses of glucocorticoids result in hormonal dependence and serious side effects. To solve this problem, a series of clinical trials were conducted by loading dexamethasone in erythrocytes, and the results demonstrated that this strategy was safe and feasible. Prolonged steady release of low doses of dexamethasone in the bloodstream maintained patients in clinical remission and allowed steroid withdrawal in the treatment of inflammatory bowel disease, chronic obstructive pulmonary disease, and pulmonary cystic fibrosis, while this effect was independent of the multi-drug resistance 1 gene polymorphism (the *MDR1* gene is responsible for the expression of P-glycoprotein, a protein that acts as an energy-dependent drug efflux pump and has the potential to influence dexamethasone bioavailability) [114–117]. In another case, primaquine-loaded RBCs exhibited a sustained drug release profile over 48 h, which was beneficial for the treatment of malaria [118, 119].

On the other hand, RBCs containing antibiotics have many advantages in combating intracellular pathogens. In this regard, various antibiotics, such as amikacin, tetracycline, gentamicin, and metronidazole, were loaded in RBCs to achieve a sustained release effect. RBC-derived carriers increased the half-life and the "area under the plasma drug concentration versus time curve" (AUC) of antibiotics (which reflects the drug concentration in the blood plasma as a function of time) and enhanced the drug accumulation in macrophages, both *in vitro* and *in vivo* [120–122]. Additionally, RBCs are also potential vectors for loading antiviral drugs. RBCs containing AZT (an analogue of thymidine), glutathione (GSH), or AZT+GSH were investigated to test their efficacy in a murine model of acquired immunodeficiency syndrome (AIDS). GSH-loaded RBCs provided additional antiviral effects, significantly decreased the pro-viral DNA content in the macrophages, and also mitigated the

GSH deficiency upon AZT treatment [123]. Similar results have been obtained with 2′,3′-dideoxycytidine 5′-triphosphate (ddCTP) and 2′,3′-dideoxycytidine-5′-phosphate (ddCMP), using RBCs as a carrier during HIV infection [124, 125]. To defeat the serious *Mycobacterium avium* complex opportunistic infection in AIDS patients, RBC carriers loaded with AZT and ethambutol were developed, which could successfully penetrate infected macrophages and inhibit replication of *M. avium in vitro* [126]. Likewise, interferon (INF)-α and ribavirin (RIBA) were simultaneously encapsulated in human RBCs for the treatment of hepatitis C. The engineered RBCs-INF-α-RIBA could target the liver macrophages, allowing for higher therapeutically efficient concentrations in the target organ, as well as (INF)-α induced autocrine stimulation of the liver and spleen to enhance cell-mediated antiviral defenses. In addition, it has been shown that RBCs-INF-α-RIBA could activate monocytes and induce surface MHC II and Fc receptors [127].

Furthermore, in addition to INF-α and ribavirin, the HCV protease inhibitor boceprevir was co-loaded in RBCs to potentiate their efficacy, which demonstrated to be an alternative combination scheme for the treatment of hepatitis C [128]. Also, erythrocytes may be ideal candidates in the field of vaccines as biomimetic carriers and adjuvants of antigens. Polvani et al. encapsulated three bacterial toxoids in RBCs using a hypotonic dialysis method. In comparison to free toxoids, significantly higher titters of specific and neutralizing antibodies were obtained via RBCs delivery after IV injection [129]. In other cases, protein antigens, glycoproteins B of herpes simplex virus type 1, or HIV-1 Tat protein were attached to the surface of autologous RBCs via the biotin–avidin–biotin bridges to induce significantly higher immunological responses in comparison to Freund's adjuvant [130, 131].

Coating nanoparticles via the erythrocyte membrane is another biomimetic strategy developed to treat infections. In a recent study, Zhang's group constructed a biomimetic toxin nanosponge, which consisted of a polymeric nanoparticle core cloaked with an RBC membrane. This nanosponge can act as a nanodecoy to adsorb and neutralize several bacterial pore-forming toxins *in vivo* (Fig. 9.5A–D) [132]. Furthermore, a redox-sensitive nanogel camouflaged by the RBC membrane was developed by the same group to combat MRSA infections.

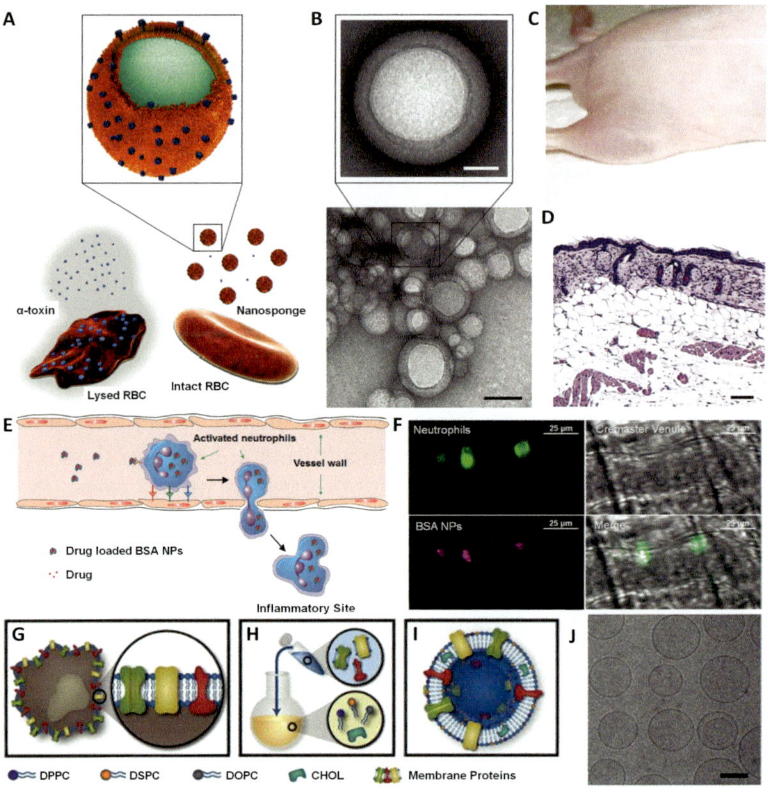

Figure 9.5 Engineering of cell-based nanocarriers. **(A)** Schematic of a toxin nanosponge and the mechanism of neutralizing pore-forming toxins. The nanosponge consists of substrate-supported RBC bilayer membranes into which pore-forming toxins can be incorporated. **(B)** TEM images of nanosponges mixed with α-toxin (scale bar, 80 nm) and the zoomed-in view of a single toxin-absorbed nanosponge (scale bar, 20 nm). **(C,D)** Mice injected with α-toxin/nanosponges. No skin lesion occurred **(C)**. No abnormality was observed in the epidermis **(D)**. **(E)** The concept of neutrophil-mediated delivery of therapeutic albumin NPs. **(F)** Intravital microscopic images of TNF-α-induced inflammation of live mouse cremaster venules, at 30 min after IV injection of Cy5-BSA NPs and Alexa Fluor-488-labeled anti-mouse Gr-1 to mark neutrophils (green). **(G–J)** Synthesis, formulation, and characterization of leukosomes. **(G)** Extraction of proteolipid materials from murine J774 macrophages. **(H)** Protein enrichment of the phospholipid film. **(I)** Vesicular formulation of leukosomes. **(J)** Cryo-TEM analysis of leukosomes. DPPC, 1,2-dihexadecanoyl-sn-glycero-3-phosphocholine; DSPC, 1,2-distearoyl-sn-glycero-3-phosphocholine; DOPC, 1,2-dioleoyl-sn-glycero-3-phosphocholine; CHOL, cholesterol. Images (A–D) are reprinted with permission from [132]. Images (E,F) are reprinted with permission from Ref. [140]. Images (G–J) are reprinted with permission from Ref. [151].

The RBC-nanogel significantly neutralized MRSA-associated toxins in the extracellular microenvironment and achieved controlled release of vancomycin in the intracellular reducing microenvironment [133]. Additionally, the incubation of RBC-membrane-coated PLGA NPs with staphylococcal α-hemolysin can generate a nanotoxoid (Hla), which can be used as an anti-virulence vaccine. Mice vaccinated with this nanotoxoid showed better protective immunity against MRSA skin infection in comparison to the control group [134]. Similarly, a study by Koo et al. demonstrated that RBC-membrane-coated PLGA NPs remarkably attenuated β-hemolysin/cytolysin-mediated hemolysis, inhibited lung epithelial and macrophage apoptosis, and suppressed macrophage inflammasome activation to diminish IL-1β production, thereby providing a novel way for adjunctive therapy of group B Streptococcal sepsis [135]. Together, this type of decoy strategies has been widely investigated for the treatment and prevention of bacterial infections [136–139].

9.3.3.2 Neutrophil-based nanocarriers

Neutrophils, the most abundant leucocyte population, can function as promising carriers for the delivery of therapeutic agents to acute inflammatory sites owing to their capacity to rapidly extravasate from the blood circulation and infiltrate infected/injured tissues. Similar to RBC carriers, there are also two types of drug delivery systems based on neutrophils, *i.e.*, neutrophils directly loaded with therapeutic agents and neutrophil membrane–derived nanocarriers.

Chu et al. utilized neutrophils loaded with albumin NPs to cross the blood vessel barrier in a mouse model of LPS-induced acute lung injury (ALI). The denatured BSA NPs were specifically endocytosed by activated neutrophils *in situ* and then migrated to the lung inflammatory site. Antibiotic-loaded albumin NPs remarkably inhibited bacterial growth. This study demonstrated the ability to hijack neutrophils to transport nanocarriers to target inflammatory sites (Fig. 9.5E,F) [140]. In another study, poly(ε-caprolactone)-b-polyethylene glycol (PCL-PEG) NPs loaded with sparfloxacin were functionalized with neutrophil membranes. This system showed prolonged circulation and effective targeting to pneumonia sites after IV injection and, therefore, significantly alleviated inflammation without any side effects [141]. In a recent study, Gao et al. first generated neutrophil EVs by nitrogen cavitation that crushed neutrophils into nanosized membrane vesicles and then encapsulated piceatannol into the EVs via pH gradient. Further studies showed that drug-loaded EVs inhibited the NF-κB pathway and efficiently attenuated LPS-induced ALI and sepsis after IV administration [142]. Employing a similar strategy, the same group co-loaded resolvin D1 (on the surface) and antibiotic (inside) in human neutrophil membrane–derived nanovesicles to treat mouse models of bacterium-induced peritonitis and lung infection [143, 144].

9.3.3.3 Macrophage-based nanocarriers

Macrophages are tissue-resident immune cells with a lifespan of several months, which play a crucial role in adaptive and innate immune responses. Owing to their immunomodulatory traits, low immunogenicity, long half-life period, and stability in blood circulation, macrophages are suitable as drug carriers for the treatment of infectious diseases. Macrophage membrane–cloaked drug delivery nanovehicles have been widely applied for the treatment of bacterial infections in recent studies. Wang et al. developed a gold–silver nanocage camouflaged by *S. aureus*–activated macrophage membrane. These NPs showed an excellent capacity to interact with the source bacteria due to the pattern-recognizing receptors on macrophages. The nanosystem could considerably accumulate at the infection sites and exhibited potent bacteria-killing ability when combined with near-infrared (NIR) laser irradiation [145]. Another study revealed that after being functionalized with macrophage membranes, microwave-responsive Fe_3O_4/Au NPs specifically bound to inflammatory cytokines, targeted *S. aureus* by Toll-like receptor 2, and could be retained in osteomyelitis infection site via a magnet. The generation of $\bullet O^{2-}$ and heat upon microwave irradiation resulted in bacterial death, macrophage polarization into the M2 phenotype, and enhanced bone regeneration [146].

Also, polymeric NPs coated with macrophage membranes were developed as decoys or vaccines to absorb and neutralize endotoxins for the management of sepsis, other bacteria-induced infections, and COVID-19 [147–150]. In addition, Molinaro et al. designed a leukosome platform by integrating proteins derived from the leukocyte membrane (J774 macrophages) with lipid NPs. It was found that leukocyte membrane proteins, including CD45, CD47, LFA-1, Mac-1, and PSGL-1, were all identified on the surface of leukosomes. Intravital microscopy analysis showed that the obtained nanovesicles preferentially accumulated in the inflamed tissue, exhibiting a five to eight-fold increase compared to liposomes. Moreover, dexamethasone-loaded leukosomes showed controlled drug release. Treatment with this biomimetic nanotherapy dramatically improved the symptoms of LPS-induced ear inflammation in mice (Fig. 9.5G–J) [151].

9.3.3.4 Platelet-based nanocarriers

Platelets are small, anuclear cell fragments that are 2–4 μm in diameter, with a lifespan of 7–10 days. Platelets present multiple features, such as high payload loading potential, immunocompatibility, pathogen adhesion, and preferential adhesion to damaged vasculature, which make them attractive drug carriers. Zhang's group reported that platelet membrane–cloaked NPs showed strong binding to platelet-adhering pathogens and that vancomycin-loaded biomimetic nanotherapy afforded much better antimicrobial efficacy, compared to an RBC membrane–coated control nanoformulation in a mouse model of systemic MRSA252 infection, due to the superior collagen-binding ability and stronger selective adhesion to damaged blood vessels [152]. Likewise, Peng et al. prepared mesoporous copper silicate microspheres camouflaged by platelet membrane (CSO@PM). Due to the formyl peptide receptor on the membrane, CSO@PM efficiently targeted bacteria and adsorbed LPS, thus promoting re-epithelialization and granulation-tissue formation as well as accelerated wound healing [153].

9.3.3.5 Other cell-based nanocarriers

In addition to the above-mentioned cell-based biomimetic nanocarriers, nanocarriers derived from other cell types have been examined for anti-infection therapy. In particular, Zhang's group developed a novel targeted nanotherapy made of antibiotic-loaded polymeric cores coated with membranes of human gastric adenocarcinoma cells (AGS cells). The resulting NPs possess inherent adhesion to *H. pylori* bacteria *in vitro*, thus demonstrating potent therapeutic efficacy against *H. pylori* infection *in vivo* [154]. The same group also found that PLGA cores cloaked with plasma membranes of $CD4^+$ T cells specifically bound with the key envelope glycoprotein of the HIV, effectively neutralized HIV, and suppressed viral infection of human peripheral mononuclear blood cells *in vitro* [155]. Recently, novel nanocarriers based on hybrid cell membranes have been developed to combat infectious diseases. A recent example of such an avenue is the development of a novel nanodecoy prepared by fusing cellular membrane nanovesicles derived from human monocytes (THP-1 cells) and cells overexpressing the angiotensin-converting enzyme II (ACE2) receptor. The nanodecoy remarkably neutralized SARS-CoV-2 and inhibited cytokine storm *in vitro* and *in vivo* [156]. Adopting a similar strategy, Wang et al. developed dual membrane–coated microfluidic microspheres via fusing membranes of macrophages and ACE2 receptor–overexpressing cells to defeat COVID-19 infection [157]. In another study, Zhang et al. engineered an effective vaccine by expressing human papillomavirus (HPV) L2 epitope peptides or H1N1 hemagglutinin (HA) glycoprotein on the membrane of 293T or Hela cells. The virus-mimetic nanovesicles (VMVs) could be retained in the liver and spleen for a long time and activate potent immune response after IV injection in mice [158].

9.4 Conclusion and Perspectives

In the recent years, various biogenic and biomimetic nanocarriers, including EVs, OMVs, and NPs derived from different types of bacteria, viruses, and mammalian cells, have been employed as versatile platforms for drug delivery (Table 9.1). The advantages of these systems are long circulating performance, stealth, immunomodulatory activity, targeted drug delivery capability, and great biocompatibility. The applications of biogenic and biomimetic nanosystems for the treatment of infectious diseases are flourishing, but there are also full of challenges in clinical translation. The main obstacle is the isolation and preparation of these nature-derived nanocarriers in good batch-to-batch reproducible, cost-efficient, and large-scalable manners. Second, on account of the biological complexity of biogenic and biomimetic systems, their molecular components and features are not fully identified. Therefore, detailed characterization and further *in vitro*/*in vivo* comprehensive studies are highly necessary. Finally, most of the currently available anti-infection nanotherapies based on biogenic and biomimetic NPs were only evaluated in preclinical studies using disease models of small animals, which are not predictive of efficacy and immunogenicity in humans. The effectiveness and long-term safety issues of these systems in multiple animal species and humans remain to be confirmed in future studies. The combination of synthetic systems such as responsive and intelligent NPs with targeted delivery capabilities of biomimetic platforms holds great potential in the treatment of infectious diseases.

Table 9.1 The current development status of biogenic or biomimetic drug delivery carriers

<table>
<tr><th colspan="3">Types of Biogenic or Biomimetic Nanocarriers</th><th>Definition</th><th>Applications</th><th>Ref.</th></tr>
<tr><td rowspan="2">Biogenic nanocarriers</td><td colspan="2">Extracellular vesicles/Exosomes</td><td>Spherical biomembrane structures secreted by most eukaryotic and prokaryotic cells</td><td>Antibacterial nanotherapies, drug delivery, vaccination</td><td>[8–10, 12, 20, 26, 27, 31, 34, 36, 37]</td></tr>
<tr><td colspan="2">Outer membrane vesicles (OMVs)</td><td>Spherical structures containing cytoplasmic components commonly secreted by gram-negative bacteria</td><td>Biomimetic therapies, drug delivery, vaccination</td><td>[40, 41, 43, 47, 51, 52, 55, 58, 61, 64, 67]</td></tr>
<tr><td rowspan="3">Biomimetic nanocarriers</td><td rowspan="3">Bacteria-based nanocarriers</td><td>Bacterial microbots</td><td>Drug-loaded nanoparticles attached to bacteria</td><td>Targeted drug delivery</td><td>[68]</td></tr>
<tr><td>Bacterial membrane–based biomimetic nanocarriers</td><td>Drug-loaded nanoparticles decorated with the membrane of bacterial extracellular vesicles</td><td>Targeted drug delivery, vaccination</td><td>[70–73]</td></tr>
<tr><td>Bacteriomimetic invasin-functionalized nanocarriers</td><td>Drug-loaded nanoparticles decorated with bacterial outer membrane invasion</td><td>Targeted drug delivery</td><td>[75–77]</td></tr>
</table>

(Continued)

Table 9.1 (*Continued*)

Types of Biogenic or Biomimetic Nanocarriers			Definition	Applications	Ref.
Biomimetic nanocarriers	Virus-based nanocarriers	Viral vectors	Viruses used for delivery of genetic materials into cells	Vaccination, gene delivery	[82, 83, 86, 94, 97]
		Virus-like particles	Assemblies of viral capsids or envelope proteins derived from viruses, but in the absence of viral genome	Vaccination, drug/gene delivery	[102, 103, 105, 106, 108]
		Virosomes	Spherical structures consisting of virus-derived proteins incorporated in phospholipid bilayer membranes	Vaccination, drug/gene delivery	[111–113]
	Mammalian cell–based nanocarriers	Erythrocyte-based nanocarriers	Drugs loaded in erythrocytes or drug-loaded nanoparticles decorated with erythrocytes membrane	Biodetoxification, vaccination, targeted drug delivery	[114, 123, 128, 132–134]
		Neutrophil-based nanocarriers	Drug-loaded nanoparticles endocytosed by activated neutrophils *in situ* or drugs loaded in neutrophil extracellular vesicles	Targeted drug delivery	[140–144]

Types of Biogenic or Biomimetic Nanocarriers			Definition	Applications	Ref.
Biomimetic nanocarriers	Mammalian cell–based nanocarriers	Macrophage-based nanocarriers	Drug-loaded nanoparticles decorated with the macrophage membrane	Biodetoxification, vaccination, targeted drug delivery	[145, 146, 149, 151]
		Platelet-based nanocarriers	Drug-loaded nanoparticles decorated with the platelet membrane	Biodetoxification, targeted drug delivery	[152, 153]
		Other cell-based nanocarriers	Drug-loaded nanoparticles decorated with membranes of AGS cells, T cells, or hybrid cell membranes	Targeted drug delivery	[154, 155, 157]

References

1. M. Xiong, Y. Bao, X. Yang, Y. Zhu, and J. Wang, Delivery of antibiotics with polymeric particles, *Adv. Drug Deliver. Rev.* **78**, 63 (2014).
2. S. EL Andaloussi, I. Mäger, X. O. Breakefield, and M. J. Wood, Extracellular vesicles: Biology and emerging therapeutic opportunities, *Nat. Rev. Drug Discov.* **12**, 347 (2013).
3. G. Raposo and W. Stoorvogel, Extracellular vesicles: Exosomes, microvesicles, and friends, *J. Cell Biol.* **200**, 373 (2013).
4. G. Camussi, M. C. Deregibus, S. Bruno, C. Grange, V. Fonsato, and C. Tetta, Exosome/microvesicle-mediated epigenetic reprogramming of cell, *Am. J. Cancer Res.* **1**, 98 (2011).
5. G. Fuhrmann, A. L. Neuer, and I. K. Herrmann, Extracellular vesicles: A promising avenue for the detection and treatment of infectious diseases? *Eur. J. Pharm. Biopharm.* **118**, 56 (2017).
6. J. S. Schorey and C. V. Harding, Extracellular vesicles and infectious diseases: New complexity to an old story, *J. Clin. Invest.* **126**, 1181 (2016).
7. J. You, Z. Fu, and L. Zou, Mechanism and potential of extracellular vesicles derived from mesenchymal stem cells for the treatment of infectious diseases, *Front. Microbiol.* **12**, 761338 (2021).

8. A. Porzionato et al., Intratracheal administration of clinical-grade mesenchymal stem cell-derived extracellular vesicles reduces lung injury in a rat model of broncho pulmonary dysplasia, *Am. J. Physiol. Lung Cell Mol. Physiol.* **316**, L6 (2019).
9. J. You et al., Human umbilical cord mesenchymal stem cell-derived small extracellular vesicles alleviate lung injury in rat model of bronchopulmonary dysplasia by affecting cell survival and angiogenesis, *Stem Cells Dev.* **29**, 1520 (2020).
10. L. O'Driscoll, Extracellular vesicles from mesenchymal stem cells as a Covid-19 treatment, *Drug Discov. Today* **25**, 1124 (2020).
11. V. Sengupta, S. Sengupta, A. Lazo, P. Woods, A. Nolan, and N. Bremer, Exosomes derived from bone marrow mesenchymal stem cells as treatment for severe COVID-19, *Stem Cells Dev.* **29**, 747 (2020).
12. J. Wang, R. Huang, Q. Xu, G. Zheng, G. Qiu, M. Ge, Q. Shu, and J. Xu, Mesenchymal stem cell-derived extracellular vesicles alleviate acute lung injury via transfer of miR-27a-3p, *Crit. Care Med.* **48**, e599 (2020).
13. Q. Hao et al., Mesenchymal stem cell-derived extracellular vesicles decrease lung injury in mice, *J. Immunol.* **203**, 1961 (2019).
14. M. Khatri, L. A. Richardson, and T. Meulia, Mesenchymal stem cell-derived extracellular vesicles attenuate influenza virus-induced acute lung injury in a pig model, *Stem Cell. Res. Ther.* **9**, 17 (2018).
15. Y. Song et al., Exosomal miR-146a contributes to the enhanced therapeutic efficacy of interleukin-1β-primed mesenchymal stem cells against sepsis, *Stem Cells* **35**, 1208 (2017).
16. M. Yao, B. Cui, W. Zhang, W. Ma, G. Zhao, and L. Xing, Exosomal miR-21 secreted by IL-1β-primed-mesenchymal stem cells induces macrophage M2 polarization and ameliorates sepsis, *Life Sci.* **264**, 118658 (2021).
17. X. Wang et al., Exosomal miR-223 contributes to mesenchymal stem cell-elicited cardioprotection in polymicrobial sepsis, *Sci. Rep.* **5**, 13721 (2015).
18. T. M. Rager, J. K. Olson, Y. Zhou, Y. Wang, and G. E. Besner, Exosomes secreted from bone marrow-derived mesenchymal stem cells protect the intestines from experimental necrotizing enterocolitis, *J. Pediatr. Surg.* **51**, 942 (2016).
19. C. J. McCulloh, J. K. Olson, Y. Wang, Y. Zhou, N. H. Tengberg, S. Deshpande, and G. E. Besner, Treatment of experimental necrotizing enterocolitis with stem cell-derived exosomes, *J. Pediatr. Surg.* **53**, 1215 (2018).

20. Y. Y. Li, Q. W. Xu, P. Y. Xu, and W. M. Li, MSC-derived exosomal miR-34a/c-5p and miR-29b-3p improve intestinal barrier function by targeting the Snail/Claudins signaling pathway, *Life Sci.* **257**, 118017 (2020).

21. D. Sun, H. Cao, L. Yang, L. Lin, B. Hou, W. Zheng, Z. Shen, and H. Song, MiR-200b in heme oxygenase-1-modified bone marrow mesenchymal stem cell-derived exosomes alleviates inflammatory injury of intestinal epithelial cells by targeting high mobility group box 3, *Cell Death Dis.* **11**, 480 (2020).

22. A. L. Nocera, M. M. Miyake, P. Seifert, X. Han, and B. S. Bleier, Exosomes mediate interepithelial transfer of functional P-glycoprotein in chronic rhinosinusitis with nasal polyps, *Laryngoscope* **127**, E295 (2017).

23. A. L. Nocera et al., Exosome swarms eliminate airway pathogens and provide passive epithelial immunoprotection through nitric oxide, *J. Allergy Clin. Immunol.* **143**, 1525 (2019).

24. C. I. Timár et al., Antibacterial effect of microvesicles released from human neutrophilic granulocytes, *Blood* **121**, 510 (2013).

25. L. Martin-Jaular, E. S. Nakayasu, M. Ferrer, I. C. Almeida, and H. A. Del Portillo, Exosomes from *Plasmodium yoelii*-infected reticulocytes protect mice from lethal infections, *PLoS One* **6**, e26588 (2011).

26. P. A. Shiekh, A. Singh, and A. Kumar, Exosome laden oxygen releasing antioxidant and antibacterial cryogel wound dressing OxOBand alleviate diabetic and infectious wound healing, *Biomaterials* **249**, 120020 (2020).

27. F. Xie et al., Engineering extracellular vesicles enriched with palmitoylated ACE2 as COVID-19 therapy, *Adv. Mater.* **33**, 2103471 (2021).

28. M. Mittelbrunn, C. Gutiérrez-Vázquez, C. Villarroya-Beltri, S. González, F. Sánchez-Cabo, M. Á. González, A. Bernad, and F. Sánchez-Madrid, Unidirectional transfer of microRNA-loaded exosomes from T cells to antigen-presenting cells, *Nat. Commun.* **2**, 282 (2011).

29. J. K. Schnitzer, S. Berzel, M. Fajardo-Moser, K. A. Remer, and H. Moll, Fragments of antigen-loaded dendritic cells (DC) and DC-derived exosomes induce protective immunity against Leishmania major, *Vaccine* **28**, 5785 (2010).

30. J. Colino and C. M. Snapper, Exosomes from bone marrow dendritic cells pulsed with diphtheria toxoid preferentially induce type 1 antigen-specific IgG responses in naive recipients in the absence of free antigen, *J. Immunol.* **177**, 3757 (2006).

31. J. Colino and M. Snapper Clifford, Dendritic cell-derived exosomes express a *Streptococcus pneumoniae* capsular polysaccharide type 14 cross-reactive antigen that induces protective immunoglobulin responses against pneumococcal infection in mice, *Infect. Immun.* **75**, 220 (2007).
32. Y. Cheng and J. S. Schorey, Exosomes carrying mycobacterial antigens can protect mice against *Mycobacterium tuberculosis* infection, *Eur. J. Immunol.* **43**, 3279 (2013).
33. C. Beauvillain, M. O. Juste, S. Dion, J. Pierre, and I. Dimier-Poisson, Exosomes are an effective vaccine against congenital toxoplasmosis in mice, *Vaccine* **27**, 1750 (2009).
34. C. Cai et al., Macrophage-derived extracellular vesicles induce long-lasting immunity against hepatitis C virus which is blunted by polyunsaturated fatty acids, *Front. Immunol.* **9**, 723 (2018).
35. W. Wu et al., Interferon-induced macrophage-derived exosomes mediate antiviral activity against hepatitis B virus through miR-574-5p, *J. Infect. Dis.* **223**, 686 (2021).
36. G. Go, J. Lee, D. S. Choi, S. S. Kim, and Y. S. Gho, Extracellular vesicle-mimetic ghost nanovesicles for delivering anti-inflammatory drugs to mitigate gram-negative bacterial outer membrane vesicle-induced systemic inflammatory response syndrome, *Adv. Healthc. Mater.* **8**, e1801082 (2019).
37. X. Yang, G. Shi, J. Guo, C. Wang, and Y. He, Exosome-encapsulated antibiotic against intracellular infections of methicillin-resistant *Staphylococcus aureus*, *Int. J. Nanomed.* **13**, 8095 (2018).
38. C. Schwechheimer and M. J. Kuehn, Outer-membrane vesicles from gram-negative bacteria: Biogenesis and functions, *Nat. Rev. Microbiol.* **13**, 605 (2015).
39. F. Micoli and C. A. MacLennan, Outer membrane vesicle vaccines, *Semin. Immunol.* **50**, 101433 (2020).
40. F. Caron et al., From tailor-made to ready-to-wear meningococcal B vaccines: Longitudinal study of a clonal meningococcal B outbreak, *Lancet Infect. Dis.* **11**, 455 (2011).
41. F. Micoli et al., Comparative immunogenicity and efficacy of equivalent outer membrane vesicle and glycoconjugate vaccines against nontyphoidal Salmonella, *Proc. Natl. Acad. Sci. U. S. A.* **115**, 10428 (2018).
42. A. L. Bishop, A. A. Tarique, B. Patimalla, S. B. Calderwood, F. Qadri, and A. Camilli, *Vibrio cholerae*-induced inflammation in the neonatal mouse cholera model, *J. Infect. Dis.* **205**, 412 (2012).

43. S. Mitra, R. Sinha, J. Mitobe, and H. Koley, Development of a cost-effective vaccine candidate with outer membrane vesicles of a tolA-disrupted *Shigella boydii* strain, *Vaccine* **34**, 1839 (2016).

44. M. J. McConnell, C. Rumbo, G. Bou, and J. Pachón, Outer membrane vesicles as an acellular vaccine against *Acinetobacter baumannii*, *Vaccine* **29**, 5705 (2011).

45. G. Bjune et al., Effect of outer membrane vesicle vaccine against group B meningococcal disease in Norway, *Lancet* **338**, 1093 (1991).

46. F. Caron, V. Delbos, E. Houivet, A.-E. Deghmane, J.-P. Leroy, E. Hong, J. Bénichou, and M.-K. Taha, Evolution of immune response against *Neisseria meningitidis* B:14:P1.7,16 before and after the outer membrane vesicle vaccine MenBvac, *Vaccine* **30**, 5059 (2012).

47. R. F. Ochoa-Azze, L. García-Imía, and V. Vérez-Bencomo, Effectiveness of a serogroup B and C meningococcal vaccine developed in Cuba, *MEDICC Rev.* **20**, 22 (2018).

48. R. F. Ochoa-Azze, Cross-protection induced by VA-MENGOC-BC® vaccine, *Hum. Vaccin. Immunother.* **14**, 1064 (2018).

49. V. G. Sierra-González, Cuban meningococcal vaccine VA-MENGOC-BC:30 years of use and future potential, *MEDICC Rev.* **21**, 19 (2019).

50. P. Oster, D. Lennon, J. O'Hallahan, K. Mulholland, S. Reid, and D. Martin, MeNZB: A safe and highly immunogenic tailor-made vaccine against the New Zealand *Neisseria meningitidis* serogroup B disease epidemic strain, *Vaccine* **23**, 2191 (2005).

51. B. J. Loring, N. Turner, and H. Petousis-Harris, MeNZB vaccine and epidemic control: When do you stop vaccinating? *Vaccine* **26**, 5899 (2008).

52. M. O'Ryan, J. Stoddard, D. Toneatto, J. Wassil, and P. M. Dull, A multi-component meningococcal serogroup B vaccine (4CMenB): The clinical development program, *Drugs* **74**, 15 (2014).

53. E. A. Semchenko, A. Tan, R. Borrow, and K. L. Seib, The serogroup B meningococcal vaccine Bexsero elicits antibodies to *Neisseria gonorrhoeae*, *Clin. Infect. Dis.* **69**, 1101 (2019).

54. A. I. Camacho, J. de Souza, S. Sánchez-Gómez, M. Pardo-Ros, J. M. Irache, and C. Gamazo, Mucosal immunization with *Shigella flexneri* outer membrane vesicles induced protection in mice, *Vaccine* **29**, 8222 (2011).

55. O. Launay et al., Safety profile and immunologic responses of a novel vaccine against *Shigella sonnei* administered intramuscularly, intradermally and intranasally: Results from two parallel randomized

Phase 1 clinical studies in healthy adult volunteers in Europe, *EBioMedicine* **22**, 164 (2017).

56. C. W. Obiero et al., A Phase 2a randomized study to evaluate the safety and immunogenicity of the 1790GAHB generalized modules for membrane antigen vaccine against *Shigella sonnei* administered intramuscularly to adults from a shigellosis-endemic country, *Front. Immunol.* **8**, 1884 (2017).

57. X. Wang, A. K. Singh, X. Zhang, and W. Sun, Induction of protective antiplague immune responses by self-adjuvanting bionanoparticles derived from engineered *Yersinia pestis*, *Infect. Immun.* **88**, e00081 (2020).

58. S. M. Baker et al., *Burkholderia pseudomallei* OMVs derived from infection mimicking conditions elicit similar protection to a live-attenuated vaccine, *NPJ Vaccines* **6**, 18 (2021).

59. M. H. Daleke-Schermerhorn et al., Decoration of outer membrane vesicles with multiple antigens by using an autotransporter approach, *Appl. Environ. Microbiol.* **80**, 5854 (2014).

60. Q. Chen, H. Bai, W. Wu, G. Huang, Y. Li, M. Wu, G. Tang, and Y. Ping, Bioengineering bacterial vesicle-coated polymeric nanomedicine for enhanced cancer immunotherapy and metastasis prevention, *Nano Lett.* **20**, 11 (2020).

61. K. Kuerban, X. Gao, H. Zhang, J. Liu, M. Dong, L. Wu, R. Ye, M. Feng, and L. Ye, Doxorubicin-loaded bacterial outer-membrane vesicles exert enhanced anti-tumor efficacy in non-small-cell lung cancer, *Acta Pharm. Sin. B* **10**, 1534 (2020).

62. V. Gujrati, S. Kim, S. H. Kim, J. J. Min, H. E. Choy, S. C. Kim, and S. Jon, Bioengineered bacterial outer membrane vesicles as cell-specific drug delivery vehicles for cancer therapy, *ACS Nano* **8**, 1525 (2014).

63. J. L. Kadurugamuwa and T. J. Beveridge, Delivery of the non-membrane-permeative antibiotic gentamicin into mammalian cells by using *Shigella flexneri* membrane vesicles, *Antimicrob. Agents Chemother.* **42**, 1476 (1998).

64. W. Huang, Q. Zhang, W. Li, M. Yuan, J. Zhou, L. Hua, Y. Chen, C. Ye, and Y. Ma, Development of novel nanoantibiotics using an outer membrane vesicle-based drug efflux mechanism, *J. Control. Release* **317**, 1 (2020).

65. W. Gao, R. H. Fang, S. Thamphiwatana, B. T. Luk, J. Li, P. Angsantikul, Q. Zhang, C.-M. J. Hu, and L. Zhang, Modulating antibacterial immunity via bacterial membrane-coated nanoparticles, *Nano Lett.* **15**, 1403 (2015).

66. Y. Zhang et al., Inhibition of pathogen adhesion by bacterial outer membrane-coated nanoparticles, *Angew. Chem. Int. Ed.* **58**, 11404 (2019).

67. G. Wu et al., Nanoparticle reinforced bacterial outer-membrane vesicles effectively prevent fatal infection of carbapenem-resistant *Klebsiella pneumoniae, Nanomedicine* **24**, 102148 (2020).

68. M. M. Stanton, B.-W. Park, D. Vilela, K. Bente, D. Faivre, M. Sitti, and S. Sánchez, Magnetotactic bacteria powered biohybrids target *E. coli* biofilms, *ACS Nano* **11**, 9968 (2017).

69. S. Fazal and R. Lee, Biomimetic bacterial membrane vesicles for drug delivery applications, *Pharmaceutics* **13**, 1430 (2021).

70. L. Qiao, Y. Rao, K. Zhu, X. Rao, and R. Zhou, Engineered remolding and application of bacterial membrane vesicles, *Front. Microbiol.* **12**, 729369 (2021).

71. F. Gao, L. Xu, B. Yang, F. Fan, and L. Yang, Kill the real with the fake: Eliminate intracellular *Staphylococcus aureus* using nanoparticle coated with its extracellular vesicle membrane as active-targeting drug carrier, *ACS Infect. Dis.* **5**, 218 (2019).

72. G. Chen, Y. Bai, Z. Li, F. Wang, X. Fan, and X. Zhou, Bacterial extracellular vesicle-coated multi-antigenic nanovaccines protect against drug-resistant *Staphylococcus aureus* infection by modulating antigen processing and presentation pathways, *Theranostics* **10**, 7131 (2020).

73. O. Y. Kim et al., Bacterial protoplast-derived nanovesicles as vaccine delivery system against bacterial infection, *Nano Lett.* **15**, 266 (2015).

74. J. M. V. Makabenta, A. Nabawy, C.-H. Li, S. Schmidt-Malan, R. Patel, and V. M. Rotello, Nanomaterial-based therapeutics for antibiotic-resistant bacterial infections, *Nat. Rev. Microbiol.* **19**, 23 (2021).

75. H. I. Labouta, S. Menina, A. Kochut, S. Gordon, R. Geyer, P. Dersch, and C. M. Lehr, Bacteriomimetic invasin-functionalized nanocarriers for intracellular delivery, *J. Control. Release* **220**, 414 (2015).

76. S. Menina, H. I. Labouta, R. Geyer, T. Krause, S. Gordon, P. Dersch, and C. M. Lehr, Invasin-functionalized liposome nanocarriers improve the intracellular delivery of anti-infective drugs, *RSC Adv.* **6**, 41622 (2016).

77. A. Castoldi, M. Empting, C. De Rossi, K. Mayr, P. Dersch, R. Hartmann, R. Müller, S. Gordon, and C. M. Lehr, Aspherical and spherical InvA497-functionalized nanocarriers for intracellular delivery of anti-infective agents, *Pharm. Res.* **36**, 22 (2018).

78. S. Menina, J. Eisenbeis, M. A. M. Kamal, M. Koch, M. Bischoff, S. Gordon, B. Loretz, and C. M. Lehr, Bioinspired liposomes for oral delivery of

colistin to combat intracellular infections by *Salmonella enterica*, *Adv. Healthc. Mater.* **8**, e1900564 (2019).

79. N. J. Sullivan, A. Sanchez, P. E. Rollin, Z. Y. Yang, and G. J. Nabel, Development of a preventive vaccine for Ebola virus infection in primates, *Nature* **408**, 605 (2000).

80. J. W. Shiver et al., Replication-incompetent adenoviral vaccine vector elicits effective anti-immunodeficiency-virus immunity, *Nature* **415**, 331 (2002).

81. A. Volkmann et al., The Brighton Collaboration standardized template for collection of key information for risk/benefit assessment of a modified vaccinia Ankara (MVA) vaccine platform, *Vaccine* **39**, 3067 (2021).

82. F. C. Zhu et al., Immunogenicity and safety of a recombinant adenovirus type-5-vectored COVID-19 vaccine in healthy adults aged 18 years or older: A randomised, double-blind, placebo-controlled, phase 2 trial, *Lancet* **396**, 479 (2020).

83. D. Y. Logunov et al., Safety and efficacy of an rAd26 and rAd5 vector-based heterologous prime-boost COVID-19 vaccine: An interim analysis of a randomised controlled phase 3 trial in Russia, *Lancet* **397**, 671 (2021).

84. J. Sadoff et al., Safety and efficacy of single-dose Ad26.COV2.S vaccine against Covid-19, *N. Engl. J. Med.* **384**, 2187 (2021).

85. K. R. W. Emary et al., Efficacy of ChAdOx1 nCoV-19 (AZD1222) vaccine against SARS-CoV-2 variant of concern 202012/01 (B.1.1.7): An exploratory analysis of a randomised controlled trial, *Lancet* **397**, 1351 (2021).

86. N. C. Salisch et al., A double-blind, randomized, placebo-controlled Phase 1 study of Ad26.ZIKV.001, an Ad26-vectored anti-Zika virus vaccine, *Ann. Intern. Med.* **174**, 585 (2021).

87. L. R. Baden et al., Safety and immunogenicity of two heterologous HIV vaccine regimens in healthy, HIV-uninfected adults (TRAVERSE): A randomised, parallel-group, placebo-controlled, double-blind, phase 1/2a study, *Lancet HIV* **7**, e688 (2020).

88. L. Coughlan et al., Heterologous two-dose vaccination with simian adenovirus and poxvirus vectors elicits long-lasting cellular immunity to influenza virus A in healthy adults, *EBioMedicine* **29**, 146 (2018).

89. I. Yacoby, M. Shamis, H. Bar, D. Shabat, and I. Benhar, Targeting antibacterial agents by using drug-carrying filamentous bacteriophages, *Antimicrob. Agents Chemother.* **50**, 2087 (2006).

90. I. Yacoby, H. Bar, and I. Benhar, Targeted drug-carrying bacteriophages as antibacterial nanomedicines, *Antimicrob. Agents Chemother.* **51**, 2156 (2007).

91. L. Vaks and I. Benhar, *In vivo* characteristics of targeted drug-carrying filamentous bacteriophage nanomedicines, *J. Nanobiotechnol.* **9**, 58 (2011).

92. C. Westwater, L. M. Kasman, D. A. Schofield, P. A. Werner, J. W. Dolan, M. G. Schmidt, and J. S. Norris, Use of genetically engineered phage to deliver antimicrobial agents to bacteria: An alternative therapy for treatment of bacterial infections, *Antimicrob. Agents Chemother.* **47**, 1301 (2003).

93. D. Bikard, C. W. Euler, W. Jiang, P. M. Nussenzweig, G. W. Goldberg, X. Duportet, V. A. Fischetti, and L. A. Marraffini, Exploiting CRISPR-Cas nucleases to produce sequence-specific antimicrobials, *Nat. Biotechnol.* **32**, 1146 (2014).

94. R. J. Krom, P. Bhargava, M. A. Lobritz, and J. J. Collins, Engineered phagemids for nonlytic, targeted antibacterial therapies, *Nano Lett.* **15**, 4808 (2015).

95. E. Alemzadeh, A. Dehshahri, K. Izadpanah, and F. Ahmadi, Plant virus nanoparticles: Novel and robust nanocarriers for drug delivery and imaging, *Colloids Surf. B Biointerfaces.* **167**, 20 (2018).

96. L. P. Shriver, K. J. Koudelka, and M. Manchester, Viral nanoparticles associate with regions of inflammation and blood brain barrier disruption during CNS infection, *J. Neuroimmunol.* **211**, 66 (2009).

97. A. M. Wen, N. Le, X. Zhou, N. F. Steinmetz, and D. L. Popkin, Tropism of CPMV to professional antigen presenting cells enables a platform to eliminate chronic infections, *ACS Biomater. Sci. Eng.* **1**, 1050 (2015).

98. G. Yang, S. Chen, and J. Zhang, Bioinspired and biomimetic nanotherapies for the treatment of infectious diseases, *Front. Pharmacol.* **10**, 751 (2019).

99. J. Zepeda-Cervantes, J. O. Ramírez-Jarquín, and L. Vaca, Interaction between virus-like particles (VLPs) and pattern recognition receptors (PRRs) from dendritic cells (DCs): Toward better engineering of VLPs, *Front. Immunol.* **11**, 1100 (2020).

100. T. Rivera-Hernandez, J. Hartas, Y. Wu, Y. P. Chuan, L. H. Lua, M. Good, M. R. Batzloff, and A. P. Middelberg, Self-adjuvanting modular virus-like particles for mucosal vaccination against group A streptococcus (GAS), *Vaccine* **31**, 1950 (2013).

101. K. L. Warfield, C. M. Bosio, B. C. Welcher, E. M. Deal, M. Mohamadzadeh, A. Schmaljohn, M. J. Aman, and S. Bavari, Ebola virus-like particles protect from lethal Ebola virus infection, *Proc. Natl. Acad. Sci. U. S. A.* **100**, 15889 (2003).

102. M. Cai et al., Virus-like particle vaccine by intranasal vaccination elicits protective immunity against respiratory syncytial viral infection in mice, *Acta Biochim. Biophys. Sin.* **49**, 74 (2017).

103. M. C. Kim et al., Microneedle patch delivery to the skin of virus-like particles containing heterologous M2e extracellular domains of influenza virus induces broad heterosubtypic cross-protection, *J. Control. Release* **210**, 208 (2015).

104. C. Mathieu, G. Rioux, M. C. Dumas, and D. Leclerc, Induction of innate immunity in lungs with virus-like nanoparticles leads to protection against influenza and *Streptococcus pneumoniae* challenge, *Nanomedicine* **9**, 839 (2013).

105. M. Lebel, J. F. Daudelin, K. Chartrand, E. Tarrab, U. Kalinke, P. Savard, N. Labrecque, D. Leclerc, and A. Lamarre, Nanoparticle adjuvant sensing by TLR7 enhances $CD8^+$ T cell-mediated protection from *Listeria monocytogenes* infection, *J. Immunol.* **192**, 1071 (2014).

106. M. Zdanowicz and J. Chroboczek, Virus-like particles as drug delivery vectors, *Acta Biochim. Pol.* **63**, 469 (2016).

107. M. T. Smith, A. K. Hawes, and B. C. Bundy, Reengineering viruses and virus-like particles through chemical functionalization strategies, *Curr. Opin. Biotechnol.* **24**, 620 (2013).

108. S. N. Crooke, J. Schimer, I. Raji, B. Wu, A. K. Oyelere, and M. G. Finn, Lung tissue delivery of virus-like particles mediated by macrolide antibiotics, *Mol. Pharm.* **16**, 2947 (2019).

109. Y. Kaneda, Virosome: A novel vector to enable multi-modal strategies for cancer therapy, *Adv. Drug Deliv. Rev.* **64**, 730 (2012).

110. M. Lakadamyali, M. J. Rust, and X. Zhuang, Endocytosis of influenza viruses, *Microbes Infect.* **6**, 929 (2004).

111. M. S. Mueller, A. Renard, F. Boato, D. Vogel, M. Naegeli, R. Zurbriggen, J. A. Robinson, and G. Pluschke, Induction of parasite growth-inhibitory antibodies by a virosomal formulation of a peptidomimetic of loop I from domain III of *Plasmodium falciparum* apical membrane antigen 1, *Infect. Immun.* **71**, 4749 (2003).

112. V. Kumar, R. Kumar, V. K. Jain, and S. Nagpal, Comparison of virosome vs. liposome as drug delivery vehicle using HepG2 and CaCo2 cell lines, *J. Microencapsul.* **38**, 263 (2021).

113. M. Amacker, O. Engler, A. R. Kammer, S. Vadrucci, D. Oberholzer, A. Cerny, and R. Zurbriggen, Peptide-loaded chimeric influenza virosomes for efficient *in vivo* induction of cytotoxic T cells, *Int. Immunol.* **17**, 695 (2005).

114. F. Bossa et al., Erythrocytes-mediated delivery of dexamethasone 21-phosphate in steroid-dependent ulcerative colitis: A randomized, double-blind sham-controlled study, *Inflamm. Bowel Dis.* **19**, 1872 (2013).

115. V. Annese et al., The polymorphism of multi-drug resistance 1 gene (*MDR1*) does not influence the pharmacokinetics of dexamethasone loaded into autologous erythrocytes of patients with inflammatory bowel disease, *Eur. Rev. Med. Pharmacol. Sci.* **10**, 27 (2006).

116. L. Rossi, S. Serafini, L. Cenerini, F. Picardi, L. Bigi, I. Panzani, and M. Magnani, Erythrocyte-mediated delivery of dexamethasone in patients with chronic obstructive pulmonary disease, *Biotechnol. Appl. Biochem.* **33**, 85 (2001).

117. L. Rossi et al., Low doses of dexamethasone constantly delivered by autologous erythrocytes slow the progression of lung disease in cystic fibrosis patients, *Blood Cells Mol. Dis.* **33**, 57 (2004).

118. F. K. Alanazi, D. Harisa Gel, A. Maqboul, M. Abdel-Hamid, S. H. Neau, and I. A. Alsarra, Biochemically altered human erythrocytes as a carrier for targeted delivery of primaquine: An *in vitro* study, *Arch. Pharm. Res.* **34**, 563 (2011).

119. N. Talwar and N. K. Jain, Erythrocyte based delivery system of primaquine: *in vitro* characterization, *J. Microencapsul.* **9**, 357 (1992).

120. E. Briones, C. I. Colino, C. G. Millán, and J. M. Lanao, Increasing the selectivity of amikacin in rat peritoneal macrophages using carrier erythrocytes, *Eur. J. Pharm. Sci.* **38**, 320 (2009).

121. H. G. Eichler, H. Rameis, K. Bauer, A. Korn, S. Bacher, and S. Gasić, Survival of gentamicin-loaded carrier erythrocytes in healthy human volunteers, *Eur. J. Clin. Invest.* **16**, 39 (1986).

122. J. R. DeLoach and G. G. Wagner, Pharmacokinetics of tetracycline encapsulated in bovine carrier erythrocytes, *Am. J. Vet. Res.* **45**, 640 (1984).

123. A. Fraternale, A. Casabianca, L. Rossi, L. Chiarantini, G. F. Schiavano, A. T. Palamara, E. Garaci, and M. Magnani, Erythrocytes as carriers of reduced glutathione (GSH) in the treatment of retroviral infections, *J. Antimicrob. Chemother.* **52**, 551 (2003).

124. M. Magnani, M. Bianchi, L. Rossi, and V. Stocchi, Human red blood cells as bioreactors for the release of 2',3'-dideoxycytidine, an inhibitor of HIV infectivity, *Biochem. Biophys. Res. Commun.* **164**, 446 (1989).
125. M. Magnani, L. Rossi, G. Brandi, G. F. Schiavano, M. Montroni, and G. Piedimonte, Targeting antiretroviral nucleoside analogues in phosphorylated form to macrophages: *In vitro* and *in vivo* studies, *Proc. Natl. Acad. Sci. U. S. A.* **89**, 6477 (1992).
126. L. Rossi, G. Brandi, G. F. Schiavano, S. Scarfi, E. Millo, G. Damonte, U. Benatti, A. De Flora, and M. Magnani, Heterodimer-loaded erythrocytes as bioreactors for slow delivery of the antiviral drug azidothymidine and the antimycobacterial drug ethambutol, *AIDS Res. Hum. Retroviruses* **15**, 345 (1999).
127. Y. Godfrin et al., International seminar on the red blood cells as vehicles for drugs, *Expert Opin. Biol. Ther.* **12**, 127 (2012).
128. M. Forouzesh and A. A. H. Zarin, A novel combinatory paradigm for chronic hepatitis C treatment using liver-targeted carrier erythrocytes co-encapsulated with interferon alpha-2b, ribavirin and boceprevir, *Iran. J. Med. Hypotheses Ideas* **4**, 0 (2010).
129. C. Polvani, A. Gasparini, U. Benatti, A. DeFlora, S. Silvestri, G. Volpini, and L. Nencioni, Murine red blood cells as efficient carriers of three bacterial antigens for the production of specific and neutralizing antibodies, *Biotechnol. Appl. Biochem.* **14**, 347 (1991).
130. L. Chiarantini, R. Argnani, S. Zucchini, L. Stevanato, P. Zabardi, M. P. Grossi, M. Magnani, and R. Manservigi, Red blood cells as delivery system for recombinant HSV-1 glycoprotein B: Immunogenicity and protection in mice, *Vaccine* **15**, 276 (1997).
131. S. Dominici et al., Red blood cell-mediated delivery of recombinant HIV-1 Tat protein in mice induces anti-Tat neutralizing antibodies and CTL, *Vaccine* **21**, 2073 (2003).
132. C. M. Hu, R. H. Fang, J. Copp, B. T. Luk, and L. Zhang, A biomimetic nanosponge that absorbs pore-forming toxins, *Nat. Nanotechnol.* **8**, 336 (2013).
133. Y. Zhang, J. Zhang, W. Chen, P. Angsantikul, K. A. Spiekermann, R. H. Fang, W. Gao, and L. Zhang, Erythrocyte membrane-coated nanogel for combinatorial antivirulence and responsive antimicrobial delivery against *Staphylococcus aureus* infection, *J. Control. Release* **263**, 185 (2017).
134. F. Wang et al., Nanoparticle-based antivirulence vaccine for the management of methicillin-resistant *Staphylococcus aureus* skin infection, *Adv. Funct. Mater.* **26**, 1628 (2016).

135. J. Koo, T. Escajadillo, L. Zhang, V. Nizet, and S. M. Lawrence, Erythrocyte-coated nanoparticles block cytotoxic effects of group B streptococcus β-hemolysin/cytolysin, *Front. Pediatr.* **7**, 410 (2019).

136. Y. Chen, Y. Zhang, M. Chen, J. Zhuang, R. H. Fang, W. Gao, and L. Zhang, Biomimetic nanosponges suppress *in vivo* lethality induced by the whole secreted proteins of pathogenic bacteria, *Small* **15**, e1804994 (2019).

137. Y. Zhang, W. Gao, Y. Chen, T. Escajadillo, J. Ungerleider, R. H. Fang, K. Christman, V. Nizet, and L. Zhang, Self-assembled colloidal gel using cell membrane-coated nanosponges as building blocks, *ACS Nano* **11**, 11923 (2017).

138. Y. Chen et al., Broad-spectrum neutralization of pore-forming toxins with human erythrocyte membrane-coated nanosponges, *Adv. Healthc. Mater.* **7**, e1701366 (2018).

139. B. Chen et al., Highly biocompatible and recyclable biomimetic nanoparticles for antibiotic-resistant bacteria infection, *Biomater. Sci.* **9**, 826 (2021).

140. D. Chu, J. Gao, and Z. Wang, Neutrophil-mediated delivery of therapeutic nanoparticles across blood vessel barrier for treatment of inflammation and infection, *ACS Nano* **9**, 11800 (2015).

141. K. Wang, Y. Lei, D. Xia, P. Xu, T. Zhu, Z. Jiang, and Y. Ma, Neutrophil membranes coated, antibiotic agent loaded nanoparticles targeting to the lung inflammation, *Colloids Surf. B* **188**, 110755 (2020).

142. J. Gao, S. Wang, and Z. Wang, High yield, scalable and remotely drug-loaded neutrophil-derived extracellular vesicles (EVs) for anti-inflammation therapy, *Biomaterials* **135**, 62 (2017).

143. J. Gao, S. Wang, X. Dong, L. G. Leanse, T. Dai, and Z. Wang, Co-delivery of resolvin D1 and antibiotics with nanovesicles to lungs resolves inflammation and clears bacteria in mice, *Commun. Biol.* **3**, 680 (2020).

144. J. Gao, X. Dong, Y. Su, and Z. Wang, Human neutrophil membrane-derived nanovesicles as a drug delivery platform for improved therapy of infectious diseases, *Acta Biomater.* **123**, 354 (2021).

145. C. Wang et al., Pretreated macrophage-membrane-coated gold nanocages for precise drug delivery for treatment of bacterial infections, *Adv. Mater.* **30**, 1804023 (2018).

146. J. Fu et al., An engineered pseudo-macrophage for rapid treatment of bacteria-infected osteomyelitis via microwave-excited anti-infection and immunoregulation, *Adv. Mater.* **33**, 2102926 (2021).

147. S. Thamphiwatana et al., Macrophage-like nanoparticles concurrently absorbing endotoxins and proinflammatory cytokines for sepsis management, *Proc. Natl. Acad. Sci. U. S. A.* **114**, 11488 (2017).

148. Y. Li et al., Coating of a novel antimicrobial nanoparticle with a macrophage membrane for the selective entry into infected macrophages and killing of intracellular staphylococci, *Adv. Funct. Mater.* **30**, 2004942 (2020).

149. X. Wei et al., Multiantigenic nanotoxoids for antivirulence vaccination against antibiotic-resistant gram-negative bacteria, *Nano Lett.* **19**, 4760 (2019).

150. Q. Zhang et al., Cellular nanosponges inhibit SARS-CoV-2 infectivity, *Nano Lett.* **20**, 5570 (2020).

151. R. Molinaro et al., Biomimetic proteolipid vesicles for targeting inflamed tissues, *Nat. Materials* **15**, 1037 (2016).

152. C. M. J. Hu et al., Nanoparticle biointerfacing by platelet membrane cloaking, *Nature* **526**, 118 (2015).

153. Z. Peng et al., Integrated endotoxin-adsorption and antibacterial properties of platelet-membrane-coated copper silicate hollow microspheres for wound healing, *J. Nanobiotechnol.* **19**, 383 (2021).

154. P. Angsantikul, S. Thamphiwatana, Q. Zhang, K. Spiekermann, J. Zhuang, R. H. Fang, W. Gao, M. Obonyo, and L. Zhang, Coating nanoparticles with gastric epithelial cell membrane for targeted antibiotic delivery against Helicobacter pylori infection, *Adv. Ther.* **1**, 1800016 (2018).

155. X. Wei, G. Zhang, D. Ran, N. Krishnan, R. H. Fang, W. Gao, S. A. Spector, and L. Zhang, T-cell-mimicking nanoparticles can neutralize HIV infectivity, *Adv. Mater.* **30**, 1802233 (2018).

156. L. Rao et al., Decoy nanoparticles protect against COVID-19 by concurrently adsorbing viruses and inflammatory cytokines, *Proc. Natl. Acad. Sci. U. S. A.* **117**, 27141 (2020).

157. Z. Wang et al., Inhaled ACE2-engineered microfluidic microsphere for intratracheal neutralization of COVID-19 and calming of the cytokine storm, *Matter* **5**, 336 (2022).

158. P. Zhang et al., Virus-mimetic nanovesicles as a versatile antigen-delivery system, *Proc. Natl. Acad. Sci. U. S. A.* **112**, E6129 (2015).

Index